INNOVATIONS IN SATELLITE COMMUNICATIONS AND SATELLITE TECHNOLOGY

INNOVATIONS IN SATELLITE COMMUNICATIONS AND SATELLITE TECHNOLOGY

The Industry Implications of DVB-S2X, High Throughput Satellites, Ultra HD, M2M, and IP

DANIEL MINOLI
Secure Enterprise Systems Inc. New York, USA

Library of Congress Cataloging-in-Publication Data:

Minoli, Daniel, 1952-
  Innovations in satellite communication and satellite technology : the industry implications
of DVB-S2X, high throughput satellites, Ultra HD, M2M, and IP / Daniel Minoli.
    pages cm
  Includes bibliographical references and index.
  ISBN 978-1-118-98405-5 (cloth)
1. Artificial satellites in telecommunication. I. Title.
  TK5104.M539 2015
  621.382'5–dc23

                                                                    2014043912

Printed in the United States of America

10 9 8 7 6 5 4 3 2 1

Typeset in 10/12 pt TimesLTStd by Laserwords Private Limited, Chennai, India

1   2015

For Anna.
And for my parents Gino and Angela.

CONTENTS

PREFACE

A number of technical and service advances affecting commercial satellite communications have been seen in the past few years. This text surveys some of these new key advances and what the implications and/or opportunities for end-users and service providers might be. Satellite communication plays and will continue to play a key role in commercial, TV/media, government, and military communications because of its intrinsic multicast/broadcast capabilities, mobility aspects, global reach, reliability, and ability to quickly support connectivity in open-space and/or hostile environments.

Business factors impacting the industry at this time include the desire for higher throughput and more cost-effective bandwidth. Improved modulation techniques allow users to increase channel datarates by employing methods such as 64APSK. High throughput is also achieved via the use of Ka (and Ku) spotbeams on High Throughput Satellites (HTS), and via the reduction of transmission latency (due to higher layer protocol stack handshakes) using Medium Earth Orbit (MEO) satellites that operate in a 5,000-mile orbit over the equator (also known as MEO-HTS), but where users must use two steerable antennas to track the spacecraft and retain signal connectivity by moving the path from one satellite in the constellation to another.

Providing services to people on-the-move, particularly for transoceanic airplane journeys is now both technically feasible and financially advantageous to the service provider stakeholders. M2M (machine-to-machine) connectivity, whether for trucks on transcontinental trips, or for aircraft real-time-telemetry aggregation, or mercantile ship data tracking, opens up new opportunities to extend the Internet of Things (IoT) to broadly-distributed entities, particularly in oceanic environments. Emerging Ultra High Definition Television (UHDTV) provides video quality that is the equivalent of 8-to-16 HDTV screens (33 million pixels, for the $7,680 \times 4,320$ resolution), compared to a maximum 2 million pixels ($1,920 \times 1,080$ resolution) for the current highest quality HDTV service – clearly this requires a lot more bandwidth. Satellite operators are planning to position themselves in this market segment, with generally-available broadcast services planned for 2020, and more targeted transmission starting at press time.

At the core-technology level, electric (instead of chemical) propulsion is being sought; such propulsion approaches can reduce spacecraft weight (and so, launch cost) and possibly extend the spacecraft life. Additionally, new launch platforms are being brought to the market, again with the goal of lowering launch cost via increased competition.

Satellite networks cannot really exist (forever) as stand-alone islands in a sea of connectivity; hence, hybrid networks have an important role to play. The widespread introduction of IP-based services, including IP-based Television (IPTV) and Over The Top (OTT) video, driven by continued deployment of fiber connectivity will ultimately re-shape the industry. In particular, Internet Protocol Version 6 (IPv6) is a technology now being deployed in various parts of the world that will allow true explicit end-to-end device addressability. As the number of intelligent systems that need direct access expands to the multiple billions (e.g., including smartphones, tablets, appliances, sensors/actuators, and even body-worn bio-metric devices), IPv6 becomes an institutional imperative, in the final analysis. The integration of satellite communication and IPv6 capabilities promises to provide a powerful networking infrastructure that can serve the evolving needs of government, military, IPTV, and mobile video stakeholders, to name just a few.

This book explores these evolving technical themes and opportunities. After an introductory overview, Chapter 2 discusses advances in modulation techniques, such as DBV-S2 extensions (DVB-S2X). Spotbeam technologies (at Ka but also at Ku) which constitute the technical basis for the emerging HTS systems and services are discussed in Chapter 3. Aeronautical mobility services such as Internet service while on-the-move are covered in Chapter 4. Maritime and other terrestrial mobility services are covered in Chapter 5. M2M applications are surveyed in Chapter 6. Emerging Ultra HD technologies are assessed in Chapter 7. Finally, new space technology, particularly Electric Propulsion and new launch platforms ultimately driving lower cost-per-bit (or cost-per-MHz) are discussed in Chapter 8.

This work will be of interest to technology investors; planners with satellite operators, carriers and service providers; CTOs; logistics professionals; engineers at equipment developers; technology integrators; Internet Service Providers (ISP), telcos, and wireless providers, both domestically and in the rest of the world.

ACKNOWLEDGMENTS

The author would like to thank Mr. William B. McDonald, President of WBMSAT Satellite Communications Consulting (Port Orchard, WA) for review, input, and guidance for this text. WBMSAT provides research, systems design, engineering, integration, testing, and project management in all aspects of commercial and military satellite communication.

The author would like to also thank Edward D. Horowitz for valuable contributions. Mr. Horowitz is Co-founder and a Director of U.S. Space LLC, a satellite services company, and Chairman of ViviSat, its in-orbit servicing venture. Recently Mr. Horowitz joined the Office of the CEO of Encompass Digital Media, a leading provider of worldwide television channel origination, live sports and news distribution, digital media and government services.

However, any pointed opinion, perspective, limitations, possible ambiguities, or lack of full clarity in this work are solely attributable to this author.

ABOUT THE AUTHOR

Mr. Minoli has many years of technical-hands-on and managerial experience in planning, designing, deploying, and operating secure IP/IPv6-, telecom-, wireless-, satellite-, and video networks for global Best-In-Class carriers and financial companies. He is currently the Chief Technology Officer at *Secure Enterprise Systems* (www.ses-engineering.us), an engineering, technology assessment, and enterprise cybersecurity firm. Previous roles in the past two decades have included General Manager and Director of Ground Systems Engineering at *SES*, the world's second largest satellite services provider, Director of Network Architecture at *Capital One Financial*, Chief Technology Officer at *InfoPort Communication Group*, and Vice President of Packet Services at *Teleport Communications Group (TCG)* (eventually acquired by AT&T).

In the recent past he has been responsible for (i) development, engineering, and deployment of metro Ethernet, IP/MPLS, and VoIP/VoMPLS networks, (ii) the development, engineering, and deployment of hybrid IPTV, non-linear, and, 3DTV video systems, (iii) deployments of a dozen large aperture antenna (7–13 m) at teleports in the U.S. and abroad; (iv) deployment of satellite monitoring services worldwide (over 40 sites); (v) development, engineering, and deployment of IPv6-based services in the M2M/Internet of Things area, in the non-linear video area, in the smartphone area, in the satellite area, and in the network security area; and (vi) the deployment of cloud computing infrastructure (Cisco UCS - 3,800 servers) for a top-line Cable TV provider in the U.S. Some Mr. Minoli's satellite-, wireless-, IP-, video-, and Internet Of Things-related work has been documented in books he has authored, including:

- *Satellite Systems Engineering in an IPv6 Environment* (Francis and Taylor 2009),
- *Wireless Sensor Networks* (co-authored) (Wiley 2007),

- *Hotspot Networks: Wi-Fi for Public Access Locations* (McGraw-Hill, 2002),
- *Mobile Video with Mobile IPv6* (Wiley 2012),
- *Linear and Non-Linear Video and TV Applications Using IPv6 and IPv6 Multicast* (Wiley 2012), and,
- *Building the Internet of Things with IPv6 and MIPv6 (Wiley, 2013).*

He also played a founding role in the launching of two companies through the high-tech incubator *Leading Edge Networks Inc.*, which he ran in the early 2000s: *Global Wireless Services*, a provider of secure broadband hotspot mobile Internet and hotspot VoIP services; and, *InfoPort Communications Group*, an optical and Giga-bit Ethernet metropolitan carrier supporting Data Center/SAN/channel extension and cloud network access services.

He has also written columns for *ComputerWorld, NetworkWorld*, and *Network Computing* (1985–2006). He has taught at *New York University* (Information Technology Institute), *Rutgers University*, and *Stevens Institute of Technology* (1984–2003). Also, he was a Technology Analyst At-Large, for Gartner/DataPro (1985–2001); based on extensive hand-on work at financial firms and carriers, he tracked technologies and wrote CTO/CIO-level technical scans in the area of telephony and data systems, including topics on security, disaster recovery/business continuity, network management, LANs, WANs (ATM, IPv4, MPLS, IPv6), wireless (LANs, public hotspot, wireless sensor networks, 3G/4G, and satellite), VoIP, network design/economics, carrier networks (such as metro Ethernet and CWDM/DWDM), and e-commerce. For several years he has been Session-, Tutorial-, and now overall Technical Program Chair for the IEEE ENTNET (Enterprise Networking) conference; ENTNET focuses on enterprise networking requirements for large financial firms and other corporate institutions (this IEEE group has now merged and has become the IEEE Technical Committee on Information Infrastructure [TCIIN]).

He has also acted as Expert Witness in a (won) $11B lawsuit regarding a VoIP-based wireless Air-to-Ground radio communication system for airplane in-cabin services, as well as for a large lawsuit related to digital scanning and transmission of bank documents/instruments (specifically, scanned checks). He has also been engaged as a technical expert in a number of patent infringement proceedings in the digital imaging, VoIP, firewall, and VPN space supporting law firms such as *Schiff Hardin LLP, Fulbright & Jaworski LLP, Dimock Stratton LLP/ Smart & Biggar LLP*, Munger, Tolles, and Olson LLP, and *Baker & McKenzie LLP*, among others.

Over the years he has advised Venture Capitalists for investments in a dozen high-tech companies. He performed extensive technical, sales, and market-ing analyses of high-tech firms seeking funding for a total of approximately $150M, developing multimedia, digital video, physical layer switching, VSATs, telemedicine, Java-based CTI, VoFR & VPNs, HDTV, optical chips, H.323 gateways, nanofabrication/QCL wireless, and TMN mediation. Included the following efforts: *MRC*: multimedia & Asynchronous Transfer Mode; *NHC:* Physical Layer switch; *CoastCom:* VSAT systems; *Cifra:* tele-medicine; *Uniforce:* Java IP-based

CTI; *Memotec:* VoFR; *Miranda*: HDTV & Electronic Cinema; *Lumenon*: optical WDMs; *Medisys*: Web-based healthcare ASP; *Tri-Link*: H.323 VoIP gateway; *Maxima*: wireless free-space optics metro networks using nanofabricated QCLs (Quantum Cascade Lasers); and, *ACE*COMM* for TMN/IPDR (financiers/VCs: *Societe' General de Financiament de Quebec; Caisse de Depot et Placement Quebec; Les Funds De Solidarite' Des Travailleurs).*

1

OVERVIEW

Satellite services, spanning the commercial arena, the military arena, and the earth sensing arena (including weather tacking), offer critical global connectivity and observation capabilities, which are perceived to be indispensable in the modern world. Whether supporting mobility in the form of Internet access and real-time telemetry from airplanes or ships on oceanic routes, or distribution of high-quality entertainment video to dispersed areas in emerging markets without significant infrastructure, or emergency communications in adverse conditions or in remote areas, or earth mapping, or military theater applications with unmanned aerial vehicles, satellites fill a void that cannot be met by other forms of communication mechanisms, including fiber optic links. Over 900 satellites were orbiting the earth as of press time. However, due to the continued rapid deployment of fiber and Internet Protocol (IP) services in major metropolitan areas where the paying customers are, including those in North America, Europe, Asia, South America, and even in Africa, tech-savvy and marketing-sophisticated approaches that organically integrate IP into the end-to-end solution are critically needed by the satellite operators to sustain growth.

Progressive satellite operators will undoubtedly opt to implement, at various degrees, some of the concepts presented here, concepts, frankly, not *per se* surprisingly novel or esoteric, since the idea of making satellites behave more than just microwave repeaters (microwave repeaters with operative functionally equivalent to the repeaters being deployed in the 1950s in the Bell System in the United States), in order to sustain market growth with vertically integrated user-impetrated

Innovations in Satellite Communications and Satellite Technology: The Industry Implications of DVB-S2X, High Throughput Satellites, Ultra HD, M2M, and IP, First Edition. Daniel Minoli.
© 2015 John Wiley & Sons, Inc. Published 2015 by John Wiley & Sons, Inc.

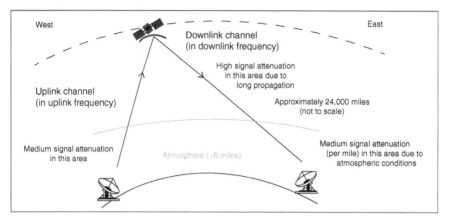

Figure 1.1 A typical satellite link.

applications, was already advocated by industry observers in the late 1970s (e.g., but certainly not only [MIN197901]) and by the early industry savants (e.g., but not only [ROS198201], [ROS198401]).

1.1 BACKGROUND

Satellite communication is based on a line-of-sight (LOS) one-way or two-way radio frequency (RF) transmission system that comprises a transmitting station utilizing an uplink channel, a space-borne satellite system acting as a signal regeneration node, and one or more receiving stations monitoring a downlink channel to receive information. In a two-way case, both endpoint stations have the transmitting and the receiving functionality (see Figure 1.1).

Satellites can reside in a number of near-earth orbits. The geostationary orbit (GSO) is a concentric circular orbit in the plane of the earth's equator at 35,786 km (22,236 miles) of altitude from the earth's surface (42,164 km from the earth's center – the earth's radius being 6,378 km). A geosynchronous (GEO) satellite[1] circles the earth in the GSO at the earth's rotational speed and in the same direction as the rotation. When the satellite is in this equatorial plane it effectively appears to be permanently stationary when observed at the earth's surface, so that an antenna pointed to it will not require tracking or (major) positional adjustments at periodic intervals of time.[2,3] Other orbits are possible, such as the medium Earth orbit (MEO) and the low Earth orbit (LEO).

---

[1] In this book whenever we use the term satellite we mean a geostationary communications satellite, unless noted otherwise by the context.

[2] In practice, the term geosynchronous and geostationary are used interchangeably.

[3] A GSO is a circular prograde orbit (prograde is an orbital motion in the same direction as the primary rotation) in the equatorial plane, with an orbital period equal to that of the earth.

Traditionally, satellite services have been officially classified into the following categories:

- *Fixed Satellite Service (FSS):* This is a satellite service between satellite terminals at specific fixed points using one or more satellites. Typically, FSS is used for the transmission of video, voice, and IP data over long distances from fixed sites. FSS makes use of geostationary satellites with fixed ground stations. Signals are transmitted from one point on the globe either to a single point (point-to-point) or from one transmitter to multiple receivers (point-to-multipoint). FSS may include satellite-to-satellite links (not commercially common) or feeder links for other satellite services such as the Mobile Satellite Service or the Broadcast Satellite Service.

- *Broadcast Satellite Service (BSS):* This is a satellite service that supports the transmission and reception via satellite of signals that are intended for direct reception by the general public. The best example is Direct Broadcast Service (DBS), which supports direct broadcast of TV and audio channels to homes or business directly from satellites at a defined frequency band. BSS/DBS makes use of geostationary satellites. Unlike FSS, which has both point-to-point and point-to-multipoint communications, BSS is only a point-to-multipoint service. Therefore, a smaller number of satellites are required to service a market.

- *Mobile Satellite Service (MSS):* This is a satellite service intended to provide wireless communication to any point on the globe. With the broad penetration of the cellular telephone, users have started to take for granted the ability to use the telephone anywhere in the world, including rural areas in developed countries. MSS is a satellite service that enhances this capability. For telephony applications, a specially configured handset is needed. MSS typically uses satellite systems in MEOs or LEOs.

- *Maritime Mobile Satellite Service (MMSS):* This is a satellite service between mobile satellite earth stations and one or more satellites.

- While not formally a service in the regulatory sense, one can add Global Positioning (Service/) System (GPS) to this list; this service uses an array of satellites to provide global positioning information to properly equipped terminals.

A number of technical and service advances affecting commercial satellite communications have been seen in the past few years; these advances are the focus of this textbook. Spectral efficiencies are being vigorously sought by end-users in order to sustain the business case for content distribution as well as interactive voice (VoIP) and Internet traffic. At the same time, to sustain sales growth, operators need to focus on delivering IP services (enterprise and Internet access), on next-generation video (hybrid distribution, caching, nonlinear/time shifting, higher resolution), and on mobility. Some of the recent technical/service advances include the following:

- Business factors impacting the industry at press time included the desire for higher overall satellite channel and system throughput. Improved modulation schemes allow users to increase channel throughput: advanced modulation and coding (modcod) techniques being introduced as standardized solutions embedded in next-generation modems provide more bits per second per unit of spectrum, and adaptive coding enables more efficient use of the higher frequency bands that are intrinsically susceptible to rain fade; extensions to the well-established baseline DVB-S2 standard are now being introduced.

- High throughput is also achieved via the use of spotbeams on High Through-put Satellites (HTSs), typically (but not always) operating at Ka-band (18.3–20.2 GHz for downlink frequencies and 28.1–30 GHz for uplink frequencies), and via the reduction of transmission latency utilizing MEO satellites. HTSs are capable of supporting over 100 Gbps of raw aggregate capacity and, thus, significantly reducing the overall per-bit costs by using high-power, focused spot beams. HTS systems and capabilities can be lever-aged by service providers to extend the portfolio of satcom service offerings. HTSs differ from traditional satellites in a number of ways, including the utilization of high-capacity beams/transponders of 100 MHz or more; the use of gateway earth stations supporting one or two dozen beams (typically with 5 Gbps capacity requirement); high per-station throughputs for all the remote stations; and advanced techniques to address rain attenuation, especially for Ka-band systems.

- Providing connectivity services to people on-the-move, for example, for people traveling on ships or airplanes, where terrestrial connectivity is lacking, is now both technically feasible and financially advantageous to the service providers. When that is desired on fast-flying planes, special antenna design considerations (e.g., tracking antennas) have to be taken into account.

- M2M (machine-to-machine) connectivity, whether for trucks on transconti-nental trips, or aircraft real-time-telemetry aggregation, or mercantile ship data tracking, opens up new opportunities to extend the Internet of Things (IoT) to broadly distributed entities, particularly in oceanic environments. With the increase in global commerce, some see increasing demands on maritime communication networks supported by satellite connectivity; sea-going communications requirements can vary by the type of vessel, type of operating company, data volumes, crew and passenger needs, and application (including the GMDSS [Global Maritime Distress and Safety System]), so that a number of solutions may be required or applicable.

- Emerging Ultra High Definition Television (UHDTV) (also known as Ultra HD or UHD) provides video quality that is equivalent to 8-to-16 HDTV screens; clearly, this requires a lot more bandwidth per channel than currently used in video transmission. So-called 4 K and 8 K versions are emerging, based on the vertical resolution of the video. Satellite operators are planning to position themselves in this market segment, with generally-available broadcast services planned for 2020, and more targeted transmission starting as of press time. UHDTV will require a bandwidth of approximately 60 Mbps for distribution services and 100 Mbps for contribution services. The use of DVB-S2 extensions (and possibly wider transponders, e.g., 72 MHz) will be a general requirement. The newly emerging H.265/HEVC (High Efficiency Video Coding) video compression standard (algorithm) provides up to 2x better compression efficiency compared with that of the baseline H.264/AVC (Advanced Video Coding) algorithm; however, it also has increased computational complexity requiring more advanced chip sets. Many demonstrations and simulations were developed in recent years, especially in 2013, and commercial-grade products were expected in the 2014–2015 time frame, just in time for Ultra HD applications (both terrestrial and satellite based). Even in the context of Standard Definition (SD)/High Definition (HD) video, upgrading ground encoding equipment by content providers to H.264 HEVC reduces the bandwidth requirements (and, hence, recurring expenditures) by up to 50%. Upgrading from DVB-S2 to DVB-S2 extensions (DVB-S2X) can reduce bandwidth by an additional 10–60%.

- Hybrid networks combining satellite and terrestrial (especially IP) connectivity have an important role to play in the near future. The widespread introduction of IP-based services, including IP-based television (IPTV) and over-the-top (OTT) video, driven by continued deployment of fiber connectivity, will ultimately reshape the industry. In particular, IP version 6 (IPv6) is a technology now being deployed in various parts of the world that will allow true explicit end-to-end device addressability. The integration of satellite communication and IPv6 capabilities promises to provide a networking hybrid infrastructure that can serve the evolving needs of government, military, IPTV, and mobile video stakeholders, to name just a few.

- At the core technology level, electric (instead of chemical) propulsion is being investigated and, in fact, being deployed; such propulsion approaches can reduce spacecraft weight (and so launch cost) and possibly extend the spacecraft life. According to proponents, the use of electric propulsion for satellite station-keeping has already changed the global satellite industry, and now, with orbit-topping and orbit-raising, it is poised to transform it.

- In addition, new launch platforms are being brought to the market, again with the goal of lowering launch cost via increased competition.

It is thus self-evident that satellite communications play and will continue to play a key role in commercial, TV/media, government, and military communications because of its intrinsic multicast/broadcast capabilities, mobility aspects, global reach, reliability, and ability to quickly support connectivity in open-space and/or hostile environments. This text surveys some of these new key advances and what the implications and/or opportunities for end-users and service providers might be. The text is intended to be generally self-contained; hence, some background technical material is included in this introductory chapter.

1.2   INDUSTRY ISSUES AND OPPORTUNITIES: EVOLVING TRENDS

1.2.1   Issues and Opportunities

Expanding on the observations made in the introductory section, it is instructive to assess some of the general industry trends as of the mid-decade, 2010s. Observations such as these partially characterize the environment and the trends:

> " ... a change of the economics of the industry [is] key to the satellite sector's long-term growth ... We have to be more relevant, more efficient, we have to push the boundaries ... the industry needs to both drive down costs and expand the market through innovation ... "

[WAI201401];

> " ... the satellite market is seeing dramatic change from the launch of new high-throughput satellites, to the dramatic drop in launch costs brought on by gutsy new entrants ... more affordable and reliable launch options [are becoming available] ... "

[WAI201401];

> " ... it is an open question as to who will see the fastest rate of growth. Will the top four continue to score the big deals and push further consolidation, or will the pendulum swing to the regional players with their closer relationships to the domestic/national client bases and launch of new "national flag" satellites? ... "

[GLO301301];

> " ... to combine with its wireless, phone and high-speed broadband Internet services as competition ramps up; the pool of pay-TV customers is peaking in the U.S. [and in Europe] because viewers are increasingly watching video online ... "

[SHE201401];

> " ... The TV and video market is experiencing a dramatic shift in the way content is accessed and consumed, that should see no turning back. Technical innovation and the packaging of new services multiply the ... interactions between content and viewers ... Four major drivers are impacting the way content is managed from its production to its distribution and monetization. They are:

> - Delinearization of content consumption, and multiplication of screens and networks to access content;
> - Faster increase in competition than in overall revenues; new sources of content, inter-mediaries and distributors challenge the value chain;
> - Shorter innovation and investment cycles to meet customer expectations; and
> - Fast growth in emerging regions opening new growth opportunities at the expense of an increasing customization to local needs ... "

[BUC201401].

> " ... seeking innovative satellite and launcher configurations is an absolute must for the satellite industry if it expects to remain competitive against terrestrial technologies ... the cost of the satellite plus the cost of the launcher will ... deliver a 36-megahertz-equivalent transponder into orbit for $1.75 million ... "

[SEL201402];

" ... Many factors can disrupt the market and have an impact on competition, demand, or pricing ... FSS cannibalization of MSS, emergence of new competitors in the earth observation arena, changes to the U.S. Government's behavior as a customer, growing competition from emerging markets, growth of government (globally) as a source for satellite financing, and adoption of a 4 K standard for DBS ... "

[SAT201401];

"Innovation and satellite manufacturing are not always words that end up in the same sentence ... Due to the expensive nature, risk aversion and technical complexity, innovation has been fairly slow in satellite communications ... "

[PAT201301];

" ... After the immediate, high-return investments have been done, new growth initiatives are either higher risk or lower return ... Is industry maturity itself a disruptor, forcing experiments that fall beyond the risk frontier? ... "

[SAT201401];

"Higher speeds, more efficient satellite communication technology and wider transponders are required to support the exchange of large and increasing volumes in data, video and voice over satellite. Moreover, end-users expect to receive connectivity anywhere anytime they travel, live or work. The biggest demand for the extensions to the DVB-S2 standard comes from video contribution and high-speed IP services, as these services are affected the most by the increased data rates ... "

[WIL201401];

"The latest market figures confirm that broadband satellites or so-called high-throughput satellite systems are on the rise. As the total cumulative capital expenditures in high-throughput satellites climbs to $12 billion, an important question must be raised: How will these new systems impact the mindset of our industry? ... The large influx of this capacity to the market has created some concerns about the risk of oversupply in regions such as Latin America, the Middle East and Africa, and Asia Pacific ... "

[DER201301];

" ... [high-throughput satellite] are designed to transform the economics and quality of service for satellite broadband ... satellites can serve the accelerating growth in bandwidth demand for multimedia Internet access over the next decade ... Current satellite systems are not designed for the high bandwidth applications that people want, such as video, photo sharing, VoIP, and peer-to-peer networking. The solution is to increase the capacity and speed of the satellite. Improving satellite service is not just about faster speeds, but about increasing the bandwidth capacity available to each customer on the network to reduce network contention ... "

[VIA201401];

"Procurement of commercial GEO communications satellites will remain stable over the next 10 years. While the industry will experience a short term decline from a high in 2013 ... it will remain driven by replacements and some extensions primarily in Ku-band and HTS. However, a number of trends will affect the growth curve and considerably change the trade-off environment for satellite manufacturing:

- New propulsion types increasingly used;
- More platforms proposed by an increasing number of suppliers;

- *Multi-beam architectures becoming more frequent;*
- *Launch services capabilities evolve toward higher masses.*

The whole industry is shaping-up; including both satellite manufacturers and launch services providers … Market shares evolved significantly in the last few years and after years of complacency, certain players were bordering on insignificance in this important space … "

[EDI201401];

" … We live in a smart, connected world. The number of things connected to the Internet now exceeds the total number of humans on the planet, and we're accelerating to as many as 50 billion connected devices by the end of the decade … the implications of this emerging "Internet of Things (IoT)" are huge. According to a recent McKinsey Global Institute report, the IoT has the potential to unleash as much as $6.2 trillion in new global economic value annually by 2025. The firm also projects that 80 to 100 percent of all manufacturers will be using IoT applications by then, leading to potential economic impact of as much as $2.3 trillion for the global manufacturing industry alone … "

[HEP201401].

" … The report has the following key findings:

- *The wireless M2M market will account for nearly $196 Billion in annual revenue by the end of 2020, following a Compound Annual Growth Rate (CAGR) of 21% during the six year period between 2014 and 2020;*
- *The installed base of M2M connections (wireless and wireline) will grow at a CAGR of 25% between 2014 and 2020, eventually accounting for nearly 9 Billion connections worldwide;*
- *The growing presence of wireless M2M solutions within the sensitive critical infrastructure industry is having a profound impact on M2M network security solutions, a market estimated to reach nearly $1.5 Billion in annual spending by the end of 2020;*
- *Driven by demands for device management, cloud based data analytics and diagnostic tools, M2M/IoT platforms (including Connected Device Platforms [CDP], Application Enablement Platforms [AEP], and Application Development Platforms [ADP]) are expected to account for $11 Billion in annual spending by the end of 2020 … "*

[SST201401].

Kevin Ashton known for coining the term "The Internet of Things" to describe a system where the Internet is connected to the physical world via ubiquitous sensors [MIN201301], recently made these very cogent observations, which given the depth are quoted here (nearly) in full:

"Yesterday [April 27, 2014], the aerial search for floating debris from Malaysia Airlines Flight 370 was called off, and an underwater search based on possible locator beacon signals was completed without success … The more than 50-day operation, which the Australian prime minister, Tony Abbott, calls "probably the most difficult search in human history," highlights a big technology gap. We live in the age of what I once called "the Internet of Things," where everything from cars to bathroom scales to Crock-Pots

*can be connected to the Internet, but somehow, airplane data systems are barely con-
nected to anything ... the plane's Aircraft Communications Addressing and Reporting
System (ACARS), which was invented in the 1970s and is based on telex, an almost
century-old ancestor of text messaging made essentially obsolete by fax machines ...
When so much is connected to the Internet, why is the aerospace industry using technology that predates fax machines to look for flash drives in the sea?*

*Because, while technology for communicating from the ground has advanced rapidly
in the last 40 years, technology for communicating from the sky has been stuck in
the 1970s. The problem starts not with planes, but with the satellites that track them.
The Sentinel-1A satellite, for example, weighs two and a half tons, costs around $400
million, and was launched on a rocket designed in Soviet Russia in the 1960s. The
Sentinel can store the same amount of data as seven iPhones. When was this relic
from the age of mainframe computers sent into orbit? On April 3. Huge, expensive,
rocket-launched satellites with little computing power may make sense for broadcasting, where one satellite sends one signal to lots of things (such as television sets) but
they are generally too expensive and not intelligent enough to be part of the Internet,
where lots of things (such as airplanes) would send lots of signals to one satellite. This is
why most satellites reflect TV signals, take pictures of the earth, or send the signals that
drive GPS systems. It is also the reason airplanes can't stream flight and location data
like they stream vapor trails: cellphone and Wi-Fi signals don't reach the ground from
30,000 feet, so airplanes need to be able to send information to satellites — satellites
that, as well as being unable to handle network data economically, are designed to talk
to rotating, dish-shaped antennas that would be impossible to retrofit to airplanes.*

*The solution to these problems is simple: We need new satellite technology. And it's
arriving. Wealthy private investors and brilliant young engineers are dragging satellites into the 21st century with inventions including "flocks" of "nanosatellites" that
weigh as little as three pounds; flat, thin antennas built from advanced substances called
"metamaterials"; and "beamforming," which steers radio signals using software. On
January 9, 2014a San Francisco-based start-up called Planet Labs sent a flock of 28
nanosatellites into space. The first application for this type of technology is taking pictures of the Earth, but it could also be used to receive data streaming from aircraft
retrofitted with those new, flat "metamaterial" antennas. There are many other possible systems. Dozens of new satellite technologies are emerging, with countless ways to
combine them. Streaming data from planes is about to become cheap and easy"*

[ASH201401].

1.2.2   Evolving Trends

The satellite[4] industry comprises spacecraft manufacturers, launch entities, satellite
operators, and system equipment developers. Major operators include *Eutelsat, Intelsat, SES, and Telesat*; a cadre of national-based providers (particularly in the context

---

[4]Services, service options, and service providers change over time (new ones are added and existing ones
may drop out as time goes by); as such, any service, service option, or service provider mentioned in
this chapter is mentioned strictly as illustrative examples of possible examples of emerging technologies,
trends, or approaches. As such, the mention is intended to provide pedagogical value. No recommendations
are implicitly or explicitly implied by the mention of any vendor or any product (or lack thereof).

of the BRICA countries: Brazil, Russia, India, China, and Africa) also exist. The industry's combined revenue had a growth at around 3–4% in 2013; emerging markets and emerging applications will be a driver for continued and/or improved growth: industry observers state that 80% of future growth in satellite services demand will be in the southern hemisphere, although the "quality" of the revenues in those areas does not compare with that of the developed countries. Generally, operators worldwide launch about two dozen commercial communications satellites a year.

As hinted above, mobility for commercial users represents a major business opportunity for satellite operators. Some industry observers see a practical convergence of what were officially MSS and FSS. Many of the evolving mobility services are based on satellites supporting principally (or in part) FSS. Many airlines are planning to retrofit their airplanes to offer in-flight connectivity services. As an illustrative example, *El Al Israel Airlines* announced in 2014 that it was outfitting its *Boeing* 737 aircraft fleet to provide in-flight satellite broadband using *ViaSat's Exede in the Air* service enabled by *Eutelsat's* KA-SAT Ka-band satellite starting in 2015. (*Eutelsat's* KA-SAT satellite covers almost all of Europe, the Middle East, parts of Russia, Central Asia, and the Eastern Atlantic.) Passengers will be offered several Internet service options, including one free service, to connect their laptops, tablets, or smartphones to the Internet; *Exede in the Air* is reportedly able to deliver 12 Mbps capacity to each passenger, a rate the company says is irrespective of the number of users on a given plane [SEL201403]. To provide a complete in-flight Internet service to the aircraft, airline companies need to add airborne terminals, tracking antennas, and radomes to the aircraft, and also subscribe to satellite-provided channel bandwidth (air time) on selected satellites.

New architectures related to how satellites are designed are also emerging. It is true that until recently satellite operators have shown tepid interest for an all-electric propulsion satellite, principally because with this type of propulsion technology it takes months, rather than weeks, for a newly launched spacecraft to reach the final GSO operating position; satellite operators have indicated they are also concerned that, when they wish to move their in-orbit satellites from one slot to another during the satellite's 15- to 20-year service lifecycle, such maneuvers will take much longer (these drifts are very common, enabling the operator to address bandwidth needs in various parts of the world, as these needs arise – in some cases up to 25% of an operator's fleet may be in some sort of drift state) [SEL201402]. However, there are ostensibly certain economic advantages to electronic-propulsion spacecraft, which can bring transponder bandwidth costs down, thereby opening up new markets and applications. Some key operators have recently announced renewed interest in the technology. A large, complex spacecraft can weigh more than 6000 kilograms; a reduction in weight will significantly reduce launch costs; electric propulsion can result in a spacecraft that weigh around 50% of what it would weigh at launch with full chemical propellant. Some industry observers expect to see the emergence of a hybrid solution that saves some of the launch mass of a satellite through electric propulsion, but retains conventional chemical propellant to speed the arrival of the satellite to final operating position.

At the end-user level, there have been a number of technological developments, including the extensions for DVB-S2, tighter filter rolloffs in modems, adaptive pre-correction for transmission equipment nonlinearity as well as for group delay. These modem developments increase the bandwidth achievable in a channel (augmenting the application's scope) or reduce the amount of analog spectrum needed to support a certain datarate (thus, reducing the cost of the application).

An area of possible new business and technical opportunities entails the "hosted payload" concept. NASA and the US Department of Defense (DoD) have begun to look at the commercial space sector for more cost-effective solutions compared with that of proprietary approaches, including the use of hosted payloads. Spacecraft constructed for commercial services can be designed for additional payload capacity in the area of mass, volume, and power. This capacity can be used to host additional (government) payloads, such as communications transponders, earth observation cameras, or technology demonstrations. These "hosted payloads" can provide government agencies with capabilities at a fraction of the cost of a dedicated satellite and also provide satellite operators with an additional source of revenue [FOU201201]. A handful of GEO communications spacecraft launched in recent years incorporated hosted payloads, and there are indications that hosted payloads are gaining broader acceptance as government agencies are increasingly challenged to do more with reduced funding. Recently the US Air Force's Space and Missiles Systems Center (SMC) formed a Hosted Payload Office to better coordinate hosted payload opportunities across the government. The Air Force previously launched the Commercially Hosted InfraRed Payload (CHIRP) as a hosted payload on a communications satellite to test a new infrared sensor for use by future missile warning systems. The Air Force is reportedly planning to build on the success of CHIRP with a follow-on program called CHIRP+, again using hosted payloads to test infrared sensors. In another example, the Australian Defense Force placed a hosted payload on a commercial satellite, Intelsat 22, to provide UHF communications for military forces. The government reportedly saved over $150 million over alternative approaches, and the hosted payload approach was 50% more effective economically than flying the payload as its own satellite, and it was 180% more efficient than leasing the capacity. Some other examples of hosted applications include *EMC-Arabsat* and *GeoMetWatch-Asiasat*. The benefits of hosted payloads are measurable, although there may be institutional challenges for both satellite operators and potential government customers to work through the procurement and integration issues.

There is a new category of space technology developing called logistics in space. Services include life extension, tug, inclination removal, hosted payloads, and perhaps fuel transfer. ViviSat is an example of a company seeking to provide on-orbit servicing for GSO satellites.

Satellite-based M2M technology and services are receiving increased attention of late. As mentioned in the previous section, there is an urgent need, for example, to modernize the global airplane fleet to reliably support multifaceted, reliable, seamless tracking of aircraft function, status, and location. It would be expected that such basic safety features would be affirmatively mandated in the future by global aeronautical regulators (e.g., International Civil Aviation Organization [ICAO]). A satellite-based

M2M antenna and modem cost around $125. While a terrestrial cellular system typically costs only $50, such a system does not have the full reach of a satellite-based solution, especially on a global scale. Work is underway to reduce the satellite-based system to $90. Providers include *Inmarsat, Iridium, Orbcomm,* and *Globalstar.* The current satellite share of the global M2M market has been estimated at 5% (each percentage point representing about 100,000 installed units); the expectation is for the growth in this segment in the near future. For example, satellite M2M messaging services provider *Orbcomm* launched a second-generation satellite constellation of 17 spacecraft on two Falcon 9 rockets operated by *Space Exploration Technologies Corporation* in 2014. The new-generation satellites will be backward compatible with existing *Orbcomm* modems and antennas, used to track the status of fixed and mobile assets, but the new satellites have six times more receivers on board than do the current spacecraft, and offer twice the message delivery speed (the second-generation constellation will have about 100 times the overall capacity of the existing satellites) [SEL201404]. Satellite M2M expands the connections enabled by terrestrial cellular networks not only domestically but also over land and sea.

Small special-purpose satellites (called smallsats and also called microsatellites or nanosatellites) are being assessed as an option by some operators. These satellites weigh in the range of 1–10 kg. Smallsats offer mission flexibility, lower costs, lower risk, faster time to orbit, and reduced operational and technical complexity. These satellites can be used for Geographic Information Systems, space science, satellite communication, satellite imagery, remote sensing, scientific research, and reconnaissance. Further along this continuum, one finds what are called picosatellites (e.g., CubeSats) that can perform a variety of scientific research and explore new technologies in space. Advances in all areas of smallsat technologies will enable such satellites to function within constellations or fractionated systems and to make cooperative measurements. Constellations of satellites enable whole new classes of missions for navigation, communications, remote sensing, and scientific research for both civilian and military purposes. The scope and affordability of such multispacecraft missions are closely tied to the capabilities, manufacturability, and technical readiness of their components, as well as the diverse launch opportunities available today [SMA201301]. Earth observations and remote sensing are expected to account for largest market share by 2019; their use in commercial communication applications is also being investigated.

In 2014, Google announced plans to deploy 180 small LEO satellites to provide Internet access to underserved regions of the globe. Google had previously invested in the O3b Networks initiative, but apparently it has been looking for another entry mechanism into the satellite space. The project may require several billion dollars to complete. Google has been seeking ways to provide Internet access to developing regions without investing in expensive ground-based infrastructure; for example, they unveiled Project Loon to deliver Internet via solar-powered, remote-controlled air balloon. Details of the project were not generally available at press time, but the satellites under discussion may be of the smallsat type or just a notch above these.

New antenna designs are also emerging. For example, Panasonics uses a phased array antenna for in-flight Internet connectivity. As another example, in general, some

MEO-HTS systems require two tracking antennas; however, some manufacturers (e.g., Kymeta) are developing a flat panel metamaterials antenna that is capable of tracking and instantaneously switching between satellites. As of press time Kymeta demonstrated the receive capability, and next the plan was to demonstrate a transmit capability. This antenna is currently Ka-band, but the vendor was reportedly planning a Ku-band version.

Some developers are assessing the opportunities of what has been called 'cognitive satellite communication systems' (also known as "CoRaSat" [Cognitive Radio for Satellite Communications]). These systems are planned to have the ability to automatically detect and respond to impairments of the transmission channel, such as, but not limited to Adjacent Satellite Interference (ASI), Adjacent Channel Interference (ACI), rain fade, and terminal RF performance variations due to a number of causes. These advancements build on the more general concept of 'cognitive radios', where dynamic spectrum management is employed to deal with the scarcity of the overall transmission spectrum. Since the allocated satellite spectrum is becoming scarce in various parts of the world due to growing demand for broadcast, multimedia, terrestrial mobility services, and interactive Internet services, applying efficient spectrum sharing techniques for enhancing spectral efficiency in satellite communication has become an important topic of late. Cognitive techniques such as Spectrum Sensing (SS), interference modeling, beamforming, interference alignment, and cognitive beamhopping are being assessed with the goal of possible implementation in the near future.

Satellite backhauling of regional 2G, 2.5G, 3G, and 4G/LTE services in underdeveloped areas (such as Africa and South America) is an emerging and evolving application. Satellite is also being utilized as the primary links in Mobile Network Operators' (MNOs') core backbones and for restoration services where fiber and cable are used as the core. The goal is to enable MNOs' mobile to reach more users while lowering overall cost to deliver these services. The addition of tower-based caching from satellite distribution can also support some video-on-demand services to smartphones in these regions of the world.

Another critical area for operators is the availability of spectrum; this relates to permits to transmit at specific frequency bands from specific orbital slots, as well as "landing rights," which are provisions for authorizing the use of signals of foreign satellite services in specified countries. A key area of discussion at the 2015 World Radiocommunication Conference (WRC) is the wireless terrestrial operators' effort to try to obtain C-band spectrum for evolving 4G/5G applications. Studies have shown that International Mobile Telecommunications (IMT) services interfere with fixed service satellites, which accounts for 38% of the satellite mix (not counting HTS) [WAI201401, SEL201401].

Satellite networks cannot continue to exist as stand-alone islands in a web of (required) anytime/anywhere/any media/any device connectivity. It follows that hybrid networks have an important role to play. The widespread introduction of IP-based services, including IP-based content distribution, will necessarily drive major changes in the industry during the next 10 years. The integration of satellite communication and IP capabilities (particularly IPv6) promises to provide a more symbiotic networking infrastructure that can better serve the evolving needs of the traveling public, the business enterprise, the government and military, and the IPTV/OTT/Content Delivery Network (CDN) players. Operators will need to become much more knowledgeable of IP to remain relevant.

At the business level, industry consolidation continues unabated. Two major consolidations in recent memory include AT&T's purchase of DirectTV and Comcast purchase of Time Warner Cable. The claim is made that the acquiring companies were trying to get ahead of key business trends: rising content costs, the need for scale, the growing importance of broadband, and the growing use of video on mobile platforms were the driving forces behind these deals. While AT&T has focused on Internet services, they view "the future [as being] about delivering video at scale," including satellite-based video and internet, all-of-the-above OTT video, and mobility [FAB201401]. The goals of these mergers relate to "synergies": more efficiency in delivering services at a lower internal cost. In the United States, Comcast's purchase of Time Warner Cable and AT&T's purchase of DirectTV are seen as a "reordering of the video landscape that would seem to set both companies in motion toward more significant competition." Other players have a need to reduce costs by re-engineering a number of functions, including nonrevenue producing ground assets, which are often excessively redundant in functionality and well beyond what would be needed to maintain business continuity. The expectation is that similar trends will affect other satellite/video operators worldwide in the next few years.

From a regulatory perspective, on May 13, 2014, the US State Department and the US Commerce Department published final rules transferring certain satellites and components from the US Munitions Import List (USMIL) to the Commerce Control List (CCL). These rules are the product of work between the Administration and Congress, in consultation with industry, to reform the regulations governing the export of satellites and related items. These changes will more appropriately calibrate controls to improve the competitiveness of American industry while ensuring that sensitive technology continues to be protected to preserve national security. The changes to the controls on radiation-hardened microelectronic microcircuits take effect 45 days after publication of the rule, while the remainder of the changes takes effect 180 days after publication. Earlier in the year, the Department of Justice published a Final Rule that revises the USMIL as part of the President's Export Control Reform (ECR) Initiative. These changes remove defense articles that were on the USMIL but no longer warrant import control under the Arms Export Control Act, allowing enforcement agencies to focus their efforts where they are most needed. This important reform modernizes the USMIL and will promote greater security [ECR201401].

Figure 1.2 depicts a general timeline for some of the key advances impacting the industry.

Advances and innovation in satellite communications and satellite are not limited to the topics listed earlier, although these are the more visible initiatives at this juncture. The topics discussed in these introductory overview and trends sections will be assessed in greater details in the chapters that follow. The rest of this chapter provides a primer on satellite communications.

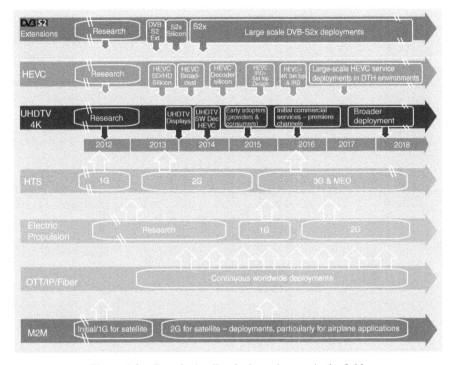

Figure 1.2   Generic timeline for key advances in the field.

1.3   BASIC SATELLITE PRIMER

This section covers some basic concepts in satellite communications; to make this treatise relatively self-contained.

1.3.1   Satellite Orbits

Communications satellites travel around the earth ("fly" in the jargon) in the well-defined orbits. Table 1.1 (loosely synthetized from references [GEO200101] and [SAT200501]) lists some key concepts related to orbits, expanding on the brief introduction to orbits provided earlier. Figure 1.3 illustrates graphically the various satellite orbits that are in common use. Most of the commercial satellites discussed in this text reside in the GSO. At the practical level, the GSO has small nonzero inclination and eccentricity, which causes the satellite to trace out a small but manageable "figure eight" in the sky. During normal operations, satellites are "station-kept" within a defined "box" around the assigned orbital slot; eventually (around the end-of-life event of a satellite – usually 15–18 years after launch) the satellite is "allowed" to enter an inclined orbit in the proximity of the assigned orbital slot (unless moved to some other maintenance position): north–south maneuvers to keep the spacecraft at the center box are not undertaken (to save fuel), but east–west

TABLE 1.1 Key Concepts Related to Orbits

Term	Description
Circular orbit	A satellite orbit where the distance between the center of mass of the satellite and of the earth is constant.
Clarke belt	The circular orbit (geostationary orbit) at approximately 35,786 km above the equator, where the satellites travel at the same speed as the earth's rotation and thus appear to be stationary to an observer on earth (named after Arthur C. Clarke who was the first to describe the concept of geostationary communication satellites).
Collocated satellites	Two or more satellites occupying approximately the same geostationary orbital position such that the angular separation between them is effectively zero when viewed from the ground. To a small receiving antenna, the satellites appear to be exactly collocated; in reality, the satellites are kept several kilometers apart in space to avoid collisions. Different operating frequencies and/or polarizations are used.
Dawn-to-dusk orbit	A special sun-synchronous (SS) orbit where the satellite trails the earth's shadow. Because the satellite never moves into this shadow, the sun's light is always on it. These satellites can, therefore, rely mostly on solar power and not on batteries; they are useful for agriculture, oceanography, forestry, hydrology, geology, cartography, and meteorology.
Geostationary orbit (GSO)	Geostationary orbits are circular orbits that are orientated in the plane of the earth's equator. A geostationary satellite completes one orbit revolution around the earth every 24 h; hence, given that the satellite spacecraft is rotating at the same angular velocity as the earth, it overflies the same point on the globe on a permanent basis (unless the satellite is repositioned by the operator). In a geostationary orbit, the satellite appears stationary, that is, in a fixed position, to an observer on the earth. The maximum footprint (service area) of a geostationary satellite covers almost one-third of the earth's surface; in practice, however, except for the oceanic satellites, most satellites have a footprint optimized for a continent and/or a portion of a continent (e.g., North America or even continental Unites States).
Geostationary satellite	A satellite orbiting the earth at such speed that it permanently appears to remain stationary with respect to the earth's surface.
Highly elliptical orbits (HEO)	HEOs typically have a perigee (point in each orbit that is closest to the earth) at about 500 km above the surface of the earth and an apogee (the point in its orbit that is farthest from the earth) as high as 50,000 km. The orbits are inclined at $63.4°$ in order to provide communications services to locations at high northern latitudes. Orbit period varies from 8 to 24 h. Owing to the high eccentricity of the orbit, a satellite spends about two-thirds of the orbital period near apogee, and during that time it appears to be almost stationary for an observer on the earth (this is referred to as apogee dwell). A well-designed HEO system places each apogee to correspond to a service area of interest. After the apogee period of

orbit, a switchover needs to occur to another satellite in the same orbit in order to avoid loss of communications. Due to the relatively large movement of a satellite in HEO with respect to an observer on the earth, satellite systems using this type of orbit need to be able to cope with Doppler shifts. An example of HEO system is the Russian Molnya system; it employs three satellites in three 12-h orbits separated by 120° around the earth, with apogee distance at 39,354 km and perigee at 1000 km.

Inclination	The angle between the plane of the orbit of a satellite and the earth's equatorial plane. An orbit of a perfectly geostationary satellite has an inclination of 0.
Inclined orbit	An orbit that approximates the geostationary orbit but whose plane is tilted slightly with respect to the equatorial plane. The satellite appears to move about its nominal position in a daily "figure-of-eight" motion when viewed from the ground. Spacecrafts (satellites) are often allowed to drift into an inclined orbit near the end of their nominal lifetime in order to conserve onboard fuel, which would otherwise be used to correct this natural drift caused by the gravitational pull of the sun and moon. North–south maneuvers are not conducted, allowing the orbit to become highly inclined.
Low earth orbit (LEO)	LEOs are either elliptical or (more commonly) circular orbits that are at a height of 2000 km or less above the surface of the earth. The orbit period at these altitudes varies between 90 min and 2 h and the maximum time during which a satellite in LEO orbit is above the local horizon for an observer on the earth is up to 20 min. With LEOs there are long periods during which a given satellite is out of view of a particular ground station; this may be acceptable for some applications, for example, for earth monitoring. Coverage can be extended by deploying more than one satellite and using multiple orbital planes. A complete global coverage system using LEOs requires a large number of satellites (40–80) in multiple orbital planes and in various inclined orbits. Most small LEO systems employ polar or near-polar orbits. Due to the relatively large movement of a satellite in LEO with respect to an observer on the earth, satellite systems using this type of orbit need to be able to cope with Doppler shifts. Satellites in LEOs are also affected by atmospheric drag that causes the orbit to deteriorate (the typical life of a LEO satellite is 5–8 years, while the typical life of a GEO satellite is 14–18 years). However, launches into LEO are less costly than to the GEO orbit and due to their much lighter weight, multiple LEO satellites can be launched at one time.
Medium earth orbits/intermediate circular orbit (MEO/ICO)	These are circular orbits at an altitude of around 10,000 km. Their orbit period is in the range of 6 h. The maximum time during which a satellite in MEO orbit is above the local horizon for an observer on the earth is in the order of a couple of hours. A global communications system using this type of orbit requires a small number of satellites in two or three orbital planes to achieve global coverage. The US GPS is an example of a MEO system.

(continued)

TABLE 1.1 *(Continued)*

Molniya orbits	See highly elliptical orbits (HEO). A Molniya orbit (named after a series of Soviet/Russian Molniya communications satellites, which have been using this type of orbit since the mid-1960s) is a type of HEO with an inclination of 63.4°, an argument of perigee of −90°, and an orbital period of one half of a sidereal day.
Orbit	The path described by the center of mass of a satellite in space, subjected to natural forces, principally gravitational attraction, but occasional low-energy corrective forces exerted by a propulsive device in order to achieve and maintain the desired path.
Orbital plane	The plane containing the center of mass of the earth and the velocity vector (direction of motion) of a satellite.
Polar orbit	Polar orbits are LEO orbits that are in a plane of the two poles. Their applications include the ability to view only the poles (e.g., to fill in gaps of GEO coverage) or to view the same place on earth at the same time each 24-h day. By placing a satellite at an altitude of about 850 km, a polar orbit period of about 100 min can be achieved (for more continuous coverage, more than one polar orbiting satellite is employed).
Sun-synchronous (SS) orbit	A special polar orbit that crosses the equator and each latitude at the same time each day is called a sun-synchronous orbit; this orbit can make data collection a convenient task. Satellites in polar orbits are mostly used for earth-sensing applications. Typically, such a satellite moves at an altitude of 1000 km. In an SS orbit, the angle between the orbital plane and sun remains constant. This orbit can be achieved by an appropriate selection of orbital height, eccentricity, and inclination that produces a precession of the orbit (node rotation) of approximately 1° eastward each day, equal to the apparent motion of the sun; this condition can only be achieved for a satellite in a retrograde orbit. As noted, SS low-altitude polar orbit is widely used for monitoring the earth because each day, as the earth rotates below it, the entire surface is covered and the satellite views the same earth location at the same time each 24-h period. All SS orbits are polar orbits, but not all polar orbits are sun-synchronous orbits.

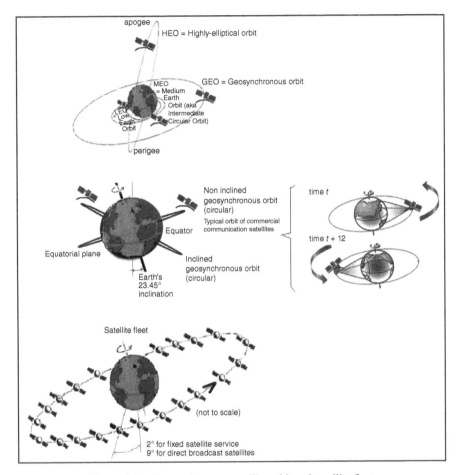

Figure 1.3   Geosynchronous satellite orbit and satellite fleet.

maneuvers are maintained to keep the orbital slot. Orbital positions are defined by international regulation and are defined as longitude values on the "geosynchronous circle," for example, 101° W, 129° W, and so on. Satellites are (now) spaced at 2° (or 9° for DBS) to allow sufficient separation to support frequency reuse, although in some applications a group of satellites can be (nearly) collocated (but each using a different frequency spectrum). In actuality, an orbital position is a "box" of about 150 km by 150 km, within which the satellite is maintained by ground control. Non-GSO satellites are used for applications such as Satellite radio services, GPS earth sensing, and military applications; however, commercial communication services are also emerging; satellites that operate in Molniya, or MEO, or LEO will add to the bandwidth supply, drive cost down, and in combination with smarter antennas on the ground provide continuous service (in fact, there is an argument to make that multiple satellites provide the ability to reconstitute quickly in the event of an anomaly).

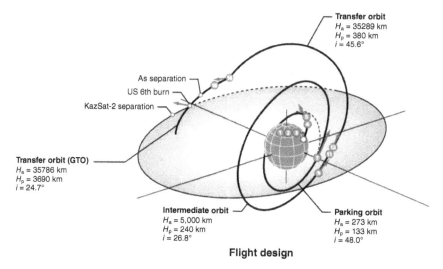

Figure 1.4   Example of a launch mission. Courtesy: International Launch Services Inc.

The major consequence of the GSO position is that signals experience a propagation delay of no less than 119 ms on an uplink (longer for earth stations at northern latitudes or for earth stations looking at satellites that are significantly offset longitudinally compared with the earth station itself[5]), and no less than 238 ms for an uplink and a downlink or a one-way, end-to-end transmission path. A two-way interactive session with a typical communications protocol, such as Transmission Control Protocol (TCP), will experience this roundabout delay twice (no less than 476 ms) since the information is making two round trips to the satellite and back. One-way or broadcast (video or data) applications easily deal with this issue since the delay is not noticeable to the video viewer or the receive data user. However, interactive data applications and voice backhaul applications typically have to accept (and adjust) to this predicament imposed by the limitations of the speed of light, which is the speed that radio waves travel. Satellite delay compensation units and "spoofing" technology have successfully been used to compensate for these delays in data circuits. Voice transmission via satellite presently accounts for only a small fraction of overall transponder capacity, and users are left to deal with the satellite delay individually, only a few find it to be objectionable.

Figure 1.4 depicts a mission overview for the launching of a mated set of satellites, based on uncopyrighted materials from International Launch Services (ILS); ILS provides mission management and launch services for the global commercial satellite industry using the premier heavy lift vehicle, the Proton Breeze M. They are

---

[5]Depending on the location of the earth station and the target satellite (which determines the look angle), the path length (and so the propagation delay) can vary by several thousand kilometers (e.g., for a satellite at 101° W and an antenna in Denver Co., the "slant" range is 37,571.99 km; for an antenna in Van Buren, ME, the range is 38,959.54 km).

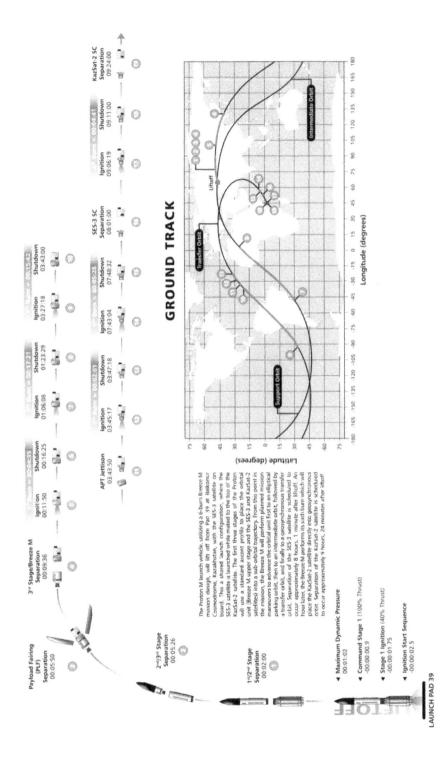

Figure 1.4 Continued.

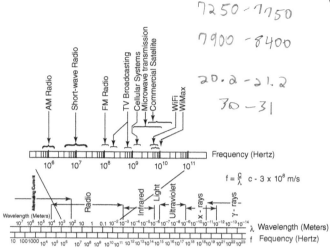

Frequency Band	Frequency Range	Propagation Modes
ELF (Extremely Low Frequency)	Less than 3 kHZ	Surface wave
VLF (Very Low Frequency)	3-30 kHz	Earth-ionosphere guided
LF (Low Frequency)	30-300 kHz	Surface wave
MF (Medium Frequency)	300 kHz - 3 MHz	Surface/sky wave for short/long distances, respectively
HF (High Frequency)	3-30 MHz	Sky wave, but very limited, short-distance ground wave also
VHF (Very High Frequency)	30-300 MHz	Space wave
UHF (Ultra High Frequency)	300 MHz - 3 GHz	Space wave
SHF (Super High Frequency)	3-30 GHz	Space wave. The "workhorse" microwave band. Line=of-sight. Terrestrial and satellite relay links
EHF (Extremely High Frequency)	30-300 GHz	Space wave. Line-of-sight. Space-to-space links, military and future use

SATELLITE FREQUENCIES (Ghz)		
BAND	DOWNLINK	UPLINK
C	3.700 - 4.200	5.925 6.425
X (Military)	7.250 -7.745	7.900 - 8.395
Ku (Europe)	FS: 10.7 - 11.7	FSS : 14.0 - 14.8
	DBS: 11.7 - 12.5	DBS: 17.3 - 18.1
	Telecom: 12.5 - 12.7	Telecom: 14.0 - 14.8
Ku (America)	FSS: 11.7 - 12.2	FSS: 14.0 - 14.5
	DBS: 12.2 -12.7	DBS: 17.3 -17.8
Ka	~18 - ~31 GHz	
V	36 - 51.4	

Figure 1.5   Electromagnetic spectrum and bands.

one of the three major launching outfits for commercial satellites. The Proton Breeze M launch vehicle launches from the Baikonur Cosmodrome, operated by the Russian Space Agency (Roscosmos) under the long-term lease from the Republic of Kazakhstan.

1.3.2   Satellite Transmission Bands

A satellite link is a radio link between a transmitting earth station and a receiving earth station through a communications satellite. A satellite link consists of one uplink and one downlink; the satellite electronics (i.e., the transponder) will remap the uplink frequency to the downlink frequency. The transmission channel of a satellite system is a radio channel using a direct-wave approach, operating in at specific RF bands within the overall electromagnetic spectrum (see Figure 1.5 [MIN199101]). Table 1.2

TABLE 1.2   Some Key Physical Parameters of Relevance to Satellite Communication

Frequency	The number of times that an electrical or electromagnetic signal repeats itself in a specified time. It is usually expressed in Hertz (Hz) (cycles per second). Satellite transmission frequencies are in the gigahertz (GHz) range.
Frequency band	A range of frequencies used for transmission or reception of radio waves (e.g., 3.7–4.2 GHz).
Frequency spectrum	A continuous range of frequencies.
Hertz (Hz)	SI unit of frequency, equivalent to one cycle per second. The frequency of a periodic phenomenon that has a periodic time of 1 s.
Kelvin (K)	SI unit of thermodynamic temperature.
Msymbol/s	Unit of data transmission rate for a radio link, equal to 1,000,000 symbol/s. Actual channel throughput is related to the modulation scheme employed.
Symbol	A unique signal state of a modulation scheme used on a transmission link, which encodes one or more information bits to the receiver.
Watt (W)	SI unit of power, equal to 1 J/s.

SI = Systeme International d'Unites (International Systems of Units)

provides some key physical parameters of relevance to satellite communication. The frequency of operation is in the super high frequency (SHF) range (3–30 GHz). Regulation and practice dictate the frequency of operation, the channel bandwidth, and the bandwidth of the subchannels within the larger channel. Different frequencies are used for the uplink and for the downlink.

Frequencies above about 30 MHz can pass through the ionosphere and, therefore, can be utilized for communicating with satellites (frequencies below 30 MHz are reflected by the ionosphere at certain stages of the sunspot cycle; however, commercial satellite services use much higher frequencies. The range 3–30 GHz represents a useful set of frequencies for geostationary satellite communication; these frequencies are also called "microwave frequencies."[6] Above about 30 GHz, the attenuation in the atmosphere due to clouds, rain, hydrometeors, sand, and dust makes a ground to satellite link unreliable (such frequencies may still be used for satellite-to-satellite links in space, although these applications have not yet developed commercially[7])

---

[6]From 30 to 300 GHz, the frequencies are referred to as "millimeter wave"; above 300 GHz optical techniques take over, these frequencies are known as "far infrared" or "quasi optical."

[7]Applications are emerging in the military arena: the Advanced Extremely High Frequency (AEHF) satellite system is a US Government joint service satellite communications system of four satellites (three of which were launched in 2014) that use the EHF spectrum; it is satellite communications system that aims at providing survivable, global, secure, protected, and jam-resistant communications for high-priority military ground, sea, and air assets. According to the US Air Force, AEHF will allow the National Security Council and Unified Combatant Commanders to control their tactical and strategic forces at all levels of conflict through general nuclear war and supports the attainment of information superiority.

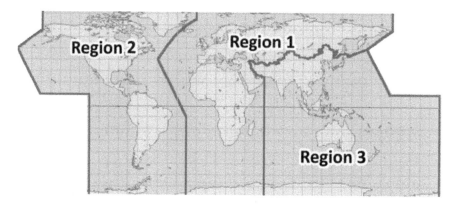

Figure 1.6 ITU regions for frequency allocations.

[JEF200401]. The actual frequencies of operation of commercial (US) satellites are[8]:

- *C-band:* 3.7–4.2 GHz for downlink frequencies and 5.925–6.425 GHz for uplink frequencies;
- *Ku-band:* 11.7–12.2 GHz for downlink frequencies and 14–14.5 GHz for uplink frequencies;
- *BSS:* 12.2–12.7 GHz for downlink frequencies and 17.3–17.8 GHz for uplink frequencies;
- *Ka-band:* 18.3–18.8 GHz and between 19.7 and 20.2 GHz for downlink frequencies and between 28.1–28.6 GHz and 29.5–30 GHz for the uplink frequencies (other specific frequencies are possible as discussed in Chapter 3 – also, some advanced services at higher frequencies, up to 40 GHz or even higher are possible in the future).

Table 1.3 depicts in a summarized form, (other) relevant satellite and/or microwave bands (higher frequencies correspond to the millimeter waves, as defined by the absolute value of the wavelength).

Note that the International Telecommunications Union (ITU) has divided the world into three regions (also see Figure 1.6):

- *Region 1:* Europe, Middle East, Russia, and Africa;
- *Region 2:* The Americas;
- *Region 3:* Asia, Australia, and Oceania.

Figure 1.7 provides some additional information on frequency bands.

---

[8]The international set of microwave bands is as follows: L-band (0.39–1.55 GHz); S-band (1.55–5.20 GHz); C-band (3.70–6.20 GHz); X-band (5.20–10.9 GHz); and K-band (10.99–36 GHz).

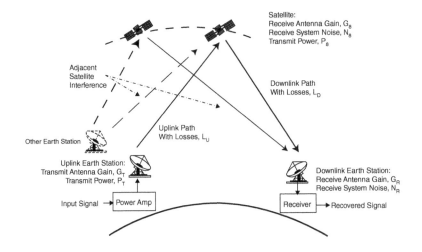

Band	Characteristics	Considerations
C-Band (6 GHz uplink and 4 GHz downlink)	• Relatively immune to atmospheric effects • Popular band, but on occasion it is congested on the ground (see note at right) • Bandwidth (~500 MHz/36 MHz transponders) allows video and high data rates • Provides good performance for video transmission • Proven technology with long heritage & good track record • Common in heavy rain zones	• Requires large antennas (3.8 - 4.5 meters or larger, especially on the transmit side) • Large footprints • Best performing band in the context of rain attenuation • Potential interference due to terrestrial microwave systems
Ku-Band (14-14.5 GHz uplink and 11.7-12.2 GHz downlink)	• Moderate to low cost hardware • Highly suited to VSAT networks • Spot beam footprint permits use of smaller earth terminals, 1-3 m wide in moderate rain zones	• Attenuated by rain and other atmospheric moisture • Spot beams generally focused on land masses • Not ideal in heavy rain zones
DBS-Band (17.3-17.8 GHz uplink and 122 -12.7 GHz downlink)	• Simplex • Multiple feeds for access to satellite neighborhoods • Small RO antennas • TV Video transmission to consumers	• Attenuated by rain and other atmospheric moisture
Ka-Band (18.3 – 18.8 GHz and 19.7 – 20.2 GHz downlink 28.1 - 28.6 GHz and 29.5 - 30.0 GHz uplink)	• Micro-spot footprint • Very small terminals, much less than 1 m • High data rates are possible 500-1000 Mbps • High Throughput Satellites	• Rain attenuation • Obstruction interference due to heavy rainfall (Black Out)

Figure 1.7   (a) Satellite transmission factors and bands. (b) Additional frequency band details for Region 2.

The Table of Frequency Allocations contained in Article 5 of the Radio Regulations (RR) allocates frequency bands in each of the three ITU regions to radiocommunication services based on various service categories as defined in the RR. There are some differences worldwide; however, the C-band and Ku-band are generally comparable.

The frequency bands are further subdivided into smaller channels that can be independently used for a variety of applications. Figure 1.8 depicts a typical subdivision of

	Channel (*)	Downlink Channel (GHz)	Description	Antenna Size (typical)	Availability
General Purpose Frequencies	C FSS (standard)	4.2 3.7	Mainstay of satellite communications Low rain fade High reliability w/ large antennas Application example: Cable TV	3 - 4.5m for cable headends	99.95
	X-Band	7.75 7.25	Military Band	Various	99.95
	Extended Ku	11.45 10.7	1 GHz split in two allocations Not in use due to medium power, regulatory restrictions & terrestrial interference 2 degree separation	75-95 cm	99.9
	Ku FSS standard	12.2 11.7	smaller antenna VSAT applications 500 MHz used by DTV/Echo for niche (int'l)services 2 degree seperation	75-95 cm	99.9
Specific Purposes Frequencies	Ku BSS US DBS	12.7 12.2	DTH applications (high power) 500 MHz by 2nd Gen Moderate rain attenuation wider spacing 4.5 - 9 degree separation	45 - 55 cm	99.85 - 99.9
	Ka BSS US RDBS	17.8 17.3	Additional Ka-band available to supplement Ku BSS for DTH applications (D/L 17.3-17.8Ghz U/L 24.75-25.25 GHz) Most countries in the Americas have this band available (no FS in the band) for DTH (uplink requires large antenna) Moderate rain attenuation ~4 degree separation to enable easier coordination with small receive dishes (FCC) Ka BSS satellites cannot operated co-located with Ku BSS satellites (transmit band same as receive band of a Ku BSS (at least 0.3 degrees of spacing is required to eliminate interference)	50 - 60 cm 4 degree spacing and high power; 50 cm 2-3 degree spacing: 60 cm and larger	99.85 - 99.9
	Ka FSS lower band Ka FSS upper band	1 GHz split into two allocations: 18.3 - 18.8 GHz 19.7 - 20.2 GHz (Main)	DIRECTV uses lower band 2 degree separation High rain attenuation small contiguous spot beams of 0.4-1 degree, covering 250-600 km with frequency reuse the capacity increase about 20-fold compared with Ku or Ka widebeam architecture supports high capacity, "bandwidth-on-demand" , Internet access and also point-to-point applications 2-way use, 70 cm dishes when s/c are 2 degrees apart (user beams connected via satellite-associated gateways) large margins needed to address rain attenuation	70-80 cm	99.8 - 99.9
Future Frequencies	Q band	50-33	Future applications	TBD	TBD
	V-band	75-50			

(*) ITU Region 2 GEO Satellite Spectrum

Figure 1.7   *(Continued)*.

the C-band into these channels, which are also called colloquially as "transponders" (transponder as a proper term is defined later in the section). The nominal subchannel bandwidth is (typically) 40 MHz with a usable (typical) bandwidth of 36 MHz. Similar frequency allocations have been established for the Ku and Ka bands. Many satellites simultaneously support a C-band and a Ku-band infrastructure (they have dedicated feeds and transponders for each band). Most communications systems fall into one of three categories: bandwidth efficient, power efficient, or cost efficient. Bandwidth efficiency describes the ability of a modulation scheme to accommodate data within a limited bandwidth. Power efficiency describes the ability of the system to reliably send information at the lowest practical power level. In satellite communications, both bandwidth efficiency and power efficiency are important [AGI200101].

Satellites usually support a number of beams, making use of frequency reuse; the use of a half-a-dozen to a dozen beams if fairly typical. These beams are implemented using different antennas and/or distinct feeds on a single antenna. Each spot beam reuses available frequencies (and/or polarizations), so that a single satellite can provide increased bandwidth. In nonoverlapping regions, the frequencies can be fully

TABLE 1.3   Satellite Bands, Generalized View, *IEEE Standard 521-1984*

Band Designator	Frequency (GHz)	Wavelength in Free Space (cm)
L-band	1–2	30.0–15.0
S-band	2–4	15–7.5
C-band	4–8	7.5–3.8
X-band	8–12	3.8–2.5
Ku-band	12–18	2.5–1.7
K-band	18–27	1.7–1.1
Ka-band	27–40	1.1–0.75
V-band	40–75	0.75–0.40
W-band	75–110	0.40–0.27

reused; in overlapping regions, nonconflicting frequencies have to be employed (see Figure 1.9 for an example). HTS support up to 100 beams in nonoverlapping geographic areas, thus greatly increasing the total usable throughput.

Figure 1.10 depicts a two-way satellite link. The end-to-end (remote-to-central point) link makes use of a radio channel, as described earlier, for the transmitting station uplink side to the satellite; in addition, it uses a downlink radio channel to the receiving station (this is also generally called the inbound link). The outbound link, from the central point to a remote, also makes use a radio channel that comprises an uplink and a downlink.

From an application's perspective, the link may be point-to-point (effectively where both ends of the link are peers) or it may be point-to-aggregation-point, for example, for handoff to a corporate network or to the Internet. Some applications are simplex, typically making use of an outbound link; other applications are duplex, using both an inbound and an outbound link.

Satellite communications now almost exclusively make use of digital modulation. Modulation is the processes of overlaying intelligence (say a bit stream) over an underlying carrier so that the information can be relayed at a distance. Demodulation is the recovery, from a modulated carrier, of a signal having the same characteristics as the original modulating signal. The underlying analog carrier is superimposed with a digital signal, typically using 4-, or 8-point Phase Shift Keying (PSK) techniques, or 16-point Quadrature Amplitude Modulation (QAM). In addition, the original signal is fairly routinely encrypted and, invariably, protected with Forward Error Correction (FEC) techniques.

As noted, different frequencies are used for the uplink and downlink to avoid self-interference, following the terrestrial microwave transmission architecture developed by the Bell System in the 1940s and 1950s. In systems using the C-band, the basic parameters are 4 GHz in the downlink, 6 GHz in the uplink, 500 MHz bandwidth over 24 transponders using vertical and horizontal polarization (a form of frequency reuse discussed later on), resulting in a transponder capacity of 36 MHz or 45–75 Mbps or more of usable throughput – depending on modulation and FEC scheme. C-band has been used for several decades and has good transmission characteristics, particularly in the presence of rain, which typically affects high-frequency

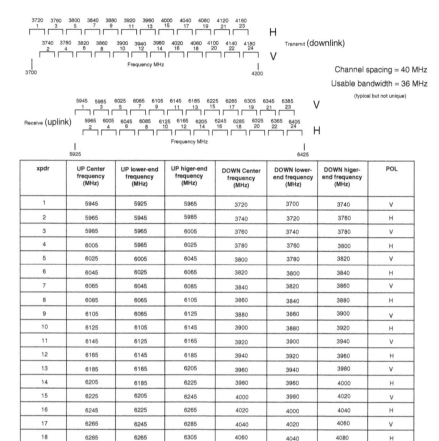

xpdr	UP Center frequency (MHz)	UP lower-end frequency (MHz)	UP higer-end frequency (MHz)	DOWN Center frequency (MHz)	DOWN lower-end frequency (MHz)	DOWN higer-end frequency (MHz)	POL
1	5945	5925	5965	3720	3700	3740	V
2	5965	5945	5985	3740	3720	3760	H
3	5985	5965	6005	3760	3740	3780	V
4	6005	5985	6025	3780	3760	3800	H
5	6025	6005	6045	3800	3780	3820	V
6	6045	6025	6065	3820	3800	3840	H
7	6065	6045	6085	3840	3820	3860	V
8	6085	6065	6105	3860	3840	3880	H
9	6105	6085	6125	3880	3860	3900	V
10	6125	6105	6145	3900	3880	3920	H
11	6145	6125	6165	3920	3900	3940	V
12	6165	6145	6185	3940	3920	3960	H
13	6185	6165	6205	3960	3940	3980	V
14	6205	6185	6225	3980	3960	4000	H
15	6225	6205	6245	4000	3980	4020	V
16	6245	6225	6265	4020	4000	4040	H
17	6265	6245	6285	4040	4020	4060	V
18	6285	6265	6305	4060	4040	4080	H
19	6305	6285	6325	4080	4060	4100	V
20	6325	6305	6345	4100	4080	4120	H
21	6345	6325	6365	4120	4100	4140	V
22	6365	6345	6385	4140	4120	4160	H
23	6385	6365	6405	4160	4140	4180	V
24	6405	6385	6425	4180	4160	4200	H

Uplink (MHz): 14000 - 14500

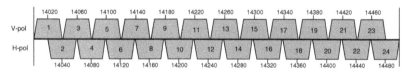

Downlink (MHz): 11700 - 12200

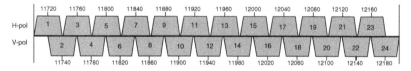

Figure 1.8 Typical subchannel ("transponder") allocations for C- and Ku-band.

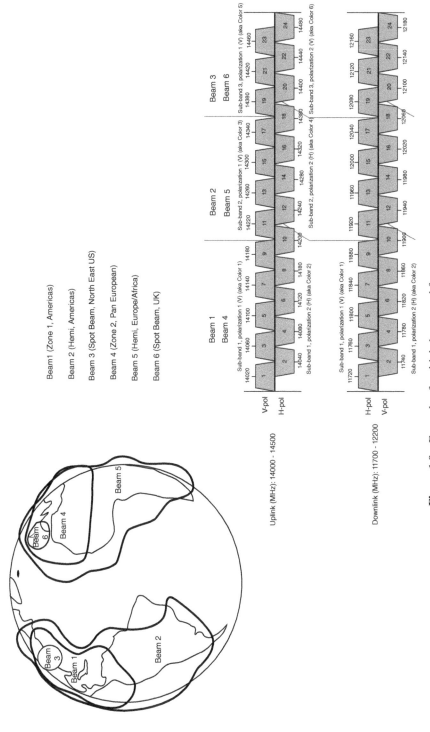

Figure 1.9 Example of multiple beams and frequency reuse.

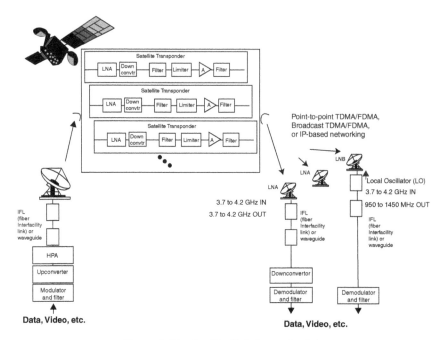

Figure 1.10   Satellite link (some details).

transmission. Generally C-band links are used for TV video distribution to headends and for military applications, among others. A number of antenna types are utilized in satellite communication, but the most commonly used narrow beam antenna type is the parabolic dish reflector antenna. C-band receive-dishes for broadcast-quality video reception are typically 3.8–4.5 m in diameter. The size is selected to optimize reception under normal (clear sky) or medium-to-severe rain conditions; however, smaller antennas of 1.5–2.4 m can also be used, depending on the intended application, service availability goals, and satellite footprint. For a two-way transmission, the same size and considerations apply (although larger antennas can also be used in some applications, especially at a major earth station); availability, acceptable bit error rate, satellite radiated power, and rain mitigation goals drive the design/size of the antenna and ground transmission power.

Enterprise applications tend to make use of the Ku-band because of the fact that smaller antennas can be employed, typically in the 0.6–2.4 m (depending on application, desired availability, rain zone, and throughput). Newer applications, typically for DBS/Direct to Home (DTH) video distribution look to make use of the Ka-band, where antenna size can range from 0.3 to 1.2 m (see Table 1.4). Spread spectrum techniques and other digital signal processing are being used in some applications to reduce the antenna size by reducing unwanted signals (either in the uplink, e.g., with spread spectrum, or in the downlink with adjacent satellite signal cancellation using digital signal processing).

TABLE 1.4   **Frequency and Wavelength of Satellite Bands**

Frequency (GHz)	Wavelength (m)	Typical Antenna Size (m)
3.7	0.081081081	1.2–4.8
4.2	0.071428571	
5.925	0.050632911	
6.425	0.046692607	
11.7	0.025641026	0.6–2.4
12.2	0.024590164	
12.7	0.023622047	
18.3	0.016393443	0.3–1.2
18.8	0.015957447	
19.7	0.015228426	
20.2	0.014851485	
27.5	0.010909091	

Related to the issue of orbits note, as stated, that there are multiple satellites in the geostationary orbit: typically every 2° on the arc, and even collocated at the "same" location, when different operating frequencies are used. Effectively, satellite systems may employ cross-satellite frequency reuse via space-division multiplexing; this implies that a large number of satellites (even neighbors) make use of the same frequency operating bands, as long as the antennas are highly directional. Some applications (e.g., direct broadcast to homes) or some (non-US) jurisdictions allow spacing at 3°; higher separation reduces the technical requirements on the antenna system but results in fewer satellites in space. DBS satellites are usually 9° apart on the orbital arc.

Unfortunately, unless the system is properly "tuned" by following all applicable regulation and technical guidelines, ASI can occur. A transmit earth station can inadvertently direct a proportion of its radiated power toward satellites that are operating at orbital positions adjacent to that of the wanted satellite. This can occur because the transmit antenna is incorrectly pointed toward the wanted satellite, or because the earth station antenna beam is not sufficiently concentrated in the direction of the satellite of interest (e.g., the antenna being too small). This unintended radiation can interfere with services that use the same frequency on the adjacent satellites. Interference into adjacent satellite systems is controlled to an acceptable level by ensuring that the transmit earth station antenna is accurately pointed toward the satellite, and that its performance (radiation pattern) is sufficient to suppress radiation toward the adjacent satellites. In general, a larger uplink antenna will have less potential for causing adjacent satellite interference, but will generally be more expensive and may require a satellite tracking system. Similarly, a receive earth station can inadvertently receive transmissions from adjacent satellite systems, which then interfere with the wanted signal. This happens because the receive antenna, although being very sensitive to signals coming from the direction of the wanted satellite, is also sensitive to transmissions coming from other directions. In general, this sensitivity reduces as the antenna size increases. As for a transmit earth station, it is also very important to accurately point the antenna toward the satellite in order to minimize ASI effects [FOC200701]. As noted, spread spectrum techniques and other digital signal processing are being used in some advanced (but not typical) applications to reduce unwanted signals (e.g., with spread spectrum or with ASI cancellation by using digital signal processing). An industry push is afoot to add an identifying ID to the signal by the modem, called Carrier ID. The Carrier ID allows one to identify the source of the transmission, and, thus, enables one to be better equipped to rectify any spurious signal reaching an (unintended) satellite.

The sharing of a channel (colloquially, a "transponder") is achieved, at this juncture, using Time Division Multiple Access (TDMA), random access techniques, Demand Access Multiple Access (DAMA), or Code Division Multiple Access (CDMA) (spread spectrum). Increasingly, the information being carried, whether voice, video, or data, is IP-based.

1.3.3   Satellite Signal Regeneration

In general, the information transfer function entails the bit transmission across a channel (medium). Since there is a variety of media in use in communication, many of the transmission techniques are specific to the medium at hand. In the context of this book, the transmission channel is a radio channel. Typical transmission problems include the following:

- Signal attenuation (e.g., free space loss or attenuation due to "rain fade");
- Signal dispersion;
- Signal nonlinearities (due, for example, to amplification or propagation phenomena);
- Internal or external noise;
- Crosstalk (e.g., spectral regrowth), intersymbol interference, and intermodulation; and
- External interference and adjacent satellite interference.

Typically, some of these impairments, but not all, can be dealt with by using a physical layer regenerator. Regeneration is the function of restoring the signal (and/or bit stream) to its original shape and power level. Physical layer regeneration techniques are specific to the medium (e.g., radio channel, fiber channel, twisted-pair copper channel, and so on). Physical layer regeneration correctively addresses signal attenuation, signal dispersion, and crosstalk; this is done via signal re-amplification, re-timing, and re-shaping. Regeneration is generally considered a Layer 1 function in the Open Systems Interconnection Reference Model (OSIRM).

A "low-end" physical layer regenerator includes only the re-amplification function. These are known as 1R regenerators. A "high-end" regenerator includes the re-amplification, re-timing, and re-shaping functionality. These are known as 3R physical layer regenerators. The functions of a 3R network element are as follows:

- Re-amplification: increases power levels above the system sensitivity;

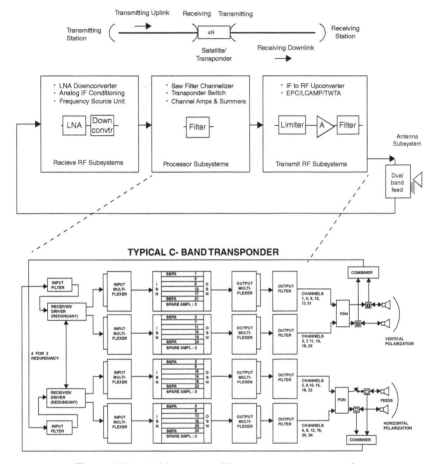

Figure 1.11 Architecture satellite regenerator, a transponder.

- Re-timing: suppresses timing jitter by optical clock recovery;
- Re-shaping: suppresses noise and amplitude fluctuations by decision stage.

Regenerators are invariably technology specific. Hence, one has LAN repeaters (even if rarely used), Wi-Fi repeaters, copper-line (T1 channel) repeaters, cable TV repeaters, and optical regenerators (of the 1R, 2R, or 3R kind). Regeneration and amplification are critical functions in satellite systems because the attenuation through space and the atmosphere is in the order of 200 dB (i.e., the power is reduced by 20 orders of magnitude). Figure 1.11 depicts a regenerator in the satellite environment: this regenerator is the satellite transponder. The term "satellite transponder" refers properly to a transmitter–receiver subsystem onboard the satellite that uses a single high-power amplification chain and processes a particular range of frequencies (the "transponder bandwidth"). There are many transponders on a typical satellite, each being capable of supporting one or more communication channels [SAT200501]. In fact, a typical satellite will have 24 transponders: 12 to regenerate the consecutive twelve 36 MHz frequency blocks that comprise the segments assigned for operation at the C-band, Ku-band, or Ka-band (500 MHz total) for use in the *vertical signal polarization* mode; and, 12 for the frequency blocks that comprise the segments assigned to the *horizontal signal polarization* mode. By utilizing transponders, commercial communication satellites perform the following functions:

- Receive signals from the ground station (uplink beam);
- Separate, shift frequency, amplify, and recombine the signals (the regeneration function);
- Transmit the signals back to (another) earth station (downlink beam).

Some advanced satellite functions include digital signal processing and packet processing (processing at a higher layer of the OSIRM). In most current cases, satellites (and their transponder apparatus) are operated in what is known as a "bent pipe" configuration, where the satellite acts as a relay as just described (possibly in a 1R, 2R, or 3R mode) to retransmit the received waveform back to points on the ground. However, there are also other designs known as "regenerative" (more specifically, "regenerative onboard processors") where the satellite demodulates the signals it receives, processes the information (e.g., packet routing and forwarding), and then remodulates these recombined signals for transmission back to points on the ground, thus performing a higher layer OSIRM function (see Figure 1.12). This is an example of onboard intelligent processing (and switching) (e.g., already advocated by this author [and others] in 1979 [MIN197901]).

1.3.4   Satellite Communication Transmission Chain

The ground segment of a satellite transmission chain consists of the earth stations (also known as ground station) that are operating within a particular satellite system or network. Earth stations are typically connected to the end-user's equipment directly with local cabling or via a terrestrial network. The ground segment supports either one or both of the following: (i) an uplink and (ii) a downlink. A satellite communications link comprises the following elements:

Link = modulation equipment + upconversion equipment + amplification equipment + uplink transmission channel + frequency shift (conversion at the satellite) equipment + downlink transmission channel + signal reception/antenna + amplification equipment + downconversion equipment + demodulation equipment.

The uplink is that portion of a satellite communications link that involves signal transmission from the ground and reception onboard the satellite. The downlink is that portion of a satellite communications link that involves signal (re-)transmission from the satellite and reception on the ground. Downconversion is the process of converting

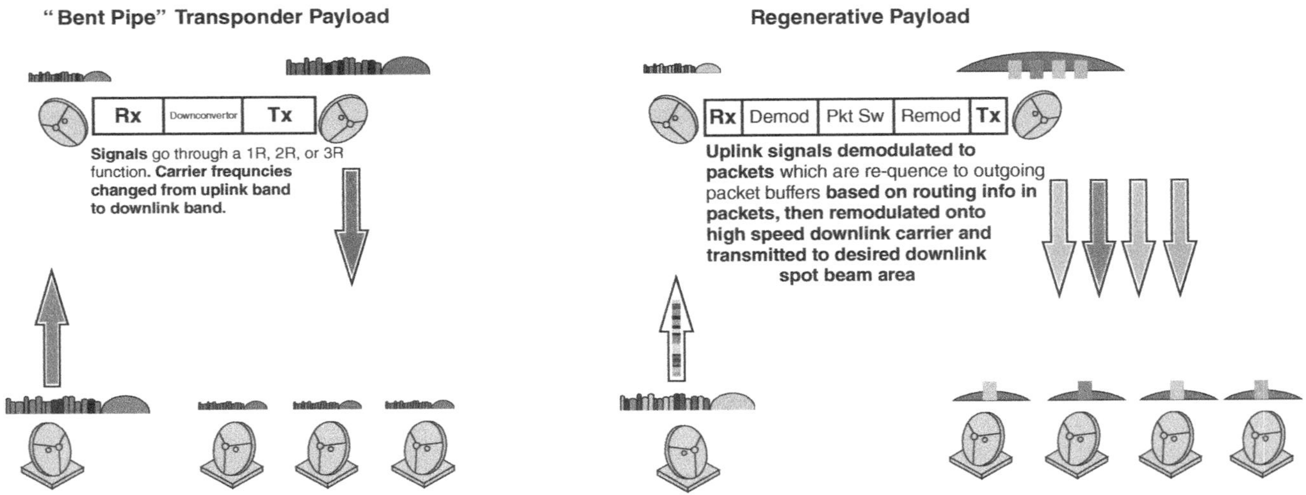

"Bent Pipe" Transponder Payload

Rx	Downconverter	Tx

Signals go through a 1R, 2R, or 3R function. Carrier frequncies changed from uplink band to downlink band.

Regenerative Payload

Rx	Demod	Pkt Sw	Remod	Tx

Uplink signals demodulated to packets which are re-quence to outgoing packet buffers based on routing info in packets, then remodulated onto high speed downlink carrier and transmitted to desired downlink spot beam area

Figure 1.12 Onboard satellite (routing) processing.

the frequency of a signal to a lower frequency; downconversion is performed at the reception point to permit the recovery of the original signal. The opposite process of upconversion is the process of converting the frequency of a signal into a higher frequency; it is done at the point of transmission. Signal management on the ground is better handled at lower frequencies; hence, the purpose/need for frequency conversion (to a higher frequency level for transmission and down to a lower frequency level at reception). The transponder is a repeater that takes in the signal from the uplink at a frequency f_1, amplifies it, and sends it back on a second frequency f_2.

An uplink system consists of the following subsystems (often in redundant mode):

- Network interface devices such as routers, encryptors, conditional access systems, and encapsulators;
- Modulators (devices that superimposes the amplitude, frequency, or phase of a wave or signal onto another wave – the carrier – which is then used to convey the original signal over the satellite link). (QPSK and 8-PSK are typical, but other more advanced methods are also used.) (FEC is typically handled at this point in the chain);
- Upconverters (devices for converting the frequency of a signal into a higher frequency; transceivers that take a 70/140/900 MHz signal and frequency converting it to either C-, Ku-, or Ka-band final frequency);
- PA (power amplifiers), specifically high-power amplifiers (HPAs), for example, solid state power amplifier (SSPA) or a klyston). Transmit power amplifiers provide amplification of signals to be transmitted to the satellite (typically 750–3,000 W); and
- Transmit antennas (the most common of which are the parabolic "dish" antennas) have a "gain" determined by the beam forming capability of the antenna, which is directly proportional to the square of the diameter of a parabolic antenna.

A downlink includes the following components:

- Receive antennas (see the discussion of the transmit antennas above);
- Downconverters (devices for converting the frequency of a signal into a lower frequency; transceivers that take a C-, Ku-, or Ka-band signal and frequency-convert it to either 70-, 140-, or 900-MHz final frequency);
- LNA (low noise amplifier), or LNB (low noise block downconverter), or LNC (low noise converter); and
- Demodulator (customer site) for recovering the data from the modulated carrier (for two-way links, both modulator and demodulator are usually combined in a single "modem."

The receive antenna provides a degree of directionality (known as antenna gain), which helps isolate the desired signal from other signals and noise when the main beam of the antenna is precisely directed toward the source of the desired signal. In the case of the common parabolic antenna, the gain is directly proportional to the square of the diameter of the antenna.

The LNA amplifies RF signal from the antenna and feeds it into a frequency converter, the output of which is typically intermediate frequency (IF) of 70/140/900 MHz. It provides 50–60 dB of amplification. An LNA is generally more precise and stable but more expensive than an LNB. The LNB amplifies the RF signal from the antenna and converts it to an L-band signal. An LNB provides 50–60 dB of amplification and also converts from one block of frequencies to another (what goes in comes out amplified and at a different frequency). The LNC is similar to an LNB, but it has a variable shift. It provides 50–60 dB of amplification and also converts block of frequencies to a specific portion of that block of frequencies (what goes into the LNC comes out amplified and at a different frequency within a given range – often it uses local oscillator to help in conversion). LNA/LNB/LNCs are typically used for one-way field antennas.

There are two LNB technologies in common use: the older dielectric resonator oscillator (DRO) LNB technology and the newer Phased Locked Loop (PLL) technology. PLLs are more stable than DROs since they utilize a more stable internal reference source (crystal oscillators) or they make use of an input from a stable external source (DROs typically have stability of ± 100 kHz or as coarse as ± 1000 kHz). It follows that for applications of low bitrates and/or small carriers (e.g., 50 kHz wide) PLL LNBs typically provide better performance. DROs are acceptable solutions when the signal entails large bitrate carriers (such as DVB-S2 or DVB-S2X video signals); a sideline advantage of the DRO technology is that it is more cost-effective than the PLL systems; this may be important in antenna with multiple feeds (e.g., simulsats). Table 1.5 provides a perspective on the use of the two LNB technologies, as per industry's best practices. The consensus is that the choice of LNBs is application dependent; generally, above 1 Msps (megasymbols per second) DROs operate in an acceptable manner. Industry best practices indicate that some applications require PLL LNB for optimal operation while other applications are equally well-served by DRO technology. Typically,

- MPEG-2/4 digital video applications operate well with high-stability DRO LNBs;
- VSAT and Point of Sale (POS) systems may use a DRO LNB, but most users prefer a PLL to ensure the highest possible system reliability;

TABLE 1.5   LNB Technologies

LNB Oscillator Type	Frequency Stability	Application
Internal reference PLL	$\pm$ 150 kHz to $\pm$ 5 kHz	News gathering VSAT
External reference PLL	0 ± 1 kHz	Narrowband data
DRO	± 1.0 MHz to ± 150 kHz frequency stability	Broadcast television wideband data broadcast

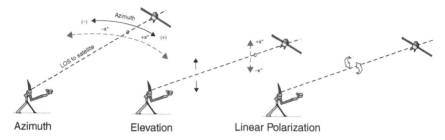

Figure 1.13 Antenna pointing.

- Satellite News Gathering (SNG) trucks typically use PLL LNBs for the most reliable performance in the worst conditions.

Antennas in a satellite environment are reflective systems, typically parabolic in shape. A highly directional antenna concentrates most of the radiated power along the antenna "boresight." It follows that a high-gain antenna is very directional and needs to be pointed with reasonable high precision. As an example, at 12 GHz the pointing accuracy needed for a 1-m diameter dish is of the order of a degree or two of arc. The antenna uses a three-axis pointing system, which one needs to adjust when pointing it to a satellite (see Figure 1.13 [LAU200701]):

- *Azimuth:* This is the magnetic compass direction (angle of sighting) at which one points the dish. It is a side-to-side adjustment. Azimuth is the angular distance from true north along the horizon to a satellite measured in degrees.
- *Elevation:* This is the angle above the horizon, at which one point the dish. This is an up-and-down adjustment.
- *Polarization:* The (linear) polarization or *skew* represents the feed's alignment needed to capture the maximum signal from the satellite consistent with the satellite signal's transmit signal orientation (polarization[9]). This is a rotational adjustment. Polarization prevents interference with signals on the same satellite at the same frequency but opposite in polarization.

A Block UpConverter (BUC) takes L-band signal and converts it to either the C- or Ku-band final frequency. This is typically used for two-way field antennas and operates in the 5–25 W range for commercial applications (but could also operate at other power levels in special circumstances).

1.4   SATELLITE APPLICATIONS

There are many traditional and emerging satellite applications. Major commercial applications include, but are not limited to, the applications that are listed as follows:

---

[9]Incorrect alignment results in picking up the undesired cross-pol signal, which will severely impact performance/quality of the intended signal.

- Video Distribution: This includes analog, SD, HD, 3D, and UHD channels.
- Direct To Home: This includes SD, HD, 3D, and UHD channels.
- Video Contribution and Occasional Use (OU): This includes analog feeds and digital feeds.
- Legacy Telephony and Carrier, including cellular backhaul.
- Enterprise Data.
- Broadband Access.
- Mobility (maritime, aeronautical).
- Government/Military Services.
- M2M Services.

Many of these applications entail IP (specifically IPv4) support; each of these may need to support IPv6 modes in the not-too-distant future.

- Two-way enterprise (Very Small Aperture Terminal) satellite communications for intranet/Internet-access connectivity (see Figure 1.14a). Enterprise customers and/or government agencies may want to use IPv6 in the future (IPv4 is common today). Some VSAT systems are configured in a "mesh" network that allows the remote sites to communicate with one another without having to go through the hub site, but this normally requires larger remote antennas.
- Video distribution to Cable Headends (see Figure 1.14b). Cable TV companies may be interested in IPv6 environments in the future.
- DTH video reception by consumers. Some ancillary applications may be able to make effective use of an IPv6 infrastructure.
- Aeronautical/mobility passenger Internet access applications (see Figure 1.15a) and M2M applications (Figure 1.15b); IPv6 is the ideal core infrastructure for evolving IoT and mobility applications.

Application acceleration is typically needed to address the one-way propagation delay of about 250 ms on an FSS satellite. In addition to the raw propagation delay, many data protocols make use of "handshakes" and/or acknowledgements, which further impact the effective user-perceived delay and the net data throughput of the connection. Therefore, VSAT systems typically employ a series of techniques to achieve end-to-end improvements. The Performance Enhancing Proxy (PEP) defined in IETF RFC 3135 provides mechanisms for TCP ACK reduction and three-way handshaking. RFC 3095 techniques can be used for header compression. Application-level acceleration is also typically used. For example, in a Web surfing application, the user will experience slow screen "fill," as Web objects are fetched one by one. To address this issue, VSAT ground systems have implemented HTTP object prefetch; the goal is to reduce the chattiness involved in fetching objects that constitute a Web page. The typical operation is as follows, as implemented in Hughes products [HUG201301]: The gateway client intercepts Web requests on the remote router and talks to a server at the data center. The normal process would involve waiting for a remote PC to parse

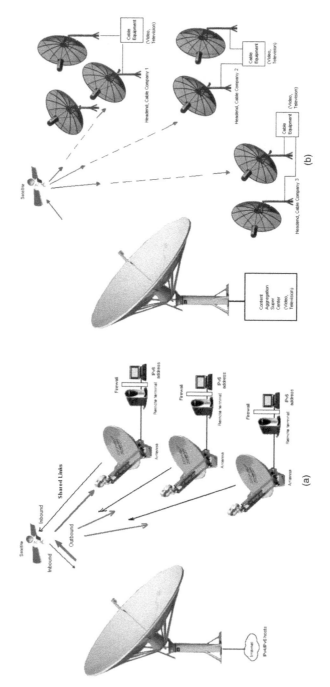

Figure 1.14 Typical applications: (a) (enterprise) (two-way) (VSAT) satellite communications. (b) satellite video distribution to Cable Headends.

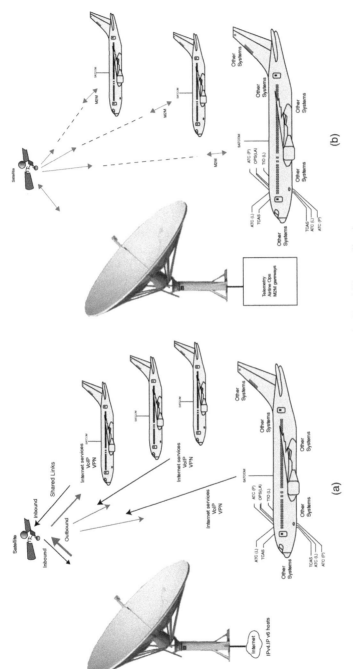

Figure 1.15 Typical aeronautical/mobility applications.

the initial HTML page, sending a DNS (domain name server) request for each server that has an object such as an image or flash file, and then initiating multiple requests to each of those other servers to retrieve each required object. Instead, the TurboPage server prefetches the objects and caches them temporarily at the terminal, providing a local delivery of the requested objects rather than requiring an end-to-end request and response. TurboPage therefore assures the freshest content from the Web server, while delivering lightning-fast performance. A two-stage payload compression scheme for HTTP traffic can also be employed, including a Byte Level Caching (BLC), which is a lossless compression algorithm that exploits duplication of byte sequences in a data stream, and a V.44 compression scheme. Figure 1.16 illustrates the data flow of the Hughes TurboPage product, including compression.

1.5 SATELLITE MARKET VIEW

In this section, an encapsulated view of the commercial communications market for satellite services is provided with data synthetized from a number of sources, including but not limited to Northern Sky Research (NSR) [GLO201301].

The combined satellite manufacturing and launch industry generated approximately US$35 billion globally in 2013; 100 new satellites were ordered (26 GEO) and 100+ satellites were launched in 2013 [EDI201401].

The total industry revenues from leased C-, Ku-, and widebeam Ka-band transponders and for the wholesale equivalent revenues from leased HTS (high-density spot beams) and MEO-HTS capacity is forecast to reach approximately US$19 billion by 2022, up from a baseline of approximately US$11 billion in 2012; this represents a CAGR of 5.3%, or 2.3% if one assumed a 3% annual inflation rate over the same period (the inflation-adjusted 2022 revenue would be US$14B in 2012 dollars). On the other hand, some industry experts view growth projections at these rates as being pollyannaishly optimistic and in being, reality, lower in a more objective assessment of the market space.

As it can be seen in Figure 1.17, the revenue growth is expected to come exclusively from Ku and HTS applications. Most of the Ku- and widebeam Ka-band revenue (and transponder demand) growth is expected to be coming from media services such as DTH and video distribution services, while the majority of the HTS and MEO-HTS revenue (and transponder demand) growth is expected to be coming from data services such as broadband access, backhaul, and VSAT networking. Table 1.6 provides an approximate global breakdown by capacity use and by revenue share for 2014.

Demand in the commercial satellite arena is generally quoted in Transponder Equivalents (TPE). The total demand for TSE is forecast to grow at a CAGR of 1.75%, taking the TSE demand level from around 5800 TPEs in 2012 to about 7000 in 2022; the growth will be in BRICA (Brazil, Russia, India, China, and Africa) countries, with near-zero growth in Europe and North America. The media market is expected to account for 60% of this increase, and the mobility markets contribute the other 40%. As it can be seen in Figure 1.17a, the number of C-band transponders

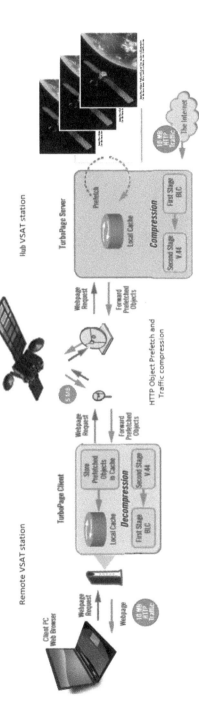

Figure 1.16 VSAT protocol/application-level acceleration.

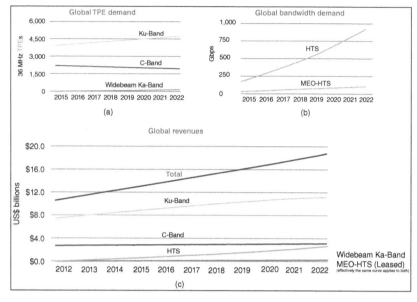

Figure 1.17 Some key forecasts about satellite capacity demand and service revenues.

TABLE 1.6 Approximate Global Breakdown by Capacity Use and by Revenue Share for 2014

	Capacity (%)	Revenue (%)
Distribution	25	30
DTH	17	26
Contribution/OU	13	9
Carriers	8	3
Enterprise/VSAT	20	17
Government	17	16
Total	100	100

is flat or slightly diminishing. Although HTS will increase, its total contribution to the TPE demand pool will remain a small fraction in this time frame (although the supported digital capacity will increase significantly – but less so for MEO-HTS, as seen in Figure 1.17b); the actual TPE increase is expected to be principally in the Ku segment. The current expectation is that mobility applications for both commercial and government clients will use an approximately equal amount of C-, Ku-, and widebeam Ka-band transponder capacity and HTS/MEO-HTS transponder capacity (mobility at C-band is more likely for large merchant and/or cruise ships where antenna sizes are less of an issue). The overall global net decline in the total C-band demand is driven by the fact that applications such C-band backhaul and IP trunking will start to make use of HTS/MEO-HTS capacity for the rest of the decade or migrate to fiber facilities.

The above-mentioned forecasts point to an interesting observation: the ratio of revenue-to-TPE in 2012 was \$1.93 M (per year), while in 2022 it would be \$2.68 M (per year), but discounting for an annual 3% inflation that ratio would be \$1.99 M (per year). This implies that overall the net per-bit-cost of satellite communications fails to follow the trend of terrestrial telecommunications services costs that generally decreases in the order of 50% per decade; therefore, in order to defend the "unchanged asking" price, the satellite operators have to become a lot more creative than they were in the 1990s and early 2000s. In turn, this means that a large amount of innovation will be required by the operators to address these ecosystem fundamentals, and staffs that have strong IP and integrated networking background over will be needed by the said operators over traditional RF-only staffers.

It should be noted that while the demand for transponder capacity is seen only growing at a CAGR of 1.75%, the new capacity entering the market in the coming 5 years (the supply side) is seen as being "unprecedented" [GLO201301]. The supply growth is expected to be such that it will continue to be a capacity (TPE) inventory of over 40% with 2700 TPEs added through 2018, while the demand base is around 6200 TPE in 2018. While this may be beneficial for the user community, a capacity glut may perhaps foster a "price war" among supplies in the next few years, especially in First World markets where prices have been kept much higher than in other parts of the world (more efficient modem technology and video encoding will also exert a downward pressure on TPE demand, unless Ultra HD comes to the rescue – but that needs consumer acceptance, which will be based on equipment costs, content availability, and subscription costs).

1.6   WHERE IS FIBER OPTIC TECHNOLOGY GOING?

Carriers will be deploying 100 Gbps ("100 G") links in their terrestrial networks in the next few years, especially in North America, Europe, Asia, and in transoceanic accessible areas around South America and Africa. Figure 1.18 depicts the progression of optical networks over time (based partially on reference [SCH201101]). These deployments will greatly increase bandwidth availability and reduce cost-per-bit. Traditionally, the carriers have quadrupled the bandwidth per link for a doubling of the cost; Ethernet-based services have fared even better with a 10-fold increase in bandwidth for a doubling of the cost.

The following observations should be of interest to satellite operators (and/or as a minimum be cognizant of these developments):

> " … Carrier preferences are trending toward 100G technology … 100G technology will grow at a faster rate than 40G did in its first years of introduction. In many ways, 40G has trailblazed for 100G, reducing risk at the component level and familiarizing service providers and test equipment vendors with coherent networking. … We believe

	Timeframe	Link speed	Relative cost/bit	Main Technologies	Applications
Initial optical networks (1st Gen)	1980–1995	155–2480 Mbps	4000–1000	SDH/SONET	Telephony, voice traffic, dial-up modem traffic
WDM/10G (2nd Gen)	1995–2005	2.4–10 Gbps	1000–500	Growing WDM, 10GbE	Internet traffic ramps
WDM/ROADM/40G (3rd Gen)	2005–2012	10–40 Gbps	500–250	Widespread 10G and introduction of 40G/OTN; SDH/SONET recodes; reconfigurable optical add-drop multiplexers (ROADMs)	FTTx broadband, video, ICPs, super data centers; Netflix, Facebook, Youtube, iTunes
Next-Gen (4th Gen)	2013–2020	40–100 Gbps	250–100	100GE, 100G, coherent optics, ROADM/OTN mesh networks, packet transport	Cloud computing; Ultra HD video, mobile broadband

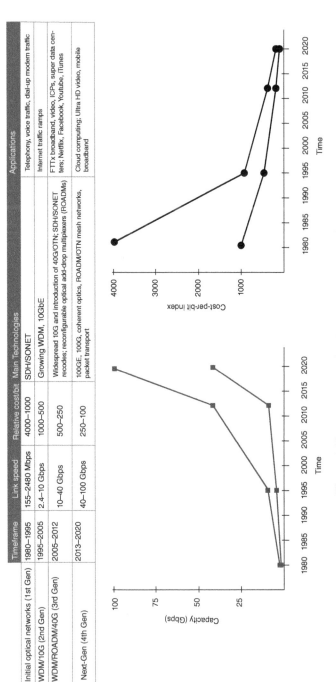

Figure 1.18 Progression of optical technology over time.

100G coherent component technology inherently should not cost more than 2x 40G technology, which when translated to equipment pricing, should result in a rapid switch over to 100G transmission ... During the last 10 years, major innovation in the network moved from the optical transport layer to the last mile: enterprise Ethernet services, cheap FTTH, widespread and cheap DSL, 100 M cable modems, and – most dramatically – mobile broadband via wireless smartphone. These technologies opened the data spigot and launched a wave of transformational businesses that exploited cheap and readily available bandwidth to consumers and enterprises. Netflix, Facebook, YouTube, and iTunes would never have happened without faster pipes to the home. The data centers of Internet content providers at the other end of these connections are virtual fire hoses of data, particularly as more applications and data storage is moved into "the cloud." ... Bit rates are increasing from 10G to 40G and 100G, but there are more changes: the use of coherent optics, Reconfigurable Optical Add Drop Multiplexors (ROADMs), and the introduction of optical transport network (OTN) switching and eventually Ethernet/MPLS as a circuit replacement for traditional SONET/SDH client interfaces. Taken together, the introduction of these technologies is another transformational change of telecom network ... 10G WDM technology was really just a sped up version of previous technology at 2.5G, 622 M, and 155 M, but the move to 40G and 100G is different, with more sophisticated modulation schemes and the introduction of coherent technology. Coherent optics bring massive spectral efficiency over greater distance, improving the central metric of carrier optical transport economics: cost per bit per kilometer transmitted. Like Wavelength Division Multiplexing (WDM) did 15 years ago, coherent optics change the economics of transport, simplifying network planning and increasing the capacity of installed fiber by at least an order of magnitude ... Survey data shows that carriers are willing to move to higher speeds at cost points higher than the traditional 4x capacity at 2.5x cost. The effect of this will be a more rapid shift to higher bit rate technologies; carriers appear to be willing to forgo large cost-per-bit reductions to move to higher bit rates ... "

[SCH201101].

On the other hand, some observers state that perhaps there could be some developments in the market whereby content providers have been charged by transport providers to increase available throughput to the consumer (for Comcast charging Netflix for a bigger pipe). If this trend continues, then the subscriber will end up paying more for content, and this might have the effect of dampening the expected growth in demand for more bandwidth [KAN201401], [EVA201401].

1.7 INNOVATION NEEDED

Chatelains may obliviously enjoy their démodé craft for yet another evanescent moment, but the proven fact in high technology over several recent decades is that the financial analytics and the end-user consumer purchases ultimately drive practically all final determinations and all outcomes. Packet-based IP services are the wave of the future for all media and applications.

The climacteric question fundamentally is: Will investor banks and investors continue to make multibillion dollar financial infusions in a utility-class, satellite-operator industry with an after-inflation growth prospect of 1–2% a year over the next decade; an industry generally reefed with superannuated middle management that often is lacking in vision for much-needed market-focused innovation; an industry swamped with superfluously redundant, zero-revenue/negative-cash-flow ground infrastructure and staff in outdated, often overbuilt, teleports; and, an industry where the growth in stable First and Second World markets is basically zero, or negative, and where the only potential tepid double-digit growth is in Third World tropical markets, many such markets in economically and politically risky environments and often saddled with difficult revenue collection prospects? In short, the answer is that the industry and the operators can and will do fine, but if and only if when innovative concepts, practitioners, and management methods are rigorously welcomed and applied, and palpably anachronistic views congealed robotically in the early 1990s are properly re-assessed and re-engineered in favor of contemporary, forward-looking business savvy and technologically-updated approaches.

REFERENCES

[AGI200101] Agilent Technologies, "Digital Modulation in Communications Systems — An Introduction", Application Note 1298, March 14, 2001, Doc. 5965-7160E, 5301 Stevens Creek Blvd, Santa Clara, CA, 95051, United States.

[ASH201401] K. Ashton, "Finding a Flash Drive in the Sea", New York Times, April 28, 2014.

[BUC201401] D. Buchs, M. Welinski, M. Bouzegaoui, *Video Content Management And Distribution, Key Figures, Concepts and Trends*, Euroconsult Report, March 2014, Euroconsult, 86 Blvd. Sebastopol, 75003 Paris, France. www.euroconsult-ec.com.

[DER201301] N. de Ruiter, "High-throughput Satellites Transforming Industry Mindset", SpaceNews online magazine, December 2, 2013.

[ECR201401] President's Export Control Reform Initiative, The ECR Blog, May 14, 2014, http://export.gov.

[EDI201401] Editor, "1000 New Satellites Needed", Advanced Television, May 22, 2014.

[EVA201401] B. Evangelista, "Comcast Accuses Netflix of Lying as Public Feud Escalates", SFGate online Magazine, April 25, 2014.

[FAB201401] D. Faber, "Here's Why We're Getting Together: AT&T, DirecTV", CNBC, May 19, 2014, www.cnbc.com.

[FOC200701] Staff, "An Introduction to Earthstations", Focalpoint 2007, http://focalpoint-consulting.com. 1 bis rue Jasmin, 31800 ST GAUDENS, France.

[FOU201201] J. Foust, "An Opening Door for Hosted Payloads", Space News, October 30, 2012.

[GEO200101] Staff, "Geostationary, LEO, MEO, HEO Orbits Including Polar and Sun-Synchronous Orbits with Example Systems and a Brief Section on Satellite History", 2001.

[GLO201301] Global Assessment of Satellite Supply & Demand, 10th Edition A Region-Specific Supply and Demand Analysis of the Commercial Geostationary Satellite Transponder Market for 2012–2022, September 2013, Northern Sky Research, www.nsr.com.

[HEP201401] J. E. Heppelmann, "The Internet of Things, How a World of Smart, Connected Products is Transforming Manufacturers", White Paper, PTC Inc., 2014, J3220–IoT-eBook–EN–214. PTC Corporate Headquarters, 140 Kendrick Street, Needham, MA 02494, USA.

[HUG201301] Staff, "The View from Jupiter: High-Throughput Satellite Stems", White Paper by Hughes, July 2013. Retrieved at www.hughes.com/resources/the-view -from-jupiter-high-throughput-satellite-systems.

[JEF200401] D. Jefferies, "Microwaves: Satcoms Applications", MSc in Satcoms Notes, University Of Surrey, Department of Electronic Engineering, School of Electronics & Physical Sciences, Guildford, Surrey, UK, GU2 7XH, 18th March 2004.

[KAN201401] C. Kang, "Netflix Strikes Deal to Pay Comcast to Ensure Online Videos are Streamed Smoothly", Washington Post, February 23, 2014.

[LAU200701] J. E. Laube, HughesNet, "Introduction to the Satellite Mobility Support Network, HughesNet User Guide" 2005–2007.

[MIN197901] D. Minoli, "Satellite On-Board Processing of Packetized Voice", ICC 1979 Conference Record, 1979, pp. 58.4.1–58.4.5.

[MIN199101] D. Minoli, "Satellite Transmission Systems", Telecommunication Technologies Handbook, First Edition, Artech House, 1991.

[MIN201301] D. Minoli, *Building the Internet of Things with IPv6 and MIPv6: The Evolving World of M2M Communications*, Wiley, New York, 2013.

[PAT201301] C. Patton, "All Electric Satellites: Revolution or Evolution?" Via Satellite, May 1, 2013.

[ROS198201] R. D. Rosner, Distributed Telecommunications Networks via Satellites and Packet Switching (Electrical Engineering), Lifetime Learning Publications, First Edition, 1982. ISBN-10: 0534979246.

[ROS198401] R. D. Rosner, *Satellites Packets and Distributed Telecommunications: A Compendium of Source Materials*, Springer, First Edition, 1984. ISBN-10: 0534979246.

[SAT200501] Satellite Internet Inc., Satellite Physical Units & Definitions, http://www. satellite-internet.ro/satellite-internet-terminology-definitions.htm.

[SAT201401] Program Guide for SATELLITE 2014 Trade show, Washington, DC, March 10–13, 2014.

[SCH201101] A. Schmitt, *"The Fast Approaching 100G Era, Infonetics Research, Inc."*, Market Report, July 2011. 900 East Hamilton Avenue, Suite 230 Campbell, CA, 95008.

[SEL201401] P. B. de Selding, "Satellite Operators On Guard Against Ground Attack at 2015 Spectrum Conclave", Space News, March 11, 2014.

[SEL201402] P. B. de Selding, "All-Electric Satellites Prove a Tough Sell for Operators Anxious for Revenue Jolt", Space News, March 11, 2014.

[SEL201403] P. B. de Selding, "El Al Israel Airlines To Offer ViaSat in-air Broadband Service", Space News, March 5, 2014.

[SEL201404] P. B. de Selding, "With 17 Satellites Slated To Launch in 2014, Orbcomm Eyes Larger Share of M2M Market", SpaceNews, March 4, 2014.

[SHE201401] A. Sherman, J. McCracken, "AT&T Said in Advanced Talks to Buy DirecTV for About $50 Billion", Bloomberg. May 12, 2014.

[SMA201301] Small Satellite Conference, Marketing Materials, 2013. Small Satellite Conference, 695 North Research Park Way, North Logan, Utah, 84341.

[SST201401] Signals and Systems Telecom, The Wireless M2M & IoT Bible: 2014–2020 – Opportunities, Challenges, Strategies, Industry Verticals and Forecasts (Report), May 2014. Reef Tower, Jumeirah Lake Towers, Sheikh Zayed Road, Dubai, UAE.

[VIA201401] Viasat On-line information on ViaSat-1. Exede/WildBlue, 349 Inverness Drive South, Englewood, CO 80112, USA. www.viasat.com.

[WAI201401] A. W. Sargent, "Satellite's 'Big Four' Weigh in on Industry's Future", Satellite TODAY News Feed, March 11, 2014.

[WIL201401] K. Willems, DVB-S2X Demystified, Newtec Whitepaper, March 2014.

2

DVB-S2 MODULATION EXTENSIONS AND OTHER ADVANCES

We noted in Chapter 1 that business factors that impact the industry at press time include the user's (and operator's) desire for higher overall satellite channel and system throughput. Evolving applications, whether in the context of the delivery of High Definition (HD), Ultra HD, or Internet access for people on the move (on airplanes or ships), drive the operator's need for higher channel throughput. Beyond the intrinsic radio channel behavior, modulation and error correction dictate, in large measure, the performance of the satellite link, the effective channel throughput, and the service availability that one is able to obtain over the satellite link. Enhanced modulation schemes allow users to increase channel throughput: refined modulation and coding (MODCOD[1]) techniques are now being introduced as standardized solutions embedded in next-generation modems that provide more bits per second per unit of spectrum. Specifically, extensions to the well-established baseline DVB-S2 standard are now being introduced in the market. Adaptive coding that is incorporated with the new modulation schemes enables more efficient use of the higher frequency bands (e.g., Ka) that are intrinsically susceptible to rain-fade. Other ways of achieving higher application-level throughput are discussed in the next chapter.

The first part of the chapter provides a foundation for the modulation and Forward Error Correction (FEC) discussion; the second part then focuses on the recent advances for ground-based elements of the satellite ecosystem, specifically DVB-S2

---

[1] MODCOD is also written as modcod.

Innovations in Satellite Communications and Satellite Technology: The Industry Implications of DVB-S2X, High Throughput Satellites, Ultra HD, M2M, and IP, First Edition. Daniel Minoli.

extensions known as DVB-S2X. The last part of the chapter briefly covers Carrier ID. The concept of Intelligent Inverse Multiplexing for satellite applications is also discussed. Brief mention is made of High Efficiency Video Coding (HEVC)/ITU-T H.265, as another aspect of ground-related advances expected to impact the industry (a topic discussed in greater details in Chapter 6).

2.1   PART 1: A REVIEW OF MODULATION AND FEC PRINCIPLES

We review some fundamental technical principles in this introductory section, before looking at the latest advancements in this arena.

2.1.1 E_b/N_0 Concepts

E_b is a measure of the bit energy; $E_b = P_{avg}/R_b$, where R_b is the bit rate. E_b/N_0, the ratio of bit energy to noise power spectral density measured in decibel, is a commonly used parameter to compare digital systems, namely, modulation outcomes. Key impairments that impact demodulation are the following (refer to the Glossary for the definition of these terms):

- Additive Gaussian noise
- Amplitude imbalance
- Carrier suppression or leakage
- Interferers
- Phase error
- Phase jitter and/or phase noise

If the input signal is distorted or greatly attenuated (e.g., see Figure 2.1), the receiver can no longer recover the symbol clock, demodulate the signal, and/or recover the information. In some cases, a symbol will fall far away from its intended constellation position that it will cross over to an adjacent position; the in-phase *(I)* and quadrature *(Q)* level detectors used in the demodulator would misinterpret such a symbol as being in the wrong location, resulting in bit errors. In this context, note that Quadrature Phase Shift Keying (QPSK) is not as spectrally efficient as, say, 8-PSK or 16-QAM (Quadrature Amplitude Modulation), but the modulation states are much farther apart and the system can thus tolerate a lot more noise before suffering symbol errors [AGI200101].

To improve the Bit Error Rate (BER) performance beyond what the basic E_b/N_0 provides (or to optimize operation at a given fixed E_b/N_0), one needs to employ various FEC techniques. FEC is a system of error control for data transmission wherein the receiving device has the capability to detect and correct, in a simplex mode (1-way communication channel), any character or code block that contains fewer than a

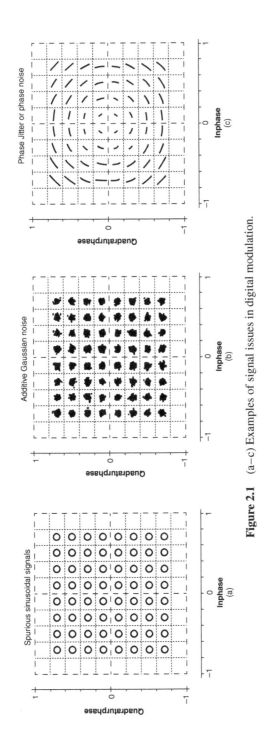

Figure 2.1 (a–c) Examples of signal issues in digital modulation.

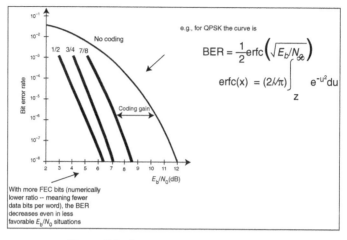

Figure 2.2   Improvements because of FEC.

predetermined number of symbols in error. FEC is accomplished by adding bits to each transmitted character or code block, using a predetermined algorithm (FEC is described in more detail in the next section). Figure 2.2 depicts the general behavior of a modulation scheme used without and with FEC; note that if there is a higher number of redundant bits, a given BER can be achieved with a lower E_b/N_0 (lower E_b or higher N_0). Also, note that the performance gradients become such that even the least improvement in E_b/N_0 can have a major impact on BER for a given FEC technique.

Some useful observations related to E_b/N_0 are as follows.

- Bit rate = (symbol rate) × (number of bits sent per symbol). For example, if two bits are transmitted per symbol (e.g., in QPSK), the symbol rate is half of the bit rate.
- N_0 is the noise density, namely, the total noise power in the frequency band of the signal divided by the bandwidth (BW) of the signal; $N_0 = P_n/B_n$ with P_n = Noise Power (the units here in Joules) and B_n = Noise BW. N_0 is measured in W/Hz. It is the noise power in 1 Hz of BW.
- 0 dB means the signal and noise power levels are equal and a 3 dB increment doubles the signal relative to the noise. Hence, if the power increases, E_b/N_0 increases; if the noise increases, E_b/N_0 decreases.
- A related concept is E_s/N_0 (also measured in dB). It represents the ratio of signal energy per symbol to noise power spectral density.

$$E_s/N_0 = (T_{sym}/T_{samp}) \times SNR$$

$$E_s/N_0 = E_b/N_0 + 10 \log_{10}(k') \text{ in } dB$$

where

SNR Signal-to-Noise Ratio
E_s Signal energy (Joules)
E_b Bit energy (Joules)
N_o Noise power spectral density (Watts/Hz)
T_{sym} Symbol period parameter of the block in E_s/N_o mode
k' Number of information bits per input symbol (which is not to be confused with Boltzmann's constant k)
T_{samp} Inherited sample time of the block, in seconds

Note that signal-to-noise ratio (SNR) is a measure of the quality of an electrical signal, usually at the receiver output. It is the ratio of the signal level to the noise level, measured within a specified BW (typically the BW of the signal). The higher the ratio, the better the quality of the signal. It is expressed in dB.

- The relationship between E_s/N_o and SNR can be derived as follows:

$$E_s/N_o(dB) = 10 \log_{10}[(S \cdot T_{sym})/N/B_n]$$

$$= 10 \log_{10}[(T_{sym}F_s) \cdot (S/N)]$$

$$= 10 \log_{10}[(T_{sym}/T_{samp}) + SNR(dB)]$$

where

T_{sym} Signal's symbol period
T_{samp} Signal's sampling period
S Input signal power, in Watts
N Noise power, in Watts
B_n Noise BW, in Hertz
F_s Sampling frequency, in Hertz

Note that $B_n = F_s = 1/T_{samp}$.

- One can show that ([MAR200201])

$$\text{probability of error} = P(e) = \frac{1}{2}\text{erfc}\left(\frac{E_b}{N_o}\right)^{1/2}$$

$$\text{erfc}\left(\frac{E_b}{N_o}\right)^{1/2} \approx (1/\sqrt{\pi}) \frac{\exp(-E_b/N_o)}{\sqrt{(E_b/N_o)}} \quad \text{when } E_b/N_o \geq 4(\sim 6 \text{ dB})$$

where erfc, that is, the complimentary error function, describes the cumulative probability curve of a Gaussian distribution, namely,

$$\text{erfc}(x) = (2/\sqrt{\pi})\int_x^\infty e^{-u^2}du$$

- Note that in an analog environment, the measures of interest are C/N (carrier power to noise power ratio) and C/N_0. These two measures are employed the same way E_b/N_0 is used in digital environments. C/N is the carrier power in the entire usable BW, C/N_0 is carrier power per unit BW. One can convert E_b/N_0 to C/N using the equation:

$$\frac{C}{N} = \frac{E_b}{N_0} \times \frac{R_b}{B}$$

where

B Channel BW
R_b Bit rate

When all the terms are in decibel, then one has

$$\frac{C}{N} = \frac{E_b}{N_0} \times R_b - B$$

2.1.2   FEC Basics

Along with modulation, FEC drives, in large measure, the performance of the satellite link, the effective channel throughput, and the service availability that one is able to obtain. FEC is required to achieve high performance in satellite links given the typical presence of high levels of noise and interference. FEC mechanisms can detect and correct a (certain) number of errors without retransmitting the data stream; this is done utilizing coded extra bits added to the transmitted data blocks.

Figure 2.3 depicts a basic diagram of FEC encoding. A FEC encoder adds redundancy to the data message at the transmitter according to certain defined rules. The FEC decoder at the receiver uses the knowledge of these rules to identify and, if possible, correct any errors that have accrued during transmission. FEC codes take a group of g data packets and generate $h = n-g$ parity packets. The total block size n consists

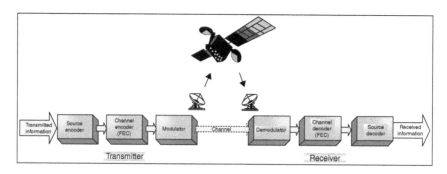

Figure 2.3   Basic diagram of FEC encoding.

of the g data and h parity packets. Once the parity packets have been computed, the block is then transmitted. The block format may vary according to the application. Typically, one has to be concerned with the following parameters:

g Number of data packets per block
h Number of parity packets per block where $h = n - k''$
n Number of total packets per block
R % of total BW used for redundancy
T_{FEC} FEC latency

There are two basic types of FEC codes: [LUB200201] and [LIT200101] (also see Table 2.1):

- Block codes. In these algorithms the encoder processes a block of message symbols and then outputs a block of codeword symbols. Broadly deployed Bose–Chaudhuri–Hocquenghem (BCH) codes belong to the family of block codes, as do the Reed–Solomon (RS) codes. The input to a FEC encoder is some number k of equal-length source symbols. The FEC encoder generates $n-k$ redundant symbols yielding an encoding block of n encoding symbols in total (composed of the k source symbols and the $n-k$ redundant symbols). Namely, a (n,k) linear block encoder takes k-bit block of message data and appends $n-k$ redundant bits algebraically related to the k message bits, producing an n-bit code block. The chosen length of the symbols can vary on each application of the FEC encoder, or it can be fixed. These encoding symbols are placed into packets for transmission. The number of encoding symbols placed into each packet can vary on a per packet basis, or a fixed number of symbols (often one) can be placed into each packet. Also, in each packet is placed enough information to identify the particular encoding symbols carried in that packet. A block FEC decoder has the property that any k of the n encoding symbols in the encoding block, which is sufficient to reconstruct the original k source symbols. On receipt of packets containing encoding symbols, the receiver feeds these encoding symbols into the corresponding FEC decoder to recreate an exact copy of the k source symbols. Ideally, the FEC decoder can recreate an exact copy from any k of the encoding symbols.
- Convolution codes. In these algorithms, instead of processing message symbols in discrete blocks, the encoder works on a continuous stream of message symbols and simultaneously generates a continuous encoded output stream. Convolutional coding is a bit-level encoding technique rather than block-level techniques such as RS coding. These codes get their name because the encoding process can be viewed as the convolution of the message symbols and the impulse response of the encoder.

The performance improvement that occurs when using error control coding is often measured in terms of coding gain. Let us suppose an uncoded communications system achieves a given BER at an SNR of 35 dB. Imagine that an error control coding

TABLE 2.1　Key FEC Families

FEC Algorithm	Definition/Characteristics
Bose–Chaudhuri–Hocquenghem (BCH)	A block code developed by the three named researchers (BCH codes were invented in 1959 by Hocquenghem, and independently in 1960 by Bose and Ray-Chaudhuri.). It is a generalization of the Hamming codes. BCH encodes k data bits into n code bits by adding $n-k$ parity checking bits. Galois Fields (GF) concepts are used. BCH codes are cyclic codes over $GF(q)$ (the channel alphabet) that are defined by a $(d-1) \times n$ check matrix over $GF(q^m)$ (the decoder alphabet) [GIL201201]: $$H = \begin{bmatrix} 1 & \alpha^b & \alpha^{2b} & \cdots & \alpha^{(n-1)b} \\ 1 & \alpha^{b+1} & \alpha^{2(b+1)} & \cdots & \alpha^{(n-1)(b+1)} \\ \vdots & \vdots & \vdots & \ddots & \vdots \\ 1 & \alpha^{b+d-2} & \alpha^{2(b+d-2)} & \cdots & \alpha^{(n-1)(b+d-2)} \end{bmatrix}$$ where • α is an element of $GF(q^m)$ of order n • b is any integer ($0 \le b < n$ is sufficient) • d is an integer with $2 \le d \le n$ ($d = 1$ and $d = n + 1$ are trivial cases) Rows of H are the first n powers of consecutive powers of α.
Reed–Solomon (RS)	Reed–Solomon codes are BCH codes where decoder alphabet = channel alphabet. The Reed–Solomon codec (coder–decoder) is a block-oriented coding system that is applied on top of the standard Viterbi coding. It corrects the data errors not detected by the other coding systems, significantly reducing the bit error rates (BERs) at nominal signal-to-noise levels ($4-8\,dB$ E_b/N_o). Bandwidth expansion is small, following the increase in code rate in the range $6-12\%$. Typically, Reed–Solomon coding is used in areas where sensitivity to transmission errors is particularly high. The "Reed–Solomon + Viterbi" option is particularly well suited to data communication applications with little or no packet acknowledgment or no packet retransmission, such as is the case in satellite communication.
Viterbi Algorithm	The term Viterbi specifically implies the use of the Viterbi Algorithm (VA) for decoding, although the term is often used to describe the entire error correction process. The encoding method is referred to as convolutional coding or trellis-coded modulation. A state diagram illustrating the sequence of possible codes creates a constrained structure called a trellis. The coded data is usually modulated; hence, the name trellis-coded modulation. The outputs are generated by convolving a signal with itself, which adds a level of dependence on past values [HEN200201].

TABLE 2.1 *(Continued)*

FEC Algorithm	Definition/Characteristics
Turbo Codes	In 1993, a new coding and decoding scheme, dubbed "Turbo codes" by its discoverers, was introduced that achieves near-capacity performance on additive white Gaussian noise channel. This technique was based in the use of two concatenated convolutional codes in parallel. After the introduction of Turbo codes the industry has seen the emergence of a plethora of codes that exploit a Turbo-like structure. The common aspect of all Turbo-like codes (TLC) is the concatenation of two or more simple codes separated by an interleaver, combined with an iterative decoding strategy. Members of the TLC family are: • Parallel Concatenated Convolutional Codes (PCCC) (this is the classical Turbo code) • Serially Concatenated Convolutional Codes (SCCC) • Low-Density Parity Check Codes (LDPC), and • Turbo Product Codes (TPC).

scheme with a coding gain of 3 dB was added to the system. This coded system would be able to achieve the same BER at the even lower SNR of 32 dB. Alternatively, if the system was still operated at an SNR of 35 dB, the BER achieved by the coded system would be the same BER that the uncoded system achieved at an SNR of 38 dB. The power of the coding gain is that it allows a communications system to either maintain a desired BER at a lower SNR than was possible without coding, or achieve a higher BER than an uncoded system could attain at a given SNR [LIT200101]. Figure 2.4 shows the advantage of combining the technologies [COM199801].

The desirable properties of an FEC algorithm are the following [TRE200401]:

• Good threshold performance – The waterfall region of an FEC algorithm's BER curve where there is a steep reduction of the error probability should occur at as low an E_b/N_o as possible. This will minimize the energy expense of transmitting information;

• Good floor performance – The error floor region of an FEC algorithm's BER curve should occur at a BER as low as possible. For communication systems employing an ARQ (Automatic Repeat Request) scheme (such as used in HDLC-like protocols) this may be as high as 10^{-6}, while most broadcast communications systems require 10^{-10} performance, and storage systems and optical fiber links require BERs as low as 10^{-15};

• Low-complexity code constraints – To allow for low-complexity decoders, particularly for high throughput applications, the constituent codes of the FEC algorithm should be simple. Furthermore, to allow the construction of high

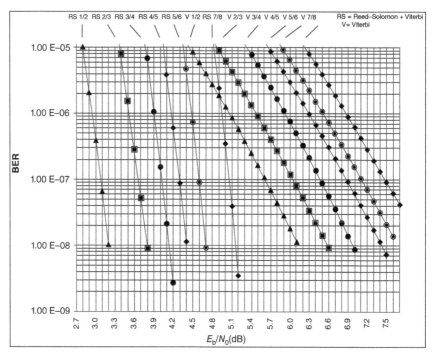

Figure 2.4 Improvement of "Reed–Solomon+Viterbi" over simple Viterbi for typical QPSK system.

throughput decoders, the code structure should be such that parallel decoder architectures with simple routing and memory structures are possible;

- Fast decoder convergence – The decoder of an FEC algorithm code should converge rapidly (i.e., the number of iterations required to achieve most of the iteration gain should be low). This will allow the construction of high through-put hardware decoders (or low-complexity software decoders);

- Code rate flexibility – Most modern communications and storage systems do not operate at a single code rate. For example, in adaptive systems the code rate is adjusted according to the available SNR so that the code overheads are minimized. It should be possible to fine-tune the code rate to adapt to varying application requirements and channel conditions. Furthermore, this code rate flexibility should not come at the expense of degraded threshold or floor perfor-mance. Some systems require code rates of 0.95 or above, but this is difficult for most FEC algorithms to achieve;

- Block size flexibility – One factor that all FEC algorithms have in common is that their threshold and floor performance is maximized by maximizing their block size. However, it is not always practical to have blocks of many thousands of bits. Therefore, it is desirable that an FEC algorithm still performs well with block sizes as small as only one or two hundred bits. Also, for flexibility, it is

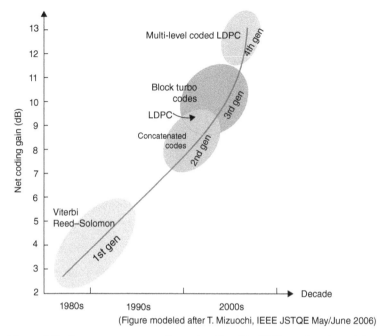

Figure 2.5   Improvements in coding gain offered by FEC as a function of time.

very desirable that the granularity of the block sizes be very small, ideally just one bit; and

- Modulation flexibility – In modern communication systems employing adaptive coding and modulation it is essential that the FEC algorithm easily support a broad range of modulation schemes.

Research activities in FEC techniques in the past 20 years have given rise to new theoretical approaches. Modern approaches are parallel or serially concatenated convolutional codes, product codes, and low-density parity check codes (LDPCs) – all of which use "turbo" (i.e., recursive) decoding techniques [MOR200401]. Satellite modems intended for common use during the past decade have employed "RS + Viterbi" encoding extensively; newer systems use either Turbo Codes or LDPC. Figure 2.5 shows the progression of the work in the recent past. In looking at a diagram that plots BER versus E_b/N_o, we can find that there typically is an initial steep reduction in error probability as E_b/N_o increases (this area is called the waterfall region), followed by a region of shallower reduction (this is called the error floor region).

Improvements have occurred in recent years. In the late 1960s, Forney introduced a concatenated scheme of inner and outer FEC codes: the inner code is decoded using soft-decision channel information, while the outer Reed–Solomon code uses errors and erasures. In 1993, a new coding and decoding scheme was reported that achieves near-capacity performance on additive white Gaussian noise channel. The technique was dubbed "Turbo codes" by its discoverers. It is based in the use of two concatenated convolutional codes in parallel. The turbo code has a floor error near 10^{-5}, but the design of the interleaver is very critical in order to achieve good results in low BER (it is possible to avoid the floor error with an adequate design of the interleaver). In 1996, a new technique was proposed, based in the same idea of Turbo codes, called Serial Concatenated Convolutional Codes (SCCCs); this new codification technique achieves better results for low BER than Turbo code. This technique avoids both problems of Turbo codes: first, the floor error disappears, and, second, the design of the interleaver is easier, because the input data of the two encoders are different [TOR199801]. After the introduction of Turbo codes. one has seen the emergence of a plethora of codes that exploit a Turbo-like structure. The common aspect of all Turbo-like codes (TLCs) is the concatenation of two or more simple codes separated by an interleaver, combined with an iterative decoding strategy. TLC algorithms are:

- Parallel Concatenated Convolutional Codes (PCCCs) (this is the classical Turbo code)
- SCCCs
- LDPC, and
- Turbo Product Codes (TPCs).

Turbo codes and TLC offer good coding gain compared to traditional FEC approaches – as much as 3 dB in many cases. The area where TLCs have traditionally excelled is in the waterfall region – there are TLCs that are within a small fraction of a dB of the Shannon limit in this region; however, many TLCs have an almost flat error floor region, or one that starts at a very high error rate, or both. This means that the coding gain of these TLCs rapidly diminishes as the target error rate is reduced. The performance of a TLC in the error floor region depends on several factors, such as the constituent code design and the interleaver design, as noted, but it typically worsens as the code rate increases or the block size decreases. Many TLC designs perform well only at high error rates, low code rates, or large block sizes; furthermore, these designs often target only a single code rate, block size, and modulation scheme, or suffer degraded performance or increased complexity to achieve flexibility in these areas [TRE200401].

Parallel Concatenated Convolutional Codes (PCCC). PCCCs consist of the parallel concatenation of two convolutional codes. One encoder is fed the block of information bits directly, while the other encoder is fed an interleaved version of the information bits. The encoded outputs of the two encoders must be mapped to the signal set used on the channel. A PCCC encoder is formed by two (or more) constituent systematic encoders joined through one or more interleavers. The input information bits feed the first encoder and, after having been scrambled by the interleaver, they enter the second encoder. A code word of a parallel concatenated code consists of the input bits to the first encoder followed by the parity check bits of both encoders.

PCCC achieves near-Shannon-limit error correction performance. Bit error probabilities as low as 10^{-5} at $E_b/N_o = 0.6$ dB have been shown. PCCCs yield very large coding gains (10 or 11 dB) at the expense of a small data reduction or BW increase [TOR199801] (Table 2.2).

Serially Concatenated Convolutional Codes (SCCC). SCCCs consist of the serial concatenation of two convolutional codes. The outer encoder is fed the block of information bits and its encoded output is interleaved before being input to the inner encoder. The encoded outputs of only the inner encoder must be mapped to the signal set used on the channel. A code word of a serial concatenated code consists of the input bits to the first encoder followed by the parity check bits of both encoders. SCCC achieves near-Shannon-limit error correction performance. Bit error probabilities as low as 10^{-7} at $E_b/N_o = 1$ dB have been shown. SCCC yields very large coding gains (10 or 11 dB) at the expense of a small data reduction or BW increase [TOR199801] (Table 2.2).

Low-Density Parity Check Codes (LDPC). LDPCs are block codes defined by a sparse parity check matrix. This sparseness admits a low-complexity iterative decoding algorithm. The generator matrix corresponding to this parity check matrix can be determined and used to encode a block of information bits; these encoded bits must then be mapped to the signal set used on the channel. The performance of LDPC is within 1 dB of the theoretical maximum performance known as the Shannon limit (Table 2.2).

Turbo Product Codes (TPC). In a TPC system the information bits are arranged as an array of equal-length rows and equal-length columns. The rows are encoded by one block code and then the columns (including the parity bits generated by the first encoding) are encoded by a second block code. The encoded bits must then be mapped onto the signal set of the channel (Table 2.2).

2.1.3 Filters and Roll-Off Factors

Newer modulation techniques utilize tighter roll-off factors. Some basic concepts related to roll-off and filters are provided in this section. High symbol rate modulation may give rise to (undesirable) signals in bands other than the assigned/intended band. Occupied BW is a measure of how much frequency spectrum is covered by the signal in question; the units are in Hertz, and measurement of occupied BW generally implies a power percentage or ratio. A typical direct conversion transmitter uses the low-pass filters (LPFs) to eliminate out-of-band signal components from the baseband signal, but these components may re-appear along the way as a result of intermodulation distortion generated within the transmitter, for example, in the downstream power amplifier (PA), such as the satellite transponder. This tendency to regain undesirable signal components is called "spectral regrowth." Adjacent Channel Power Ratio (ACPR) is a popular measure of spectral regrowth in digital transmitters and is often a design specification [CHE200701], [AMO199701]. Nonlinearities in the transmitter introduce harmonics of ω_m about the carrier fundamental. Given these observations on the possibility for spectral regrowth, a key question is whether a

TABLE 2.2   Key Aspects of Various Turbo Codes

Coding Type	Characteristics (loosely based on [TRE200401])
Parallel concatenated convolutional codes (PCCC)	• Good threshold performance – Among the best threshold performance of all TLCs; • Poor floor performance – The worst floor performance of all TLCs. With 8-state constituent codes BER floors are typically in the range 10^{-6}–10^{-8}, but this can be reduced to the 10^{-8}–10^{-10} range by using 16-state constituent codes. However, achieving floors below 10^{-10} is difficult, particularly for high code rates and/or small block sizes. There are PCCC variants that have improved floors, such as concatenating more than two constituent codes, but only at the expense of increased complexity; • High complexity code constraints – With 8- or 16-state constituent codes, the PCCC decoder complexity is rather high; • Fast convergence – Among the best convergence of all TLCs. Typically, only 4–8 iterations are required; • Fair code rate flexibility – Code rate flexibility is easily achieved by puncturing the outputs of the constituent codes. However, for very high code rates the amount of puncturing required is rather high, which degrades performance and increases decoder complexity (particularly with windowed decoding algorithms); • Good block size flexibility – The block size is easily modified by changing the size of the interleaver, and there exist many flexible interleaver algorithms that achieve good performance; and • Good modulation flexibility – The systematic bits and the parity bits from each constituent code must be combined and mapped to the signal set.
Serially concatenated convolutional codes (SCCC)	• Medium threshold performance – The threshold performance of SCCCs is typically 0.3 dB worse than PCCCs; • Good floor performance – Among the best floor performance of all TLCs. Possible to have BER floors in the range 10^{-8}–10^{-10} with 4-state constituent codes, and below 10^{-10} with 8-state constituent codes. However, floor performance is degraded with high code rates; • Medium complexity code constraints – Code complexity constraints are typically low, but are higher for high code rates. Also, constituent decoder complexity is higher than for the equivalent code in a PCCC because soft decisions of both systematic and parity bits must be formed;

TABLE 2.2   *(Continued)*

Coding Type	Characteristics (loosely based on [TRE200401])
	• Fast convergence – Convergence is even faster than PCCCs, with typically 4–6 iterations required; • Poor code rate flexibility – As for PCCCs, code rate flexibility is achieved by puncturing the outputs of the constituent encoders. However, because of the serial concatenation, this puncturing must be higher for SCCCs than for PCCCs for an equivalent overall code rate; • Good block size flexibility – As for PCCCs, block size flexibility is achieved simply by changing the size of the interleaver. However, for equivalent information block sizes the interleaver of an SCCC is larger than for a PCCC, so there is a complexity penalty in SCCCs for large block sizes. Also, if the code rate is adjusted by puncturing the outer code, then the interleaver size will depend on both the code rate and block size, which complicates reconfigurability; and • Very good modulation flexibility – As the inner code on an SCCC is connected directly to the channel, it is relatively simple to map the bits onto the signal set.
Low-density parity check codes (LDPC)	• Good threshold performance – LDPCs have been reported that have threshold performance within a tiny fraction of a dB of the Shannon Limit. However, for practical decoders their threshold performance is usually comparable to that of PCCCs; • Medium floor performance – Floors are typically better than PCCCs, but worse than SCCCs; • Low-complexity code constraints – LDPCs have the lowest complexity code constraints of all TLCs. However, high throughput LDPC decoders require large routing resources or inefficient memory architectures, which may dominate decoder complexity. Also, LPDC encoders are typically a lot more complex than other TLC encoders; • Slow convergence – LDPC decoders have the slowest convergence of all TLCs. Many published results are for 100 iterations or more. However, practical LDPC decoders will typically use 20–30 iterations; • Good code rate flexibility – LDPCs can achieve good code rate flexibility;

(continued)

TABLE 2.2   *(Continued)*

Coding Type	Characteristics (loosely based on [TRE200401])
	• Poor block size flexibility – For LDPCs to change block size, they must change their parity check matrix, which is quite difficult in a practical, high throughput decoder; and • Good modulation flexibility – As with PCCCs the output bits of an LDPC must be mapped onto the signal sets of different modulation schemes.
Turbo product codes (TPC)	• Poor threshold performance – TPCs have the worst threshold performance of all TLCs; they can have thresholds that are as much as 1 dB worse than PCCCs. However, for very high code rates (0.95 and above) they will typically outperform other TLCs; • Medium floor performance – Their floors are typically lower than PCCC, but not as low as SCCC; • Low-complexity code constraints – TPC decoders have the lowest complexity of all TLCs, and high throughput parallel decoders can readily be constructed; • Medium convergence – TPC decoders converge quickly, with 8–16 iterations being typically required; • Poor rate flexibility – The overall rate of a TPC is dictated by the rate of its constituent codes. There is some flexibility available in these rates, but it is difficult to choose an arbitrary rate; • Poor block size flexibility – The overall block size of a TPC is dictated by the block size of its constituent codes. It is difficult to choose an arbitrary block size, and especially difficult to choose an arbitrary code rate and an arbitrary block size; and • Good modulation flexibility – As with PCCCs and LDPCs, the output bits must be mapped onto the signal sets.

modulated signal fits "neatly" inside a transponder, or whether there may be a leakage of signal into another transponder. Figure 2.6 depicts the relative power spectral density (in W/Hz, measured in dB) of a digitally modulated carrier that uses BPSK and QPSK without applying any filtering (the x-axis shows normalized frequency $(f-f_c)/$(bit rate) where f_c is the carrier frequency; the y-axis is the relative level of the power density with respect to the maximum value at carrier frequency f_c). Two issues are worth considering:

1. The width of the principal lobe of the spectrum of the modulated signal, which affects the required/utilized BW of the channel; and

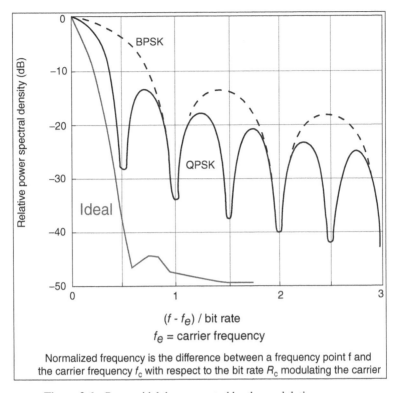

Figure 2.6 Power sidelobes generated by the modulation process.

2. The secondary sidelobes, which affect adjacent carriers by generating interference when the spectral decay with respect to frequency is not as sharp as one would hope.

The figure shows that QPSK is "better" than BPSK, having a lower main lobe. This is because, in general, constellation diagrams show that transition to new states could result in large amplitude changes (e.g., going from 010 to 110 in 8-PSK), and a signal that changes amplitude over a very large range will exercise amplifier nonlinearities, which cause distortion products. In continuously modulated systems large signal changes will cause "spectral regrowth" or wider modulation sidebands (a phenomenon related to intermodulation distortion). The problem lies in nonlinearities in the circuits. If the amplifier and associated circuits were perfectly linear, the spectrum (spectral occupancy or occupied BW) would be unchanged [AGI200101]. As stated, any fast transition in a signal, whether in amplitude, phase, or frequency, results in a power-frequency spectrum that requires a wide occupied BW. Any technique that helps to slow down transitions narrows the occupied BW. To deal with this spectral regrowth issue and limit unwanted interference to adjacent channels *filtering* is used at the transmitter side. In addition to the intrinsic phenomenon of spectral spreading of the modulation sidelobes, an attempt to transmit at maximum power, either at the remote terminal or at the satellite, produces amplitude and phase distortions. To avoid operating in the nonlinear region of the amplifier, one typically needs to backoff (from using) the maximum transponder power.

The rest of this section focuses on filtering. Filtering allows the transmitted BW to be reduced without losing the content of the digital data, thereby improving the spectral efficiency of the signal. The most common filtering techniques are: raised cosine filters, square-root raised cosine filters, and Gaussian filters. Filtering smoothes transitions (in I and Q), and, as a result, reduces interference because it reduces the tendency of one signal or one transmitter to interfere with another in a Frequency Division Multiple Access (FDMA) system. At the receiver's end, reduced BW improves sensitivity because more noise and interference are rejected. It should be noted, however, that some trade-offs must be taken into consideration: carrier power cannot be limited (clipped) without causing the spectrum to spread out once again; because narrowing the spectral occupancy was the reason the filtering was inserted in the first place; the designer must select the choice carefully [AGI200101]. One trade-off is that some types of filtering cause the trajectory of the signal (the path of transitions between the states) to overshoot in many cases. This overshoot can occur in certain types of filters (such as Nyquist filters). This overshoot path represents carrier power and phase. For the carrier to take on these values more output power from the transmitter amplifiers is required; specifically, it requires more power than would be necessary to transmit the actual symbol itself. Another consideration is Inter-Symbol Interference (ISI). ISI is the interference between adjacent symbols often caused by system filtering, dispersion in optical fibers, or multipath propagation in radio system. This occurs when the signal is filtered enough so that the symbols blur together and each symbol affects those around it. This is determined by the time-domain response or impulse response of the filter. There are different types of filtering. As noted, the most common are[2]: Raised cosine, Square-root raised cosine, and Gaussian filters.

Raised cosine filters are a class of Nyquist filters. Nyquist filters have the property that their impulse response rings[3] at the symbol rate. The time response of the filter goes through zero with a period that exactly corresponds to the symbol spacing. Adjacent symbols do not interfere with each other at the symbol times because the response equals zero at all symbol times, except the center (desired) one. Nyquist filters filter the signal heavily without blurring the symbols together at the symbol times; this is important for transmitting information without errors caused by ISI. Note that ISI does exist at all times except the symbol (decision) times. Usually, the filter is split, half being in the transmit path and half in the receiver path; in this case root Nyquist filters (also commonly called root raised cosine) are used in each part, such that their combined response is that of a Nyquist filter. The raised cosine filter is an implementation of a low-pass Nyquist filter; its spectrum exhibits odd symmetry about $1/2T$ where T is the symbol period of the communications system. The frequency-domain

---

[2] The rest of this subsection is based on Reference [AGI200101].
[3] That is, it has the impulse response of the filter cross through zero.

description of the raised cosine filter is a piecewise function characterized by two values: α, the *roll-off factor*, and T, specified as:

$$H(f) = \begin{cases} 1.0, & |f| \leq \dfrac{1-\alpha}{2T} & \text{with } 0 \leq \alpha \leq 1 \\[3mm] \dfrac{1}{2}\left[1 + \cos\left(\dfrac{\pi T}{\alpha}\left[|f| - \dfrac{1-\alpha}{2T}\right]\right)\right], & \dfrac{1-\alpha}{2T} < |f| \leq \dfrac{1+\alpha}{2T} & \text{with } 0 \leq \alpha \leq 1 \\[3mm] 0, & |f| > \dfrac{1-\alpha}{2T} \end{cases}$$

The roll-off factor, α, is a measure of the excess BW of the filter, that is, the BW occupied beyond the Nyquist BW of $1/2T$. If we denote the excess BW as Δf, then

$$\alpha = \frac{\Delta f}{\left(\frac{1}{2T}\right)} = \frac{\Delta f}{R_s/2} = 2T\Delta f$$

where $R_s = 1/T$ and T the symbol period (symbol energy expands into the $[-T, T]$ interval in a time-amplitude domain). The time-domain ripple level increases as α decreases (from 0 to 1). The excess BW of the filter can be reduced, but only at the expense of an elongated impulse response. As α approaches 0, the roll-off zone becomes infinitesimally narrow; hence, the filter converges to an ideal or brick wall filter. When $\alpha = 1$, the nonzero portion of the spectrum is a pure raised cosine. The BW of a raised cosine filter is most commonly defined as the width of the nonzero portion of its spectrum, that is:

$$\text{BW} = \frac{1}{2}R_s(1 + \alpha)$$

α is sometimes called the "excess BW factor" as it indicates the amount of occupied BW that will be required in the case of excessive, ideal occupied BW (which would be the same as the symbol rate).

The sharpness of a raised cosine filter is described by α (Figure 2.7). α provides a measure of the occupied BW of the system:

$$\text{occupied BW} = \text{symbol rate} \times (1 + \alpha)$$

with symbol rate $= 1/2T = (1/2)\,R_s$.

If the filter had a perfect (brick wall) characteristic with sharp transitions and $\alpha = 0$, the occupied BW would be

$$\text{occupied BW} = \text{symbol rate} \times (1 + 0) = \text{symbol rate}.$$

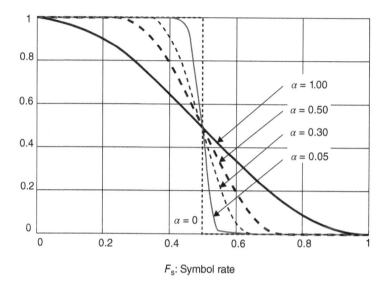

F_s: Symbol rate

Figure 2.7   Filter bandwidth parameters "α".

In a perfect world, the occupied BW would be the same as the symbol rate, but this is not practical. An $\alpha = 0$ is impossible to implement. At the other extreme, consider a broader filter with an $\alpha = 1$, which is easier to implement. The occupied BW will be

$$\text{occupied BW} = \text{symbol rate} \times (1 + 1) = 2 \times \text{symbol rate}.$$

An $\alpha = 1$ uses twice as much BW as an $\alpha = 0$. In practice, it is possible to implement an α below 0.2 and make good, compact, practical devices, though some video systems use an α as low as 0.10. Typical values range from 0.35 to 0.5.[4]

A designer seeks to maximize the power and BW efficiency of satellite terminals, in particular, for Very Small Aperture Terminals (VSATs), where a key limitation to communication capacity is a nonlinear transmitting PA. It is highly undesirable to waste the RF spectrum by using channel bands that are too wide. Therefore, filters are used to reduce the occupied BW of the transmission. The data pulse stream is band-limited for BW efficiency; however, the effect of filtering (in the time domain) is to cause envelope excursions outside the linear region of the PA characteristic. The traditional approach is to reduce the average power at the input of the amplifier such that the maximum signal excursions are within the linear range of the amplifier characteristic. This reduction is termed power backoff and, as defined earlier, can be significant for BW-efficient systems [AMB200301].

---

[4]The corresponding term for a Gaussian filter is BT (bandwidth time product). Occupied bandwidth cannot be stated in terms of BT because a Gaussian filter's frequency response does not go identically to zero, as does a raised cosine. Common values for BT are 0.3–0.5.

Narrow filters with sufficient accuracy and repeatability are somewhat difficult to build. Smaller values of α increase ISI because more symbols can contribute to the demodulated signal; this tightens the requirements on clock accuracy. These narrower filters also result in more overshoot and, therefore, more peak carrier power. The PA must then accommodate the higher peak power without distortion. Larger amplifiers cause more heat and thus cause more electrical interference to be produced because the RF current in the PA will interfere with other circuits.

Predistortion methods in the uplink station can be used to minimize the effect of transponder nonlinearity. As discussed in Appendix A, PAs are inherently nonlinear and when operated near saturation cause intermodulation products that interfere with adjacent and alternate channels. This interference affects the adjacent channel leakage ratio (ACLR). Linearization techniques allow PAs to operate efficiently and at the same time maintain acceptable ACLR levels (in effect, linearization affords the use of a lower-cost more efficient PA in place of a higher-cost less-efficient PA). PAs in the field today are predominately linearized by some form of feed-forward technology. In recent years, however, designers have stated to employ digital predistortion (also known as preemphasis); compared to feed-forward architectures, designs based on digital predistortion approaches enjoy higher efficiency at lower cost [SIL200201]. Predistortion is a system process designed to increase, within a band of frequencies, the magnitude of some (usually higher) frequencies with respect to the magnitude of other (usually lower) frequencies, in order to improve the overall SNR by minimizing the adverse effects of such phenomena as attenuation differences, or saturation of the PA [ANS200001]. In a group of waves that have slightly different individual frequencies, the group delay time is defined as the time required for any defined point on the envelope (i.e., the envelope determined by the additive resultant of the group of waves) to travel through a device or transmission facility. Group delay is thus the rate of change of the total phase shift with respect to angular frequency, $d\theta/d\omega$, through a device or transmission medium, where θ the total phase shift can be found, ω the angular frequency is equal to $2\pi sf$, and f is the frequency. Predistortion can be employed to deal with group delay.

2.2 PART 2: DVB-S2 AND DVB-S2 EXTENSIONS

Because of the nature of the satellite link (noise, attenuation), 8-PSK is the traditional technical "sweet spot"; however, more complex modulation schemes are being introduced. It should be noted that it is useful to employ standards to deploy systems so that equipment from multiple suppliers can be used in a large network (although generally for a variety of other reasons – particularly network management – operators tend not to comingle equipment for the same element of an RF link in any large measure).

2.2.1 DVB-S2 Modulation

DVB-S2 is a second-generation framing structure, channel coding and modulation system for broadcasting, interactive services, news gathering and other broadband satellite applications. DVB-S2 is a specification developed by the Digital Video Broadcasting (DVB) Project adopted by European Telecommunications Standards Institute (ETSI) standards is now used worldwide. As the name implies, DVB-S2 is second-generation specification for satellite broadcasting; it was developed in 2003. Extensions were being proposed and standardized at press time. This section discusses DVB-S2 (Section 2.2.1) followed by an analysis of extensions, which build on and extend DVB-S2 (Section 2.2.2).

The DVB-S2 standard has been specified with the goals of: (i) achieving best transmission performance, (ii) embodying flexibility, and (iii) requiring reasonably low receiver complexity (using existing chip technology). The basic documents defining the DVB-S2 standard are:

- ETSI: Draft EN 302 307: *DVB; Second generation framing structure, channel coding and modulation systems for Broadcasting, Interactive Services, News Gathering and other broadband satellite applications (DVB-S2).*
  - EN302307v1.1.1 in 2005-03
  - EN302307v1.1.2 in 2006-06
  - EN302307v1.2.1 in 2009-08
  - EN302307v1.3.1 in 2013-03
- ETSI: EN 300 421: *DVB; Framing structure, channel coding and modulation for 11/12 GHz satellite services*
- ETSI: EN 301 210: *DVB: Framing structure, channel coding and modulation for Digital Satellite News Gathering (DSNG) and other contribution applications by satellite.*

The DVB-S2 system has been designed for satellite broadband applications such as broadcast services for MPEG-2/MPEG-4 Standard Definition Television (SDTV) and High Definition Television (HDTV) (including single or multiple MPEG Transport Streams (TSs), continuous bit streams, and IP packets), and interactive services, namely, Internet access. The DVB-S2 system may be used in "single-carrier-per-transponder" or in "multi-carriers-per-transponder" configurations. When introduced in the 2000s, the DVB-S2 standard was a quantum leap in terms of BW efficiency compared to the former DVB-S and DVB-DSNG standards. This improvement is not only because of the use of LDPC for FEC, but also of new modulation schemes and new modes of operations, specifically, Variable Coding and Modulation (VCM) and Adaptive Coding and Modulation (ACM) [BRE200501]. AMC is a link adaptation method (typically used in 3G mobile wireless communications) that provides the flexibility to match the modulation-coding scheme to the average channel conditions for each user. When using AMC, the modulation and coding format is changed dynamically to match the current-received signal quality or channel conditions (however, the power of the transmitted signal is held constant over a frame interval). A DVB-S2 system can operate at C/N from −2.4 dB (using QPSK with FEC at 1/4 rate) to 16 dB (using 32-APSK with FEC at 9/10 rate), assuming an Additive White Gaussian Noise (AWGN) channel and ideal demodulator. The distance from the Shannon limit ranges from 0.7 to 1.2 dB. On AWGN, the result is typically a 20–35% capacity increase over DVB-S and DVB-DSNG under the same transmission conditions and 2–2.5 dB more robust reception for the same spectrum efficiency.

DVB-S is the original DVB specification for satellite-based television distribution; DVB-S was standardized in 1994. The broadcast industry adopted the format in the late 1990s because it established a universal framework for MPEG-2 based digital television services to be broadcast over satellite using Viterbi concatenated with RS (Viterbi + RS) FEC and QPSK modulation. The performance of this code is such that a coding rate of at least 2/3 QPSK is required (to achieve a BER $< 10^{-7}$); a typical E_b/N_o of 6.5 dB is needed. The actual (payload) bit rate for 30 Mbaud with QPSK 2/3 is $1.229 \times 30 = 36.87$ Mbps. In 1999, the DVB-S standard was extended and became the DVB-DSNG standard. This new standard allowed for more efficient modes of modulation (such as 8-PSK and 16-QAM) to be utilized. Introducing the higher-order modulation resulted in overall savings in BW and required an increase of power to achieve similar E_b/N_o results [RAD200501].

We discuss DVB-S2 (and DVB-S) features briefly herewith to counterpoint these features with the capabilities of DVB-S2X. As noted above, to achieve the optimal performance complexity trade-off (about 30% capacity gain over DVB-S), DVB-S2 makes use of recent advancements in channel coding and modulation. Specifically, DVB-S2 uses LDPC concatenated with BCH coding. The chosen LDPC FEC codes utilize large blocks (64,800 bits for applications not too critical for delays, and 16,200 bits for application that are more critical). Code rates 1/4, 1/3, 2/5, 1/2, 3/5, 2/3, 3/4, 4/5, 5/6, 8/9, and 9/10 are available, depending on the selected modulation and the system requirements. Coding rates 1/4, 1/3, and 2/5 have been introduced to operate, in combination with QPSK, under poor link conditions, where the signal level is below the noise level. Four modulation modes can be selected for the transmitted payload as shown in Figure 2.8: QPSK, 8-PSK, 16-APSK, and 32-APSK. QPSK and 8-PSK are used for broadcast applications and can be used in nonlinear satellite transponders driven near saturation. While the 8-PSK scheme has seen extensive implementation and deployment, the 16-APSK and 32-APSK modes have been targeted at professional applications (they can also be used for broadcasting); these modulation schemes require a higher level of available *C/N* and the adoption of advanced predistortion methods in the uplink station to minimize the effect of transponder nonlinearity. Although these modes are not as power efficient as the other modes, the spectrum efficiency is greater. The 16-APSK and 32-APSK constellations have been optimized to operate over a nonlinear transponder by placing the points on circles. Nevertheless, their performance on a linear channel is comparable with those of 16-QAM and 32-QAM, respectively. By selecting the modulation constellation and code rates, spectrum efficiencies from 0.5 to 4.5 bit per symbol become available and can be chosen on the basis of the capabilities and restrictions of the satellite transponder used [MOR200401]. It should be noted that, in general, DVB-S2 is optimized for nonlinear operations [e.g., for Direct to Home (DTH) applications].

Figure 2.9 depicts these efficiencies for the modulation schemes supported. DVB-S2 has three "roll-off factor" choices to determine spectrum shape: 0.35 as

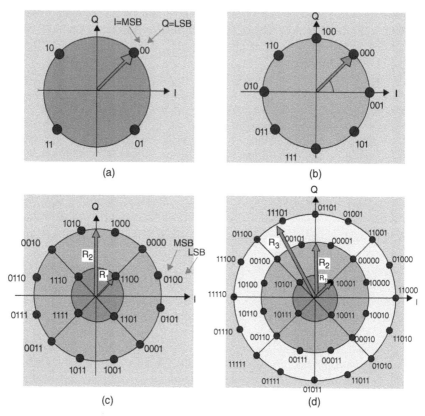

Figure 2.8 (a–d) DVB-S2 modulation constellations: 2-PSK, 3-PSK, 16-APSK, and 32-APSK.

in DVB-S, and 0.25 and 0.20 for tighter BW restrictions (the use of the narrower roll-off $\alpha = 0.25$ and $\alpha = 0.20$ allows transmission capacity to increase, but may also produce larger nonlinear degradations by satellite for single-carrier operation). (As seen above, DVB-S2 standard allows up to 5 bits per symbol with a 32-APSK constellation.)

Constant Coding Modulation (CCM) is the simplest mode of DVB-S2; it is similar to the DVB-S, in the sense that all data frames are modulated and coded using the same fixed parameters. The LDPC code has a performance such that a coding rate of 4/5 is sufficient for the same channel conditions. For example, a minimum E_s/N_o of 5.4 dB supports QPSK 4/5. The useful bit rate is then given by 30 Mbaud × 1.587 = 47.61 Mbps. Compared to DVB-S, this is an efficiency improvement of around 30%. This represents a 2–3 dB improvement compared to DVB-S [BRE200501].

Figure 2.10 compares DVB-S and DVB-S2 performance for QPSK and 8-PSK modulations with a variety of FEC rates. When one compares 8-PSK 2/3 FEC rate for a 1×10^{-9} bit error rate, the required E_b/N_o level for DVB-S is 6.10 dB versus 4.35 dB for DVB-S2, equating to a 1.75 dB reduction in the required power for

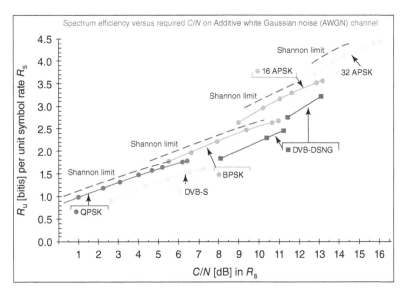

Figure 2.9 Efficiencies for the modulation schemes supported in DVB-S2.

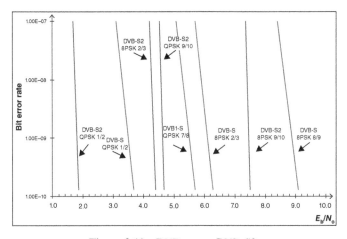

Figure 2.10 DVB versus DVB-S2.

the same E_b/N_o, reducing the required spectral power by 28.7%. For a data rate of 6 Mbps using 8-PSK 2/3 FEC rate, the DVB-S would have a symbol rate of 3.26 Msps, assuming a carrier spacing factor of 1.3, and would require 4.23 MHz of BW on the satellite. In comparison, the DVB-S2 would have a symbol rate of 3.03 Msps requiring 3.94 MHz of BW on the satellite, resulting in a BW savings of approximately 6.9% [RAD200501].

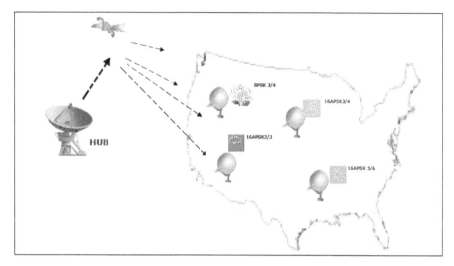

Figure 2.11   DVB-S2 variable coding and modulation.

A feature of the DVB-S2 standard is that different services can be transmitted on the same carrier, each using their own modulation scheme and coding rate (Figure 2.11). This "multiplexing" at the physical layer is known as VCM. VCM is useful when different services do not need the same protection level or different services are intended for different stations with different average receiving conditions. Using VCM, different Coding/Modulation (CM) mechanisms (also known as MODCODs) can be selected for each station. Typically, these CMs range between QPSK 4/5 and 16-APSK 2/3. If the total baud rate (30 Mbaud) is distributed equally over each station, each station will use a 30/20 = 1.5 Mbaud part of the carrier (assuming there were 20 stations in the system). The corresponding bit rates vary between 2.38 and 3.96 Mbps according to the selected CM. The total available rate is then 61.09 Mbps. This is an improvement of around 65% compared to DVB-S [BRE200501]. When a return channel is available from each receiving site to the transmit site, DVB-S2 offers an even more versatile feature known as ACM. With ACM, it is possible to dynamically modify the coding rate and modulation scheme for every single frame, according to the measured channel conditions where the frame must be received. The return channel is used to dynamically report the receiving conditions at each receiving site. In this scenario, the CMs range between 8-PSK 3/4 and 16-PSK 5/6. If distributed equally over each station, each station will use a 30/20 = 1.5 Mbaud part of the carrier (assuming there were 20 stations in the system). The corresponding bit rates vary between 3.34 and 4.95 Mbps to each site. The total available rate is then 86.3 Mbps. This is more than 130% higher than DVB-S.

To address a transition from DVB-S to DVB-S2, optional backward-compatible modes have been defined in DVB-S2. These modes allow one to send two TSs on a single satellite channel. The first (High Priority, HP) stream is compatible with DVB-S receivers as well as with DVB-S2 receivers, while the second (Low Priority, LP) stream is compatible with DVB-S2 receivers only.

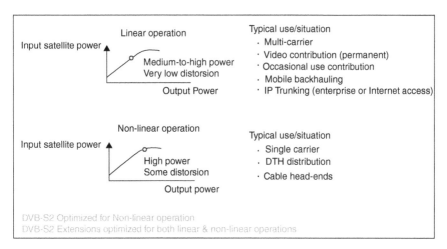

Figure 2.12   Positioning of DVB-S2X.

When DVB-S2 is transmitted by satellite, quasi-constant envelope modulations such as QPSK and 8-PSK are power efficient in the single-carrier-per-transponder configuration, because they can operate on transponders driven near saturation. 16-APSK and 32-APSK, which are inherently more sensitive to nonlinear distortions and would require quasi-linear transponders [i.e., with larger Output Back Off (OBO)], may be improved in terms of power efficiency by using nonlinear compensation techniques in the uplink station. In FDM configurations, where multiple carriers occupy the same transponder, the latter must be kept in the quasi-linear operating region (i.e., with large OBO) to avoid excessive intermodulation interference between signals.

2.2.2   DVB-S2 Extensions

During[5] the past few years, work has been underway to further enhance the capabilities of DVB-S2. As mentioned already, higher throughput is needed for evolving applications such as Ultra HD and commercial aeronautical Internet access. Technical work was being wrapped up in 2014, and the expectation was that the new standard would be adopted; most likely it will be published as Annex to the DVB-S2 standard. On February 27, 2014, the DVB steering committee approved the specifications for the new standard, to be called DVB-S2X. DVB-S2X is a superset of DVB-S2. The E_s/N_o range extends from -10dB to $+24$dB. DVB-S2X provides increased availability. While DVB-S2 is optimized for nonlinear operations, DVB-S2X handles both environments equally well (Figure 2.12).

---

[5]Services, service options, and service providers change over time (new ones are added and existing ones may drop out as time goes by); as such, any service, service option, or service provider mentioned in this chapter is mentioned strictly as illustrative and possible examples of emerging technologies, trends, or approaches. As such, the intention is to provide pedagogical value. No recommendations are implicitly or explicitly implied by the mention of any vendor or any product (or lack thereof).

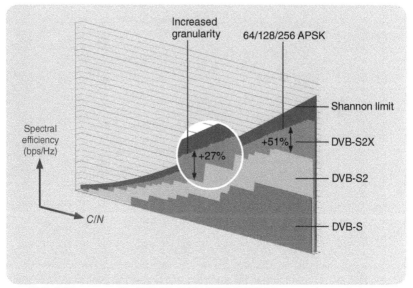

DVB-S2X compared to DVB-S2 (64/128/256 APSK & increased granularity)

Figure 2.13   Relative improvements of DVB-S2 extensions. Courtesy Newtec Cy N.V.

DVB-S2X allows increased granularity in MODCODs and higher-order modulation schemes (e.g., 64-APSK). This allows one to achieve up to 37% gain in spectral efficiency; also, generally, there are gains at lower MODCODs. Figure 2.13 depicts, in general terms, the relative improvements of the extensions. Compared with existing commercial implementations, the combination of technologies incorporated in the new standard results in a gain of up to 20% for DTH networks and 51% for other professional applications compared to DVB-S2 [WIL201401].

Modulation improvements are because of the ability to do more processing (continued advancements under "Moore's Law"). The features are as follows [NEW201401]:

- Lower roll-off factor of 5%, 10%, and 15%;
- Advanced filtering;
- Increased number of FEC choices (increased granularity MODCODs) providing the highest resolution for optimal modulation in all circumstances with E_s/N_0 between -3 and 19 dB dynamic range;
- Higher modulation scheme of 64-APSK (6 bits-per-Hertz), 128-APSK (7 bits-per-Hertz), and 256-APSK (8 bits-per-Hertz) are included in the standard to work with improved link budget because of larger antenna and more powerful satellite;

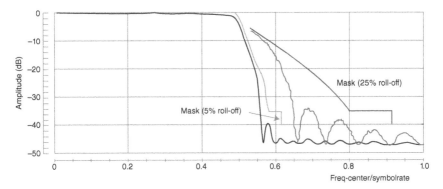

Figure 2.14 Impact of roll-off.

- Ability to support different network configurations/environments, such as Very Low SNR MODCODs to support mobile (land, sea, air) applications;
- A better matching between constellations and FEC rates for linear and nonlinear channels;
- Wideband 72 Mbaud carrier implementation; and
- Bonding of TV streams.

The tighter roll-off (smaller value of α) is a function of the filtering technology and could be applicable to any scheme; however, it is an intrinsic feature of DVB-S2X. Figure 2.14 depicts graphically the advantage of lower roll-off, in allowing less energy to spill into an adjacent band, leading to operations without excessive guard band. Figure 2.15 depicts how a tighter filtering would benefit the existing DVB-S2 standard.

To come back to DVB-S2X, the advanced filtering utilized in DVB-S2X allows elimination of side lobes next to carriers; thus the carrier spacing is reduced to only 1.05 times their symbol rates; in turn, this supports the possibility of putting carriers closer to each other, which results in BW gain (Figure 2.16).

DVB-S2X allows additional MODCODs compared with DVB-S2; it doubles the existing MODCOD points. More granularity means having more choices, which leads to more efficiency. Better FECs implies a lower E_s/N_o threshold, which leads to better efficiency. Higher Modulations of 64-APSK leads to higher efficiency, up to 5 bits/Hz (see Figure 2.17 for the constellations). Typical modulation symbol rates are: 5–15, 36, 54, and 72 Mbaud. Better implementation by the use of more sophisticated receivers, namely, a linear adaptive equalizer, or even more advanced and complex recursive decoding/demapping technologies, can significantly improve the efficiency of S2 and S2X, ranging from 5% to 10% [NEW201401]. Figure 2.18 pulls all these concepts together to show the advantage of DVB-S2X in optimizing spectral efficiency.

DVB-S2X allows the service user to differentiate between linear operation (for a nonsaturated transponder) and nonlinear operation (for a saturated transponder). It

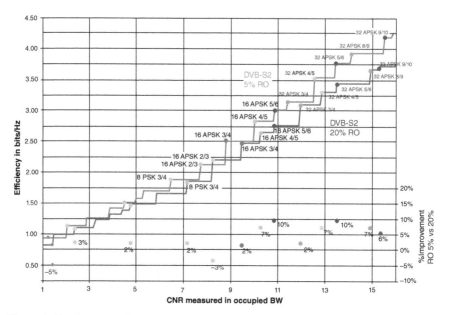

Figure 2.15 Example of different MODCOD comparing DVB-S2 with 5% roll-over and DVB-S2 with 20% roll-over (for traditional MODCOD).

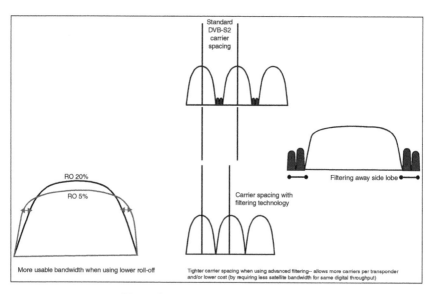

Figure 2.16 Improved carrier spacing.

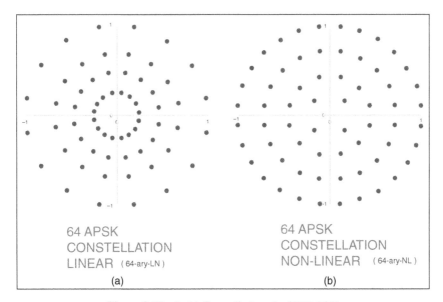

Figure 2.17 (a, b) Constellations for DVB-S2X.

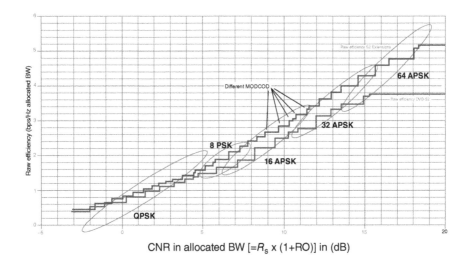

Figure 2.18 CNR comparison S2 extensions with 5% roll-over to DVB-S2with 20% roll-over for traditional and new MODCOD.

introduces MODCODs that are less sensitive to distortion for saturated transponders by allowing the use of nonlinear MODCOD for saturated transponders and linear MODCOD for multicarrier operation. There are 87 MODCOD in total. One can select the optimal MODCOD for full transponder operation by looking at the required E_s/N_o, the value for the optimal OBO, and the resulting nonlinear degradation.

The following 58 nonlinear MODCODs are supported:

QPSK: 45/180, 60/180, 72/180, 80/180, 90/180, 100/180, 108/180, 114/180, 120/180, 126/180, 135/180, 144/180, 150/180, 160/180, 162/180

8-PSK: 80/180, 90/180, 100/180, 108/180, 114/180, 120/180, 126/180, 135/180, 144/180, 150/180

16-APSK: 80/180, 90/180, 100/180, 108/180, 114/180, 120/180, 126/180, 135/180, 144/180, 150/180, 160/180, 162/180

32-APSK: 100/180, 108/180, 114/180, 120/180, 126/180, 135/180, 144/180, 150/180, 160/180, 162/180

64-APSK: 90/180, 100/180, 108/180, 114/180, 120/180, 126/180, 135/180, 144/180, 150/180, 160/180, 162/180

The following 58 linear MODCODs are supported:

QPSK: 45/180, 60/180, 72/180, 80/180, 90/180, 100/180, 108/180, 114/180, 120/180, 126/180, 135/180, 144/180, 150/180, 160/180, 162/180

8-PSK-L: 80/180, 90/180, 100/180, 108/180, 114/180, 120/180

8-PSK: 126/180, 135/180, 144/180, 150/180

16-APSK-L: 80/180, 90/180, 100/180, 108/180, 114/180, 120/180, 126/180, 135/180, 144/180, 150/180, 160/180, 162/180

32-APSK: 100/180, 108/180, 114/180, 120/180, 126/180, 135/180, 144/180, 150/180, 160/180, 162/180

64-APSK-L: 90/180, 100/180, 108/180, 114/180, 120/180, 126/180, 135/180, 144/180, 150/180, 160/180, 162/180

DVB-S2X can be utilized in cases where one has high SNRs (which utilize the high-order modulation schemes) and low or very low SNR environments, using the low-order MODCODs. Table 2.3 [NEW201401] depicts possible MODCODs and roll-off arrangements for various applications.

Figure 2.19 shows a comparison of the throughput achievable on a standard transponder using the three DVB techniques: DVB-S, DVB-S2, and DVB-S2X.

Importantly, the DVB-S2X standard supports wideband transponders that are becoming available on satellites [especially High Throughput Satellites (HTSs)] to support high-speed data links. The wideband implementation in DVB-S2X typically addresses satellite transponders with BWs from 72 MHz (e.g., on C-band systems) up to several hundred MHz (e.g., on Ka-band HTS systems, where a typical HTS has a significant number of ultrawide band [UWB]) transponders distributed among the beams, each with a BW of 100 MHz or more). The DVB-S2X demodulator can receive the complete wideband signal up to 72 Mbaud, resulting in a high data rate, along with an additional 20% efficiency gain.

High-end satellite modem and modulator manufacturers (e.g., Newtec) have been conducting "record breaking demos" using production equipment. For example, in

TABLE 2.3   DVB-S2X Possible MODCODs and Roll-off Arrangements for Various Applications

DVB-SX NL (RO 10%) versus DVB-S2 (RO 20%)	Nonlinear	
	Efficiency (%)	#modcods
Mobile (−10 − -3)	26.77	6
Low (−3 − 5)	11.50	11
Broadcast (5 − 12)	9.53	15
Professional (12 − 24)	17.16	18
Full (− 10 − 24)	16.24	50
DVB-SX L (RO 10%) versus DVB-S2 (RO 20%)	Linear	
	Efficiency (%)	#modcods
Mobile (−10 − -3)	10.01	6
Low (−3 − 5)	6.10	14
Broadcast (5 − 12)	12.19	20
Professional (12 − 24)	29.71	13
Full (− 10 − 24)	16.50	53

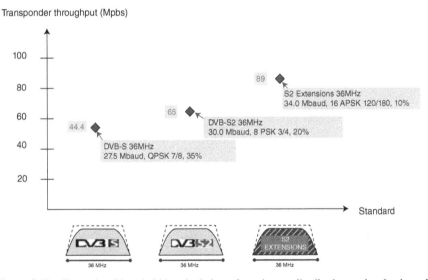

Figure 2.19   Example of bandwidth calculations for primary distribution to headends and towers.

June 2012, in a two-way high-speed backbone test, Newtec achieved 506 Mbps (2 × 253 Mbps) over a 72 MHz transponder on a Eutelsat W2A. For this test, a 32-APSK 135/180 with 5% roll-off modulation MODCOD was used [NEW201401]. An earlier test between Newtec and PSSI Global Services, which was conducted in preparation for the UHD demonstration, achieved 140 Mbps over a 36 MHz transponder on Galaxy 13 to a 4.6-meter antenna using to be standardized DVB-S2X technology

(32-APSK 150/180, 5% roll-off, 34.285 Mbaud). Other press time DVB-S2X demos are as follows:

- Yahsat – 310 Mbps over a 36 MHz Ka-band transponder;
- Intelsat – 485 Mbps over a 72 MHz Ku-band transponder with BWC;
- Eutelsat – 377.5 Mbps over 72 MHz one-way in 64-APSK 162/180; and
- Newtec MDM6000 – 2× 380 Mbps

This technology may be useful in Ultra HD applications. For example, Intelsat and Ericsson demonstrated true UHD 4K, end-to-end video transmission via satellite in June 2013 at Turner Broadcasting in Atlanta. Intelsat's Galaxy 13 satellite delivered a 100 Mbps, 4:2:2, 10-bit, 4K UHD signal at 60 frames per second.

Broadcast modulator, demodulator, and modem platform equipment reaching the market at press time (e.g., Newtec's M6100 and MDM6100) was already capable of supporting DVB-S2X, DVB RF-CID, and transmodulation (DVB-S2X to DVB-S2 or to DVB-S).

The implications for satellite operators are that users may, in principle, require less transponder BW to support the same needed throughput. Or, that, considering the ever-increasing BW growth driven by evolving applications, users will simply benefit from the increased throughput thus achievable on a specified transponder BW.

Note, in conclusion, that as of press time, not all service providers (e.g., DirecTV) and terminal manufacturers have shown immediate interest (especially in consideration of any large embedded base). Also, the extension is much more complicated than DVB-S2 itself, and the high-order constellations are only for the forward link.

2.3   PART 3: OTHER GROUND-SIDE ADVANCES

2.3.1   Carrier ID

Another recent advance is the inclusion of a Carrier Identifier (CID) in a transmitted signal. According to a recent report, 93% of operators and users who responded to a questionnaire by the Satellite Interference Reduction Group (sIRG) suffer from interference at least once a year and 27% suffer on a weekly basis [SOE201301]. Embedding a CID into the carrier is a robust mechanism being developed for expeditious pinpointing of the source of satellite interference. Without automated mechanisms, manual searches are needed, which may involve expensive air-borne (e.g., helicopter-based) triangularization or other radiolocation-based technologies. Looking at the received (interfering) signal itself is the most cost-effective and reliable mechanism (when appropriate technologies are used) to identify interference issues. There is a need for standardization to support global interoperability. Two approaches have evolved of late:

- Approach (Phase) 1: Embed Carrier ID in DVB Network Information Table (NIT) (by the World Broadcasting Unions' International Satellite Operations Group (WBU-ISOG), as proposed in 2009); supports MPEG TS-based video carriers. It entails changes to the NIT in the original stream. It is a low robustness approach; for example, the CID is not recoverable if the main carrier is inoperative. This approach requires the insertion of CID Information in the MPEG Stream from WBU-ISOG (2011-11-09). The Encoder/Mux inserts descriptors into the NIT, and the carrier ID can be edited on the modulator. The 80 character ID includes Equipment Manufacturer; Equipment Serial Number; Telephone Number; CID; Longitude and Latitude; and User Information. In multiuplink and multihop configurations, adaptation is required. This effort has been supported by the WBU-ISOG Forum, the Global VSAT Forum (GVF), and the sIRG.

- Approach (Phase) 2: DVB Carrier ID standard (DVB-CID - 2013). With this standard, the DVB-CID signal is an overlay to original carrier; this approach is agnostic to traffic carrier or transport mechanism (TS video and IP data). It provides high robustness: it can be decoded even if the main carrier is jammed. The RF carrier ID standard was released by DVB as Bluebook A164 on February 28, 2013. It was adapted as ETSI TS 103 129 v1.1.1 in May 2013. The CID was injected by modulator into carrier with fixed source ID such as MAC address and user configurable data such as GPS coordinates, carrier name, and contact information. The WBU-ISOG Forum (May 15–16, 2013) issued a resolution in support of the ETSI TS 103 129 v1.1.1 (2013 – 05).

The DVB Carrier ID standard is the preferred method in the long term. It will enable the operators and users to rapidly identify interfering carriers and respond to Radio Frequency Interference (RFI), shortening the duration of each event.

The DVB-CID data resides under the noise floor for the Rx. The impact on overall link budget degradation of main carrier is typically less than 0.28 dB. The BW used is 224 kHz if symbol rate (SR) $\geq$ 512 kBaud; or 112 kHz if SR < 512 kBaud. The carrier is 220 Hz offset compared with the main signal carrier (Figure 2.20). The minimum content of the Carrier ID is the DVB-CID Global Unique Identifier (GUI), fixed by the equipment manufacturer. In addition, it may contain other information that is configurable by the user, such as GPS coordinates, contact phone numbers, and so on, to simplify and speed up the process to stop the interference event. For the Carrier ID to be compatible with all carriers used in satellite today and easy to be included in all satellite modulators, the CID waveform is superimposed on the Host Data Carrier. The system uses BPSK spread spectrum modulation, differential encoding, scrambling, and a concatenated error protection strategy on the basis of repetition, cyclic redundancy check (CRC), and BCH codes. The CID carrier is assigned a power spectrum density level well beneath the data carrier level, thus allowing for a negligible degradation of the Data Carrier performance (typically below 0.1 dB). At the same time, the adoption of Spread Spectrum technique, together with the Differentially Encoded BPSK modulation and a BCH FEC protection, allows for a very robust Carrier ID system. It should, in fact, be possible, in most practical cases, to identify the interferer without switching off the wanted signal, as particularly required by broadcast services [DVB201301]. Manufacturers were in the process of bringing the technology to market in their modem products as of press time. Figure 2.21 depicts WBU-ISOG resolutions regarding Carrier ID made in 2013.

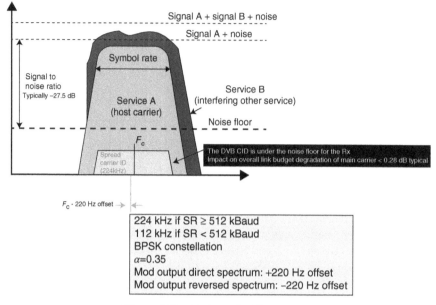

Figure 2.20   Basic carrier ID concepts.

WBU-ISOG Resolutions July 2013	
Now	• All uplinkers of SCPC and MCPC Video and Data, fixed and mobile systems, shall include Carrier ID functionality in the required specifications of all current and future proposals and quotes.
January 1 2015	• All new model modulators and codecs with integrated modulators purchased by end users for video uplinking contain a Carrier ID that meets the ETSI TS 103 129 standard (issued May 29,2013).
	• All satellite operators start the transition to use Carrier ID that meets WBU-ISOG NIT or the ETSI standard for all SNG, DSNG and any other new uplink transmission services.
January 1 2018	• All uplinkers shall ensure that Carrier ID is included for all their respective SCPC and MCPC Video and Data transmissions.
	• CID NIT shall be phased-out in preference to the ETSI standard.

Figure 2.21   Carrier ID implementation goals (WBU-ISOG).

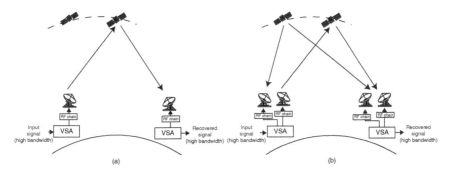

Figure 2.22 (a, b) Inverse multiplexing on satellites.

2.3.2 Intelligent Inverse Multiplexing

Inverse multiplexing, the ability to provide transparently higher BW to the endpoints by "stitching together" a group of lower-speed facilities within the transport network, has been around commercially in terrestrial networks for at least 20 years, and longer than that for internal-provider-network applications. Recently, this technology has become available to satellite applications. Not only disaggregated BW on a single or multiple transponders on a satellite but also BW across multiple satellites can be "bonded" by an intelligent inverse multiplexer (I^2M); in this case sophisticated timing management mechanisms (particularly buffer management and synchronization algorithms) have to be implemented in the modems (Figure 2.22). Commercial products typically combine the I^2M technology with a modem capability in the same device. I^2M technology may be employed in the areas of VSAT networks, cellular backhaul trunking, and possibly for Ultra HD. The challenge is to develop a modem with an acceptable price point; currently, the products are about 5–10x the cost of a normal satellite modem.

Traditionally, when information (e.g., data, voice, or video) is transported via satellite, a carrier signal is assigned a specific frequency range into which the carrier signal is placed, using FDMA principles. A satellite operator's transponders accommodate multiple customers with each assigned a specific-sized block or BW that meets their needs. However, over the course of time and as customer requirements change, BW gaps are created, and these gaps represent unused capacity to the satellite operator. BW aggregation enables the customer to utilize these gaps as if they were one contiguous block; this is achieved by combining each block's information-carrying capability to form a higher capacity channel. For the satellite operator this provides more opportunities to sell unused fragmented satellite capacity, and the customers get a more flexible network to improve their services (Figure 2.23).

There are some additional advantages to using inverse multiplexing techniques, in addition to BW aggregation: intelligent satellite BW management technology can deliver significant additional advantages, namely, interference mitigation, hitless BW adjustment, and virtualization of physical layers to incorporate fiber and satellite aggregation [PLA201301]. Another such advantage is security (confidentiality,

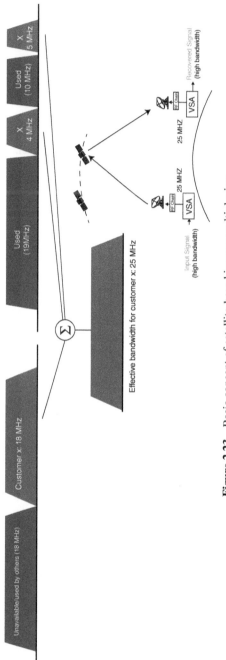

Figure 2.23 Basic concept of satellite-based inverse multiplexing.

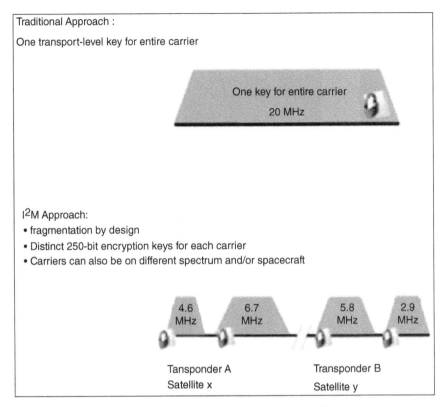

Figure 2.24 Improved security when using I^2M.

integrity, and availability): in the traditional approach, a customer's content is typically encrypted and transmitted via a single carrier; an I^2M allows the information to be spread across multiple carriers, each with its own and unique encryption key, making it more difficult to compromise the content by an intruder (Figure 2.24).

Another benefit relates to interference mitigation and antijamming: when interference occurs, it typically appears in a section of the (transponder) spectrum. In the traditional single-carrier situation, a hit by an interfering signal causes the information channel to either be rendered completely useless or be severely degraded. However, if the information is sent over multiple carriers, which have been intentionally separated in frequency using I^2M technology, the impact of interference is greatly reduced; in many cases it may be completely transparent to the customer. The same methodology is applicable to the intentional jammer scenario (Figure 2.25).

Recently (at the Satellite 2013 trade show), Alcatel-Lucent demonstrated a satellite modem platform, the Virtual Spectrum Aggregator (VSA) that supports I^2M technology; the multimodem provides BW aggregation capabilities, namely,

Traditional approach

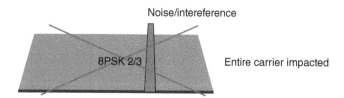

Noise/intereference

Entire carrier impacted

I²M Approach

Noise/intereference

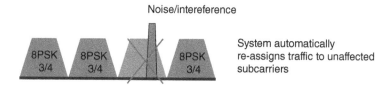

System automatically
re-assigns traffic to unaffected
subcarriers

Figure 2.25   Inference mitigation.

the functionality described above [PLA201301]. The VSA expands the possibilities to how networks are designed for efficiency, flexibility, and robustness, and how satellite BW can be offered, especially in capacity-constrained situations. A throughput in the Gbps range is possible via the aggregation of multiple transponders and/or the seamless inclusion of terrestrial links for hybrid (satellite-terrestrial) networks.

As noted earlier, the concept of inverse multiplexing (also known as channel bonding) is not new. There are traditional ways of doing inverse multiplexing, such as

Ethernet Link Aggregation, that rely on layer 2/3 header fields for the classification of traffic into flows that are mapped to randomly chosen physical channels. The drawback of this approach is its inability to deal with traffic comprising just a single flow – such as a video streaming session or a large file transfer – all of which maps onto a single physical link and keeps other links unused. That same scheme is also incapable of dealing with traffic such as TSs over DVB-ASI that does not natively carry any layer 2/3 information: broadcast video requires adaptation over a layer 2/3 transport mechanism before it can be inverse multiplexed using traditional inverse multiplexing, and this approach wastes BW. The VSA technology addresses these limitations, according to the vendor, thus allowing satellite operators to utilize their transponder capacity for both video and data traffic even in the presence of fragmented spectrum. Support for up to 64-QAM constellations allows setup of high throughput links with a high degree of resiliency following the multiple levels of interference mitigation options. For example, an IP trunk can be set up using multiple disjoint carriers such that a complete loss of one of the carriers – either because of interference or equipment failure – is hitless, from a packet delivery perspective. Another option is the rapid and automatic reconfiguration of the IP trunk in the event of a complete loss of one of the carriers. Being able to increase the spectral BW (for increased throughput) without interrupting the service is another advantage of this technology. The VSA technology is flexible; press time releases supported aggregation of up to four links with aggregate throughput of up to 1 Gbps – higher aggregate BW and number of links is possible [PLA201301].

2.3.3   Implications of H.265 Coding

HEVC is a newly developed standard. It has the same general structure as MPEG-2 and H.264/AVC, but it affords more efficient compression; namely, it can provide the same picture quality as the predecessor but with a (much) lower data rate (and storage requirement), or it can provide better picture quality than its predecessor but at the same data rate (and storage requirement). The operative feature is that video can be compressed into a smaller size or bit rate; savings ranging from 30% to 50% have been cited in the literature (i.e., up to 2 × better compression efficiency compared to the baseline H.264/AVC algorithm). Many demonstrations and simulations were developed in recent years, especially in 2013, and commercial-grade products are expected in the 2014–2015 timeframe, just in time for Ultra HD applications [ANG201301], [BRO201301], [MAR200301], [SHA201301], [SUL201201], [RIC201301], [X26201401]. The implications for satellite operators are that users may, in principle, require less transponder BW to support the same needed video quality or, that, considering the ever-increasing BW growth driven by evolving applications such as Ultra HD, users will either increase the video quality and retain the same transponder BW or possibly need more BW when Ultra HD services are implemented. The topic is revisited in greater detail in Chapter 6.

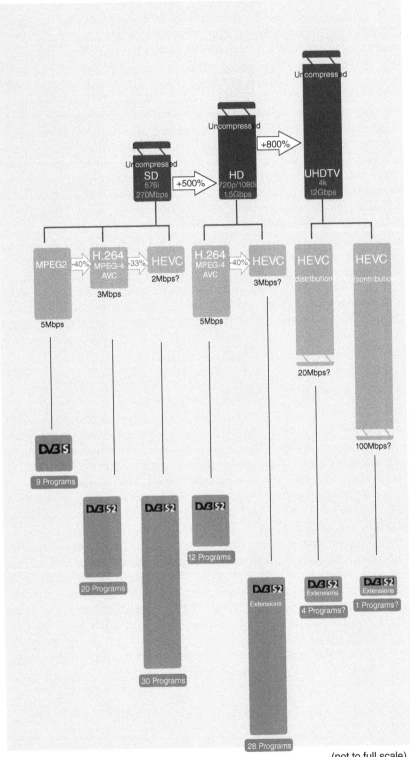

Figure 2.26 The changing scene for bandwidth use/consumption by end users: a satellite operator's dilemma.

Figure 2.26 provides a summary of the implications of the technologies discussed (or alluded to) in this chapter.

REFERENCES

[AGI200101] Agilent Technologies, "Digital Modulation in Communications Systems – An Introduction", Application Note 1298, March 14, 2001, Doc. 5965-7160E, 5301 Stevens Creek Blvd, Santa Clara, CA, 95051, United States.

[AMB200301] A. Ambroze, M. Tomlinson and G. Wade, "Magnitude Modulation for Small Satellite Earth Terminals using QPSK and OQPSK", Proceedings of IEEE ICC2003 conference in Alaska, USA, May 2003. IEEE, Piscataway, NJ. Faculty of Technology, University of Plymouth, Drake Circus, Plymouth PL4 8AA, United Kingdom.

[AMO199701] F. Amoroso, R.A. Monzingo, "Analysis Of Data Spectral Regrowth From Non-linear Amplification", IEEE, ICPWC'97, 1997, pg 142 ff.

[ANG201301] C. Angelini, "Next-Gen Video Encoding: x265 Tackles HEVC/H.265", Tom's Hardware online magazine, July 23, 2013.

[ANS200001] ANS T1.523-2001, Telecom Glossary 2000, American National Standard (ANS), an outgrowth of the Federal Standard 1037 series, *Glossary of Telecommunication Terms*, 1996.

[BRO201301] B. Bross, et al., "High Efficiency Video Coding (HEVC) text specification draft 10 (for FDIS & Last Call)", Output Document of JCT-VC, 12th Meeting: Geneva, CH, 14–23 Jan. 2013, Joint Collaborative Team on Video Coding (JCT-VC) of ITU-T SG 16 WP 3 and ISO/IEC JTC 1/SC 29/WG 11.

[BRE200501] D. Breynaert, M. d'Oreye de Lantremange, "Analysis of the Bandwidth Efficiency of DVB-S2 in a Typical Data Distribution Network", Newtec, CCBN2005, Beijing, March 21–23 2005

[CHE200701] J. Chen, "Extracting and Using J-models to Estimate ACPR in Direct Conversion Transmitters", Candence White Paper, 2655 Seely Avenue, San Jose, CA 95134.

[COM199801] CM701/DT7000 Reed–Solomon (DVB Version), Option Card, Com-Stream Corporation, A Spar Company, 6350 Sequence Drive, San Diego, California 92121–2724, 1998.

[DVB201301] DVB Carrier Identification, DVB's Caller-ID solution to stop satellite interference, DVB Fact Sheet - August 2013

[GIL201201] J. Gill, "Definition of BCH Codes", Stanford University Lecture Notes, November 2012

[HEN200201] H. Hendrix, "Viterbi Decoding Techniques for the TMS320C54x DSP Generation", Texas Instruments, Application Report SPRA071A - January 2002. Texas Instruments

[LIT200101] L. Litwin, "Error Control Coding in Digital Communications Systems", RF Design, Jul 1, 2001.

[LUB200201] M. Luby, L. Vicisano, J. Gemmell, L. Rizzo, M. Handley, J. Crowcroft, "The Use of Forward Error Correction (FEC) in Reliable Multicast", Request for Comments: 3453, December 2002.

[MAR200201] G. Maral, M. Bousquet, "*Satellite Communications Systems: Systems, Techniques and Technology*", 2002, Wiley, New York.

[MAR200301] D. Marpe, H. Schwarz, and T. Wiegand, "Context-Based Adaptive Binary Arithmetic Coding in the H.264/AVC Video Compression Standard", IEEE Transactions on Circuits and Systems for Video Technology, VOL. 13, NO. 7, JULY 2003.

[MOR200401] A. Morello and V. Mignone, "DVB-S2— Ready for Lift Off", Digital Video Broadcasting, EBU TECHNICAL REVIEW – October 2004, RAI Radiotelevisione Italiana.

[NEW201401] Newtec Technical Materials, Newtec Cy N.V., Laarstraat 5, B-9100 Sint-Niklaas, Belgium.

[PLA201301] C. Placido, "Vinay Purohit, CTO Alcatel-Lucent Ventures", Satcom Post, August 2, 2013.

[RAD200501] Radyne ComStream Staff, "DVB-S2 and the Radyne ComStream DM240", White Paper, WP017, Rev. 1.3, January 2005, Radyne ComStream, Inc., 3138 E. Elwood St., Phoenix, AZ 85034.

[RIC201301] I. Richardson, "HEVC: An introduction to High Efficiency Video Coding" White Paper. Vcodex Limited, 35 Regent Quay, Aberdeen AB11 5BE

[SHA201301] M. S. Sharabayko, N. G. Markov, "Iterative Intra Prediction Search for H.265/HEVC", 2013 International Siberian Conference on Control and Communication (SIBCON), IEEE.

[SIL200201] J. Sills, "Improving PA Performance with Digital Predistortion", CommsDesign, Oct 02, 2002.

[SOE201301] S. Soenens, K. Roost, M. Coleman, "Are You Ready to Turn Carrier ID On?", Newtec/sIRG Webinar Material, November 14, 2013.

[SUL201201] G. J. Sullivan, J.-R. Ohm, W.-J. Han, and T. Wiegand, "Overview of the High Efficiency Video Coding (HEVC) Standard", IEEE Transactions on Circuits and Systems for Video Technology, VOL. 22, NO. 12, December 2012.

[TOR199801] A. Torres, V. Demjanenko, "Inclusion of Concatenated Convolutional Codes in the ANSI T1.413 Issue 3", Contribution to Standards Committee T1-Telecommunications, Plano, Texas T1E1.4/98-301R1, November 30-December 4, 1998, VoCAL Technologies Ltd.

[TRE200401] TrellisWare Technologies Staff, "FlexiCodes: A Highly Flexible FEC Solution", April 2004, TrellisWare Technologies, Inc. 16516 Via Esprillo, Ste. 300, San Diego, CA 92127–1708.

[WIL201401] K. Willems, "DVB-S2X Demystified", Newtec White Paper, March 2014. Newtec Cy N.V., Laarstraat 5, B-9100 Sint-Niklaas, Belgium.

[X26201401] http://x265.org/hevc.html. "x265 is an Open-Source Project and Free Application Library for Encoding Video Streams into the H.265/HEVC Format".

3

HIGH THROUGHPUT SATELLITES (HTS) AND KA/KU SPOT BEAM TECHNOLOGIES

This chapter[1] looks at High Throughput Satellite (HTS) technology, applications, and key design issues. Here, we discuss multiple access schemes and frequency reuse principles, followed by a discussion on the HTS spot beam approach, the frequency colors, the frequency bands of operation (specifically, the Ka band), the losses and rain considerations, and the typical HTS applications. A comparison between approaches, namely, a discussion on Ku-based versus Ka-based HTS systems, and some HTS design factors are provided. After some detailed spot beam antenna design basics information for the on-board satellite antenna, some examples of commercially deployed HTSs are provided.

A HTS can be defined as a satellite system that makes use of a large number of geographically confined spot beams distributed over a specified service area, offering a contiguous (or noncontiguous) covering[2] of that service area and providing high system capacity and user throughput at a lower net cost per bit. While most satellites in orbit operating at the Ku band currently provide large contoured coverages for

---

[1] Services, service options, and service providers change over time (new ones are added and existing ones may drop out as time goes by); as such, any service, service option, or service provider mentioned in this chapter is mentioned strictly as possible examples of emerging technologies, trends, or approaches. As such, it is intended to provide pedagogical value. No recommendations are implicitly or explicitly implied by the mention of any vendor or any product (or lack thereof).

[2] The intuitive concept of coverage of an area with a set of spot beams is formalized in the mathematical concept of a "cover" of a topological space (e.g., see [MAC199401], [GAM199901]).

Innovations in Satellite Communications and Satellite Technology: The Industry Implications of DVB-S2X, High Throughput Satellites, Ultra HD, M2M, and IP, First Edition. Daniel Minoli.
© 2015 John Wiley & Sons, Inc. Published 2015 by John Wiley & Sons, Inc.

video broadcast applications, there is an evolving market opportunity for satellites that provide broadband data services utilizing spot beams technology that employ Ka-band frequencies. Proponents make the argument that new high-capacity satellite systems are transforming the economics and quality of services that satellite broadband can provide; they see HTS as the future of satellite communications, "the wave of the future." In fact, the economics of HTSs enable service providers to offer Internet access services with data rates and monthly data download "gigabyte quotas" that are generally competitive with 4G/LTE wireless services, where these are available. The intrinsic concept of satellite spot-beam-based architectures goes back to at least 30 years, and a large number of patents have been issued on various aspects of the technology (see Appendix A for an abbreviated listing). This decade, however, is seeing advancement in the implementation and deployment of this technology on a scaled-up basis, specifically regarding the number of on-board spot-supporting feeds and the number of spacecraft with these designs. These advancements are driven by the growing worldwide demand for Internet services, especially in mobility environments, and by the orbital spectrum availability offered by the Ka band (although the Ku band is also being used to some extent for this and other related applications).

Traditional satellites have provided regional coverage at the C band and Ku band using one or a handful of service-supporting beams. The need for increased bandwidth – whether to support Direct to Home (DTH) video for High Definition (HD) channels (or eventually Ultra HD TV), or Internet access (including streaming video reception) – is driving operators to "new solutions" that combine high-density frequency reuse and new, less orbit-congested bands – specifically the Ka band – along with wider spectrum transponders. The reader should keep in mind that a traditional satellite that supports a payload of 24 C-band transponders (36 MHz each) and 24 Ku-band transponders (36 MHz each), only delivers a total of 1.7 GHz of capacity on the 48 transponders (48×36) in overlapping geographies, and, perhaps, two times as much if, say, two nonoverlapping regions are served (e.g., with a Ku-beam Continental US [CONUS] coverage from a satellite at 45 W along with a Ku-beam Western Europe coverage on the same spacecraft, for example, as was shown in Figure 1.9). Assuming that a 36 MHz transponder supports about 70 Mbps of data with DVB-S2, then this 1.7 GHz of capacity would achieve about 3 Gbps of total digital capacity (6 Gbps if two nonoverlapping regions are served). Even with DVB-S2X, the total bandwidth would be approximately 10 Gbps, which is the equivalent of a SONET (Synchronous Optical Network) OC-192 (STM 64). These data rates are mundane in the terrestrial fiber network environment, where OC-192 links are invariably common, as noted in the last section of Chapter 1. If a frequency reuse of 24 times can be achieved with the 500 MHz allocated to the satellite, this can result in an effective capacity of 12 GHz or, say, 48 Gbps (OC-768, STM-256) on the spacecraft, which is a lot more desirable. In a somewhat parallel fashion, the utilization of small cells/frequency reuse is what enabled the terrestrial cellular industry to greatly increase their system-level throughput and support the avalanche of new 3G/4G smartphone users.

The primary currently-envisioned application for HTSs is high-speed/broadband Internet connectivity for mass markets; Figure 3.1 depicts a few examples of press

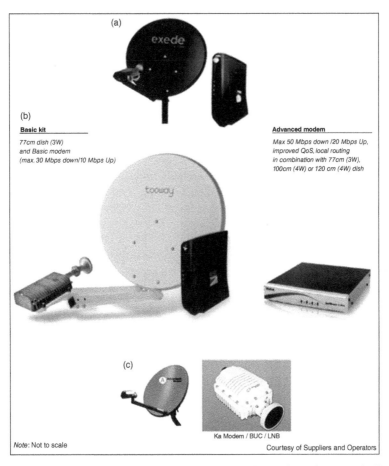

Figure 3.1 HTS gear, examples. (a) Hardware with satellite service (ViaSat-1). (b) Eutelsat KA-SAT. (c) Advantech wireless equipment.

time Ka HTS user terminals. Observers state that the total global bandwidth consumption is doubling every 2–3 years; hence, the deployment of HTS capabilities is a critical business imperative for satellite providers to remain competitive. The percentage of the world's population using Internet was around 33% at press time[3], which suggests that there is a growth opportunity for service providers. For example, Africa is a continent where approximately 90% of the citizens do not have Internet access – the continent has fiber optic cables spanning its coast, but little has been laid inland so far; hence, there is a dearth of infrastructure for Internet access, often limited to urban areas, and even then users have wired and wireless voice access that are barely capable of providing dial-up Internet access [ITU201201]. The latest HTSs

[3] An estimated 2.5 billion people out of 7 billion have Internet access: about 80% in North America, 70% in Australia, 65% in Europe, 45% in Latin America, 40% in the Middle East, 30% in Asia, and 10–15% in Africa.

typically deliver download speeds of more than 10 Mbps to individual customers and a single third-generation Ka-band HTS can support 2–3 million subscribers. Hence, HTSs can play an important role in the next few years, as terrestrial infrastructure develops. Since 2004, when HTSs were first put into service, many satellite operators have launched or are planning to launch Ka-based HTS. There have already been several generations of HTS in recent years (see Figure 3.2). The vast majority of HTS systems utilize Ka band, with Thaicom, Insat, and Intelsat being the major exceptions that use Ku band. Platform vendors that supply the intrinsic on-the-ground technology include but are not limited to ViaSat, HNS, Gilat, NewTec, and Idirect.

For global HTS services, market research firms forecast that HTS capacity demand will increase to over 900 Gbps by 2022, while the emerging Medium Earth Orbit (MEO) HTS (MEO-HTS) segment will add another 100 Gbps of leased capacity demand as of 2022. For an example of a bandwidth driver, note that the European Union has set the goal of establishing 30 Mbps satellite coverage available to all EU citizens by 2020. The large majority of the forecasted bandwidth growth will be for HTS capacity to serve broadband access services (about 75%); this is a "high-volume/low-margin" business that requires large amounts of low-cost bandwidth to maintain an acceptable service quality for subscribers – the per-Mbps price of HTS capacity is estimated to fall by over 40% between 2013 and 2022, partly driven by increased price competition and gains in spectral efficiency through technological advancements [RUI201301]. Beyond that, for the other 25% of the bandwidth growth, a combination of cell backhaul, Very Small Aperture Terminal (VSAT) networking, and IP trunking services will drive approximately similar quantities of HTS and MEO-HTS bandwidth demand globally, within the enterprise data segment. HTS and MEO-HTS capacity is also expected to compete for commercial and government mobility applications [GLO201301]. In this context, a market-savvy approach for satellite service providers (as well as spacecraft manufacturers and satellite infrastructure providers) is to deliver a menu (and/or include mechanisms in the platforms to be able to deliver a menu) of service plans for different market segments, for example, in terms of different availability; the concept of different Service Level Agreements (SLAs) is intrinsically supportable with HTS-based approaches.

3.1   OVERVIEW

HTSs are an evolution of conventional spacecraft. When a satellite provides coverage of the entire region of the earth visible from the satellite utilizing a single beam, the gain of the satellite antenna is limited by the beamwidth. For a geostationary orbit (GSO) satellite, global coverage implies a 3 dB beamwidth of 17.5°, which in turn implies an antenna gain of about 20 dB; it follows that each user must be equipped with a relatively large aperture antenna to support a high traffic rate, impacting the cost and the practicality of deployment. Furthermore, only one stream of (multiplexed) information can be supported with a given frequency range (e.g., 36 MHz

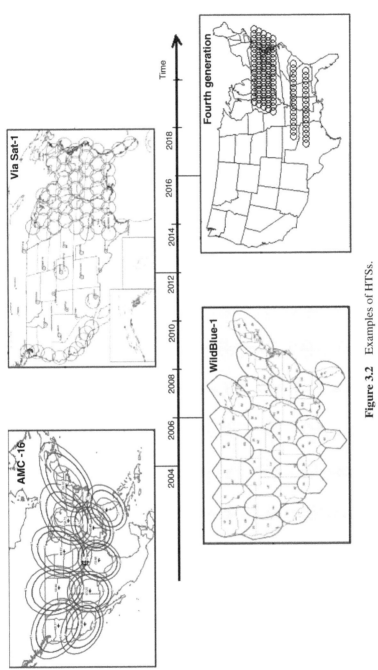

Figure 3.2 Examples of HTSs.

or even 500 MHz). To address this predicament, the spot beam[4] approach has been widely utilized for satellite communication systems, but until recently typically only with a few beams (usually half-a-dozen to a dozen beams).

Spot beams are areas of discrete signal reception on the ground and discrete transmission reception in the spacecraft as implemented by supporting antenna structures. In spot beam satellite communication environments, the satellite provides coverage of only a portion of the earth, usually a nation or subcontinent, by using shaped narrow beams pointed to different geographic areas. The advantage of this approach is a higher satellite antenna gain following a reduction in the aperture angle of the antenna beam, which in turn implies that the user can employ a small aperture antenna.[5] In addition, the multibeam technique supports the reuse of frequencies for different beams, thereby effectively increasing the total system capacity. However, it is well known that satellite power is a scarce resource; moreover, the traffic requirements of each beam may be different, perhaps because of time-zone differences (driving usage patterns); furthermore, it must be noted that when two (or more) beams utilize the same frequency, interbeam interference is introduced following the nonzero gain of the antenna sidelobes. Therefore, when there is interbeam interference between the beams, the capacity allocated to each beam is determined not only by the power allocated to the specific beam, but also by the power allocated to the other beams [WAN201401]. As a result, it is critical to optimize the power allocation to each beam to meet the specific traffic demands. In spite of these general limitations, HTSs are designed to provide acceptable performance and system capacity.

Further along this topological coverage progression, with HTS, the service area is typically covered by many (up to 200 or more) edge-overlapping high-gain spot beams with frequency reuse to support interactive broadband services using small terminals. A high-count spot beam satellite, say, for a CONUS or Western European coverage, can provide about 10 times higher capacity than classical satellites using the same input power and similar receive-station antenna size. To enable increased user-accessible bandwidth and system throughput, HTS systems make use of the efficiency achievable with space division multiplexing as implemented with the use of spot beam. Thus, HTS achieve spectrum efficiency and enhanced performance through the use of spot beam antennas (often) in conjunction with ultrawideband (UWB) satellite transponders. The narrow beamwidths intrinsic with spot beams are achieved by employing antenna structures that focus downlink energy to distinct (but small) areas on the earth's ground surface and have large areas for collecting uplink energy. User-to-Internet or user-in-beam-A-to-user-in-beam-B communication is then achieved with one (or more) terrestrial gateway node(s).

---

[4]The term multibeam (or multi-beam) is also used by some in lieu of the term spot beam (although that term can also be used – and is often used – for systems with just a handful of spot beams); we generally use the term spot beam (or spot-beam) in the context of HTSs.

[5]The physics of antenna design dictate that for a given frequency, to generate higher gain, the antenna must be larger (which, of course, results in the smaller aperture angle); the smaller aperture angle is due to the fact that spacecraft utilize a few larger diameter reflectors, with a separate feed horn (or multiple feed horns) for each beam. Hence, in practical terms, it is the resulting higher gain of the antenna on the satellite that allows the use of smaller antennas by the user.

HTS spacecraft can operate at the Ku band, but recent implementations are principally at the Ka band (Ku-band HTSs include but are not limited to Intelsat 29E/EPIC, with 25–60 Gbps throughput); they can be in the nongeosynchronous orbit (non-GSO), but principally they are in the GSO. In general terms, the reason why most HTS operate at Ka band is because (i) it is relatively difficult for global operators to secure Ku-band orbital slots in various parts of the world, and (ii) the Ka band offers more assigned spectrum bandwidth (upward of 1.1 GHz).

3.2 MULTIPLE ACCESS SCHEMES AND FREQUENCY REUSE

Before proceeding with additional discussion of HTS systems, we provide a quick review of multiplexing and multiple access schemes used in satellite communication. Readers who are familiar with these concepts may opt to skip to the next section.

Multiplexing, the ability to support multiple users over a single facility (e.g., a radio channel, a satellite band, a transponder) is as fundamental to communications as modulation. Multiple access protocols are channel allocation schemes that have desirable performance characteristics. Key types of multiple access schemes are:

- *Space Division Multiple Access* (SDMA) (also known as *Space Division Multiplexing*).
- *Frequency Division Multiple Access* (FDMA) (also known as *Frequency Division Multiplexing*).
- *Time Division Multiple Access* (TDMA) (also known as *Time Division Multiplexing*).
- *Code Division Multiple Access* (CDMA) (also known as *Code Division Multiplexing*).
- *Random Access.* The use of algorithms that support randomized transmission in a distributed system without central control.

At the highest level of the multiple access protocols classification, one has two categories: *conflict-free* protocols and *contention* protocols. Conflict-free protocols are those ensuring a transmission, whenever initiated, which is a successful one, that is, will not be interfered by another transmission. In conflict-free protocols the channel is allocated to the users without any overlap between the portions of the channel allocated to different users. The channel resources can be viewed from a time, frequency, or mixed time-frequency perspective. The channel is allocated to the users either by static or dynamic techniques. The channel can be "divided" at the Physical Layer (of the Open Systems Interconnection Reference Model) by giving the entire frequency range (bandwidth) to a single user for a fraction of the time as done in TDMA, or by giving a fraction of the frequency range to every user all of the time as done in FDMA, or by providing every user access to the bandwidth at varying frequency or time "minislots" by utilizing spread spectrum mechanisms such as CDMA. Dynamic allocation (in the conflict-free protocols context) assigns the channel on the basis of demand (and/or reservation) such that a user who happens to be idle uses only little of the channel resources, leaving the majority of the bandwidth to the other, more active users. Such an allocation can be done by statistical multiplexing at the Data Link Layer (of the Open Systems Interconnection Reference Model).

One of the drawbacks of conflict-free protocols is that idle users do consume a(n assigned) portion of the channel resources, unless a more complex Data Link Layer mechanism is used; this becomes a major issue when the number of potential users in the system is large. In systems with contention schemes, a transmitting user is not guaranteed to be instantaneously successful (retransmissions may be needed); however, they are more bandwidth-efficient. The best way to actually achieve dynamic allocation in the "modern era" would be to assign the satellite bandwidth (say a saturated transponder) as a high-capacity IP trunk where different users contend for resources at the packet layer, rather than nailing down bandwidth to users whether they needed it (at that instantaneous moment) or not; good traffic management techniques would (effectively) guarantee that all users receive adequate service levels (in throughput, latency, packet loss, and so on). An increasing number of broadcast-grade video these days is in fact being transmitted using Multiprotocol Label Switching (MPLS) IP-based services (at least terrestrially).

Space division multiplexing (also known as frequency reuse) is the ability to use the same frequencies (and/or polarizations) repeatedly across a system, but where there is geographical and/or physical separation. Space division multiplexing is a method of allowing multiple users to share a single communications channel by arranging the users so that they are not in one another's communications range. Spectrum is a valuable resource, both in the satellite arena as well in the terrestrial wireless arena. In any frequency band (C, Ku, Ka), there is a limited amount of spectrum available, hence that spectrum must be utilized efficiently. A well-known approach is to employ frequency reuse, where a single satellite uses the same frequency multiple times simultaneously (frequency reuse is also intrinsically employed in satellite communications when multiple/all C-band satellites around the world use the C band, and multiple/all Ku satellites use the Ku band, etc.). It is a method whereby a number of entities use a single service capability by having amounts of physical space within the service facility dedicated for their individual use. Thus, it should be noted that the method is used both across multiple satellites (at different orbital locations) as well by a single satellite illuminating discrete nonadjacent geographies with the same frequency. Frequency reuse is a concept that is used both terrestrially (e.g., cellular networks) and on satellites. For example, the radio FM band and the over-the-air TV band are reused in localities that are 75+ miles apart. Cellular phone frequencies also make use of this technique. The use of a C-band frequency by two earth stations using two C-band satellites is another example of space division multiplexing.

To go back to the single satellite use of space division multiplexing, the more frequency reuse supported by that single satellite, the greater the total system-level throughput that spectrum can achieve with that satellite. Satellite operators have implemented frequency reuse in their satellites for several decades in traditional (widebeam) applications. HTSs make extensive endogenous use of space division multiplexing, as implemented via fractional-degree spot beam coverage (and supporting antenna structures). As discussed already, HTSs support high overall throughput: the design goal is for satellite capacity to be in hundreds of gigabits per second (Gbps) compared to a few Gbps for Fixed Satellite Services (FSS) satellites. For example, ViaSat-1 launched in early 2012 has a total throughput capacity of 140 Gbps. Following the argument used in the introductory section, a satellite with 60 beams has an equivalent throughput of 30 GHz (30 times the amount of spectrum of a single beam satellite − 500 MHz on each polarity on a given band), or $30 \times 4 = 120$ Gbps (or, 150 Gbps, if 5-bit MODCODs are used).

TDMA is a communications technique that uses a common channel (multipoint or broadcast) for communications among multiple users by allocating unique time slots to different users. TDMA is used extensively in satellite systems, although it is employed in other environments as well. TDMA is multiple access technique whereby users share a transmission medium by being assigned and using (each user at a time and for a limited time interval) a number of time division multiplexed slots (subchannels); several transmitters use the same overall channel for sending several bit streams [ANS200001]. In TDMA a channel (of a given bandwidth − typically a transponder) is shared by all the active remote stations, but each is permitted to transmit only in predefined short bursts of time (slots) allocated by a predetermined sequencing mechanism, thus sharing the channel between all the remote stations by dividing it over time (hence, time division). Therefore, in TDMA, the entire bandwidth is used by each user for a fraction of the time.

FDMA makes use of frequency division to provide multiple and simultaneous transmissions to a single transponder. Frequency division multiplexing (FDM) is the approach of deriving two or more simultaneous, continuous channels from a transmission medium by assigning a separate portion of the available frequency spectrum to each of the individual channels. Hence, in FDMA, the bandwidth of the available spectrum is divided into separate channels, with each individual channel frequency being allocated to a different active remote station for transmission. FDMA splits the available frequency band into smaller fixed frequency channels, and each transmitter or receiver uses a separate frequency. In FDMA, a fraction of the frequency bandwidth is allocated to every user all the time. Transmitters are narrowband or frequency-limited. A narrowband transmitter is used along with a receiver that has a narrowband filter so that it can demodulate the desired signal and reject unwanted signals, such as interfering signals from adjacent radios [AGI200101]. In satellite communications, FDM is used to define the 12 (24 with polarization) transponders; however, the use of FDMA *within* a transponder is on the decline (with TDMA being the leading approach, as noted). (Some call this arrangement Frequency-Time Division Multiple Access [FTDMA].)

Satellites make use of all three of these techniques effectively: space division multiplexing is supported when multiple beams are employed over different regions while using the same frequency band among the different spot beams; frequency division multiplexing is used when allocating different transponders to different users or channels or functions; time division multiplexing is used when saturating a transponder with a single large carrier that comprises a stream of different dedicated digital channels.

CDMA is a coding scheme where multiple channels are independently coded for transmission over a single wideband channel. In some communication systems, CDMA is used as an access method that permits carriers from different stations to use the same transmission medium by using a wider bandwidth than the individual carriers. On reception, each carrier can be distinguished from the others by means of a specific modulation code, thereby allowing for the reception of signals that were originally overlapping in frequency and time. Thus, several transmissions can occur simultaneously within the same bandwidth, with the mutual interference reduced by the degree of orthogonality of the unique codes used in each transmission [ANS200001]. CDMA is an access method where multiple users are permitted to transmit simultaneously on the same frequency. The channel (probably derived via FDM methods), however, is larger than otherwise would be the case. Channelization is added, in the form of coding. In a CDMA system, all users access the same bandwidth, but they there are distinguished (separated) from one another by a uniquely different coding parameter. Each user is assigned a code that is utilized to transform the user's signal into spread-spectrum-coded version of the user's data stream, typically using Direct Sequence Spread Spectrum (digital) techniques. The receiver then uses the same spreading algorithm and code parameter to transform the spread-spectrum signal back into the original user's data stream. Users time share a higher-rate digital channel by overlaying a higher-rate digital sequence on their transmission. A different sequence is assigned to each terminal so that the signals can be discerned from one another by correlating them with the overlaid sequence [AGI200101].

In static channel allocation strategies channel allocation is predetermined (typically at system design time) and does not change during the operation of the system. For both the FDMA and the TDMA protocols, because of the static and fixed assignment, parts of the channel can be idle even though some other users in the system may have data to transmit. As noted, dynamic channel allocation protocols attempt to overcome this by changing the channel allocation on the basis of the current demands of the users.

Static conflict-free protocols such as the FDMA and TDMA protocols do not utilize the shared channel very efficiently, especially when the system is lightly loaded or when the loads of different users are asymmetric. This is one of the motivations for random access schemes. In the satellite context, this approach is typically used in VSAT environments. Most random access schemes (contention-based multiple access protocols) are noncentralized protocols where there is no single node coordinating the activities of the others (although all nodes behave according to the same set of rules). Contention schemes do not automatically guarantee successful transmission.

When contention-based multiple access protocols are used, the necessity arises to resolve the conflicts, whenever they occur. Both static and dynamic resolutions exist. Static resolution means that the actual behavior is not influenced by the dynamics of the system. A static resolution can be based, for example, on user IDs or any other fixed priority assignment, meaning that whenever a conflict arises the first user to finally transmit a message will be the one with, say, the smallest ID (this is done in some tree-resolution protocols). A static resolution can also be probabilistic, meaning that the transmission schedule for the interfering users is chosen from a fixed distribution that is independent of the actual number of interfering users, as is done in Aloha-type protocols and the Carrier Sense Multiple Access (CSMA) protocol used in classical Ethernet systems. Dynamic resolution, namely, taking advantage and tracking system changes, is also possible in contention-based protocols. For example, resolution can be based on the time of arrival, giving the highest (or lowest) priority to the oldest message in the system. Alternatively, resolution can be probabilistic but such that the statistics change dynamically according to the extent of the interference [ROM198901].

The Aloha family of protocols is popular because of its seniority, as it is the first random access technique introduced; also, many of these protocols are simple and the implementation is straightforward. The Aloha family of protocols belongs to the contention-type or random retransmission protocols in which the success of a transmission is not guaranteed in advance. The reason is that whenever two or more users are transmitting on the shared channel simultaneously, a collision occurs and the data cannot be received correctly. This being the case, packets may have to be transmitted and retransmitted until eventually they are correctly received. Transmission scheduling is therefore the focal concern of contention-type protocols. Because of the great popularity of Aloha protocols, analyses have been carried out for a large number of variations (e.g., unslotted Aloha, slotted Aloha, and many others).

3.3   SPOT BEAM APPROACH

HTS systems can be classified as follows: Ka-band small spot beam systems, Ka-band large spot beam (widebeam) systems, and Ku-band spot beam systems. In the 1990s, some satellite operators endeavored to offer Ka-band capacity using a small Ka-band payload on satellites that were primarily C- or Ku band focused; the Ka-band beam was typically a wide-area beam with little or no frequency reuse – this wide-area beam approach had limited commercial appeal. New approaches were used in the decade that followed. This came about (under the thrust of market interest in interactive Internet access) because the total cost to design, construct, and launch a satellite is approximately the same whether the satellite is optimized for capacity (many beams) or for coverage (a few beams). To achieve a high degree of frequency reuse, the broader geography of interest is no longer covered by a large single beam, but by a high number of slightly overlapping, high-gain spot beams (however, user-to-user communications requires a dual-hop intervention of the gateway). As a result, the cost per delivered bit for the HTS design is significantly lower than for a satellite optimized for large broadcast coverage.

Each spot beam reuses available frequencies (and/or polarizations), so that a single HTS spacecraft can provide up to approximately 10 times the capacity of traditional satellites – channel frequencies are reused numerous times in geographically nonoverlapping spots, but the spectrum of the individual beams is constrained within the available satellite bands, typically 500 MHz (1100 MHz in some circumstances) in total. HTS payloads commonly have 5–10 GHz of aggregate internal transponder bandwidth to support the combined throughput. These spot beams provide high signal strength and signal gain (Effective Isotropically Radiated Power [EIRP] and G/T), allowing the satellite to close links to small aperture earth stations at high data rates with reasonable rain fade margin, to provide acceptable overall link availability [CAP201201]. Spot beams may be steered or may be fixed relative to the satellite.

There is no theoretical limit to the number of times that the spectrum can be reused (by achievable polarization discrimination purity and beam isolation); in practical terms, however, the number is limited by the size, weight, and complexity of the satellite antenna [AGR198601]. A HTS spacecraft may have beams of different sizes; for example, it might have 60 0.40-degree beams and 40 0.80-degree beams. HTS leverages frequency reuse across multiple narrowly focused spot beams as compared to traditional satellite technology, which is based on a wide-coverage single beam (a few thousand miles), or at most a handful of beams (e.g., 6 beams). Each HTS spot beam covers an area 1–2% the size of a conventional satellite beam; beams are typically 70–100 miles in diameter (each beam generally covers a few thousand square miles), but the size can vary as needed. Because the spot beams have limited geographic coverage, HTSs typically have gateway beams and transponders dedicated to supporting connections with gateway nodes on the ground, where the traffic is handed over to the Internet or some other network.

The first commercial HTS was IPSTAR (also known as Thaicom 4), which operated in the Ku band. The first wave of Ka HTSs was launched in the mid-2000s with spacecraft such as AMC 15/16, WildBlue-1 (first dedicated HTS in North America), and Telesat's Anik F2, among others (Generation 1 and 2). A second wave was launched in the mid-2010s with spacecrafts such as ViaSat-1 and -2, KA-SAT (the first dedicated HTS in Europe), YahSat-1A and -1B, and RSCC, among others (Generation 3 and emerging Generation 4). These newer satellites tend to use Ka-band frequencies in the FSS mode. There are additional HTS projects either under construction or planned as of press time. Each of these succeeding generations of spacecraft has increased the number of beams and, thus, the derivable system-level throughput. The HTSs placed in service at the beginning of the decade have a download speed for the individual customer of more than 10 Mbps (examples of these are ViaSat-1 and KA-SAT platforms – these have also been called third-generation broadband VSAT systems, in contrast to second-generation VSAT systems that typically have a download speed to the individual customer up to 3–5 Mbps) [CAS201201] (see Figure 3.3).

The satellite antenna system is a critical element for the realization of a multiple spot beam coverage environment with overlapping spots. A basic HTS approach is to use a specified number of feed horns in space (say 24) fitted on a geometric matrix and illuminating a single reflector (a spacecraft may, in fact, have multiple such arrangements of feeds and reflectors, say three or four or more). Beam-forming techniques are more sophisticated approaches than static arrangements, whereby multiple feed horns are energized with signals constructed on a "network," called beam-forming network (BFN), to alter the phase of the distinct feeds such that constructive interference can be used to generate either a large shaped beam (shaped to a specific

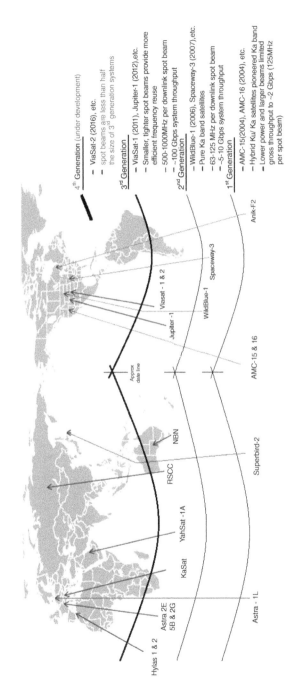

4th Generation (under development)
  - ViaSat-2 (2016), etc.
  - spot beams are less than half the size of 3rd generation systems

3rd Generation
  - ViaSat-1 (2011), Jupiter-1 (2012),etc.
  - Smaller, tighter spot beams provide more efficient frequency reuse
  - 500-1000MHz per downlink spot beam
  - ~100 Gbps system throughput

2nd Generation
  - WildBlue-1 (2006), Spaceway-3 (2007),etc.
  - Pure Ka band satellites
  - 63-125 MHz per downlink spot beam
  - ~5-10 Gbps system throughput

1st Generation
  - AMC-15(2004), AMC-16 (2004), etc.
  - Hybrid Ku/ Ka satellites pioneered Ka band
  - Lower power and larger beams limited gross throughput to ~2 Gbps (125MHz per spot beam)

Anik-F2

Viasat - 1 & 2

Spaceway-3

WildBlue-1

Jupiter -1

AMC-15 & 16

Approx date line

Superbird-2

NBN

RSCC

YahSat -1A

KaSat

Astra - 1L

Astra 2E
5B & 2G

Hylas 1 & 2

Figure 3.3   Generations of HTS.

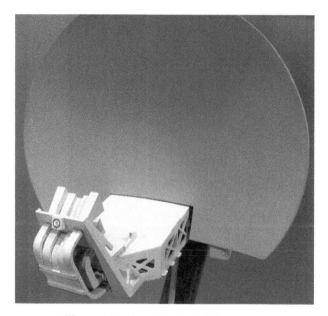

Figure 3.4   Example of an MFB antenna.

geography) or a pattern of spot beams. If a dynamically reconfigurable pattern is desired, then BFNs can be used as a core component of the multiple beams antennas; however, BFNs add weight and complexity.[6]

Specifically, two implementation approaches are possible for the HTS satellite antenna(s): (i) single feed per beam (SFB) and (ii) multiple feeds per beam (MFB) [SCH201101]. SFB designs use one feed horn for each spot; this implementation enjoys hardware simplicity and achieves a modestly better electrical performance compared with the MFB, but at the expense of an increased number of reflectors (apertures). To provide overlapping spots, several reflector apertures, typically four, are required (in some instances it is also possible to create a four-color scenario using only three reflectors). SFB antennas represent the state of the art for spot beam systems (see Figure 3.4 for a model of such antenna with 10 CONUS beams, e.g., AMC 15/16 – non-CONUS beams supported with other antennas). MFB designs use small subarrays for each spot; adjacent spots share some of the array elements. In this case, overlapping feed arrays are created, which allow producing overlapping spots using a single reflector aperture. The elements of the array are energized by a complex orthogonal BFN mentioned earlier.

Technologies such as phased array antennas and ground-based beam forming (GBBF) allow altering the coverage and the beam shape; this implies that satellite

---

[6]BFNs can and have been used to on spacecraft antennas to design geographic-specific coverage; shaped reflectors are a simpler way to achieve the desired geographic patterns at lower cost/complexity, but once designed they cannot be altered; this is a drawback if the operator at some point desires to drift the satellite to a new (distant) orbital position to cover a new geography.

operators who consider deploying these technologies will be able to increase the capacity allocated to a beam. Phased array antennas are widely deployed in military systems, and GBBF are used in MSS constellations; however, significant additional costs on the spacecraft (phased array) or ground segment (for GBBF capabilities) have led to limited deployment of these technologies in the HTS market so far [RUI201301] (we cover GBBF in Chapter 4). Additional antenna information is provided in Section 3.11.

The salient aspects of HTS, as synthesized from the discussion above, are: the satellite antenna of these HTSs are characterized by many small beams with high gain (up to about 200 in the most recent systems), which allows for closing the link to relatively small user terminals; these satellite antennas also allow for multiple frequency reuse, resulting in a throughput in the range of hundreds of Gbps. The small diameter earth station antennas require adequate spacing of the GSO satellites to avoid interference. The saturated power (EIRP) at beam edge is typically in the 45–50 dBW range; the nominal bandwidth is 72- or 108 MHz at Ku band, and possibly even higher at Ka band where a typical HTS has a number of UWB transponders distributed among the various beams. UWB transponders on HTS may, in fact, operate at 100-, 250-, or 500 MHz; they can support a single carrier or multiple carriers. For example, Anik F2, a Boeing 702 spacecraft, uses a half dozen wideband 492-MHz transponders to transmit traffic to gateways in the United States and Canada. Wideband modulators supporting up to 475 Msymbols/s (MSps) (up to 1.85 Gbps throughput) have appeared on the market, enabling one to maximize throughput on single beam, without requiring any additional backoff. For the end user, HTS provides the capability to increase data rates, upward of 100 Mbps to a single site in some cases, and to improve application performance compared to traditional satellite services [LAN201301].

3.4   FREQUENCY COLORS

As noted, HTSs have a large number of beams making heavy use of frequency reuse, employing frequency reuse concepts partially similar to small-cell cellular telephony. The beams can be arranged in a honeycomb to provide coverage of a service region.

Figure 3.5 depicts, in general, the concept of multibeam for a CONUS coverage. A number of distinct frequencies are employed; in fact, both frequency differentiation and polarization differentiation can be utilized among neighboring beams. For example, a system could use f1 = first quarter of the Ka bandwidth, f2 = second quarter of the Ka bandwidth, f3 = third quarter of the Ka bandwidth, f4 = fourth quarter of the Ka bandwidth; or, the system could use f1 = first half of the Ka bandwidth and right-hand circular polarization, f2 = second half of the Ka bandwidth and right-hand circular polarization, f3 = first half of the Ka bandwidth and left-hand circular polarization, f4 = second half of the Ka bandwidth and left-hand circular polarization). Other frequency reuse schemes are possible, on the basis of the amount of spectrum available and the amount of spectrum serving a given area. The concept of the Four Color Theorem (4CT) is used (e.g., see [MIN197501], [MIN197501], and [MIN197701], among any number of references on this topic). The term "color" is

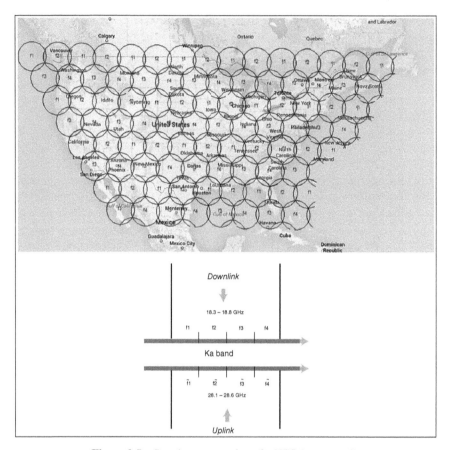

Figure 3.5 Spot-beam operation of a HTS (conceptual).

used to discuss different nonoverlapping (or noninterfering by using different polarizations) frequency bands, clearly with a reference to the mathematical concept of the 4CT.[7] As noted, the "four colors" can be supported with two different frequency subbands within the full Ku or Ka band and two orthogonal polarizations (usually

---

[7] The 4CT states that on a planar map, four colors are always sufficient to color a map such that no two adjacent "countries" sharing an edge boundary are of the same color – in fact, three colors are generally sufficient, but some complex/synthetic maps may need four (natural maps requiring four colors are rare). (The coloring has to be done "right"; however, situations may arise where more colors are needed.) For a long time this was a "known empirical fact" and it remained a mathematical conjecture; it took mathematicians over a hundred years to prove the theorem (from 1852 to 1976). The proof remains somewhat controversial even today because it entails having to use computer-assisted mechanisms for the analysis (more recently, theorem-proving software was used) in showing that there is a particular set of 1,936 maps, each of which cannot be part of a smallest-sized counterexample to the 4CT; any map that could be a counterexample must have a portion that looks like one of these maps. Since no smallest counterexamples exist, because any such example must contain one of these 1,936 maps, the contradiction in the logical argument implies there are no counterexamples and that the theorem is thus true.

right-hand and left-hand circular at Ka, vertical, and horizontal at Ku) (e.g., refer to Figure 1.9 for a view of subbands).

Adjacent spots are of different "colors," differing in frequency or polarization; hence, they can support the transmission of different information without mutual interferences. Spot beams with the same color use the same frequency and the same polarization, but they are spatially isolated from one another; no spot has a neighbor with the same color: because of the spatial separation, spot beams with the same color can support the transfer of different information. In most cases, the four-color approach is the best compromise between system capacity and performance; however, other frequency reuse schemes can also be utilized. Some HTS systems have optimized the total capacity at the expense of wide area coverage, because of the on-board antenna geometries used. Newer HTS systems are designed to double the bandwidth economics of earlier generations *and simultaneously* provide much larger coverage areas (e.g., ViaSat-2 will increase *seven times* the coverage area of ViaSat-1).

Figure 3.6 illustrates some footprint degradation, with decreased EIRP, because of satellite-borne antenna (optical) distortion; in some cases, these distortions can be addressed, but not all satellites have capabilities to do so. Figure 3.7 shows a possible example of a spot-beam coverage of the Indian and Far East region. As mentioned, to support a (large) set of spot beams that offer a covering of a region, a number of distinct frequency subbands ("colors") are needed. If, for example, the 500 MHz band is divided into two, there will be four colors (considering the polarizations). Figure 3.8 depicts an abstract spot beam system over Europe with 48 spot beams; the top shows the use of four colors, the bottom the use of three colors.

The *frequency reuse factor* of a multi-spot-beam antenna is the number of spots it can support divided by the number of colors; however, because of the overlap of

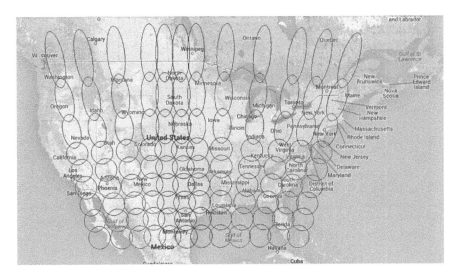

Figure 3.6   Spot-beam operation of a HTS (typical because of antenna distortion).

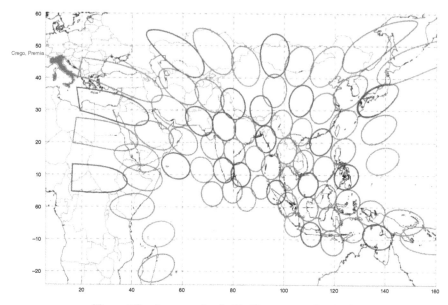

Figure 3.7   An example of a Far East–supporting HTS.

the beams and other technical constrains the *actual frequency reuse factor* is usually reduced to about 50–75% of the theoretical value. The higher the frequency reuse factor, to the degree possible, the better (note that for one zone contour beam covering, say, CONUS, the factor is 1). For example, a four-color HTS with 48 spot beams has a frequency reuse factor of 12, but because of the practical considerations just cited, the *actual frequency reuse factor* is 8; still, this means a multi-spot-beam satellite with 48 spots can provide about 8 times more capacity than a satellite with a large contour. A three color HTS with 48 spot beams has a frequency reuse factor of 16. This increase in satellite capacity is achieved without an increase in RF or DC power, without an increase in mass, and only minor increase in cost for the satellite [SCH201101].

The design desideratum for HTS tends to fall in one of the two broad categories, as hinted earlier: optimization of geographic coverage area and optimization of RF link performance. One can thus categorize the systems that have emerged of late, especially those based on the Ka-band HTS systems into two classes: those principally optimized to achieve high availability links, and those optimized mostly for large geographic coverage. The former is characterized by antennas that have beamwidths of fractional degrees; the latter relaxes link performance to use larger spot beams. HTS spot beams typically have 3 dB beamwidths between 0.4° and 1.5°; 0.5° is typical. The developers of HTS-based services (e.g., satellite operators procuring new spacecraft from manufacturers) must balance their geographic coverage needs against the improved link performance that small spot beams can provide. For example, such a trade-off must take into account the fact that the number of transponders, the payload complexity, and the spacecraft power requirements all increase linearly with the number of beams on the satellite; hence, the reliance on very small beams also limits

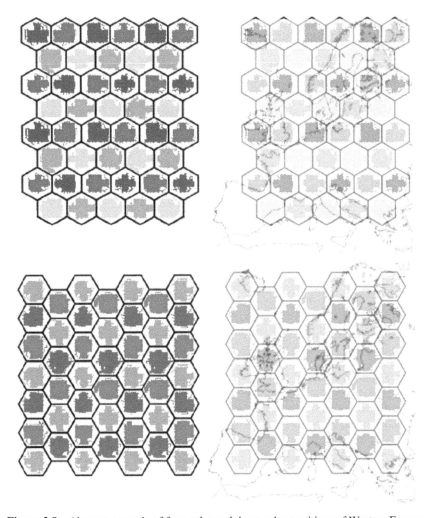

Figure 3.8 Abstract example of four-color and three-color partitions of Western Europe.

the available service area of the HTS [PAW201301]. As just noted, multi-spot-beam antennas provide beams with a typical diameter of 0.4° or less; hence, a very accurate pointing of the spacecraft antenna is required. This can be achieved by the use of an active pointing system, where a beacon signal is received by a special RF sensing feed chain (one RF sensing feed chain for each reflector is required); typically, the beacon station is located inside the user geographic area. It follows that the RF sensing capability has to be provided by the feed chain on the spacecraft in addition to the Tx/Rx user and gateway function [SCH201101].

From a macro perspective, Figure 3.9 shows a conceptual view of feed horn array/reflectors on a HTS. The satellite antenna is an offset-type antenna. For

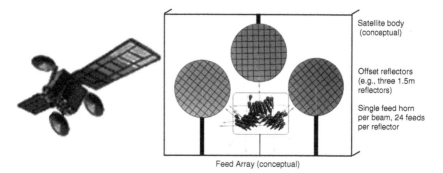

Figure 3.9 Conceptual view of feed horn array/reflectors on a HTS (three-antennas approach).

example, from a macro perspective, Figure 3.10 shows the utilization of dual polarization from the feed array. Figure 3.11 depicts the frequency assignments to the beams; note that if the exercise is done correctly not only will neighboring beams have different frequencies, but each spacecraft antenna will support a full complement of the frequencies (i.e., will use all of the subbands). The spreading of the signals from different locations to different spacecraft antenna reduces the interbeam interference; Figure 3.12 shows the polygonal antenna contours for the overall operation of an antenna trio and the beam (feed) assignments on the spacecraft antenna.

Figures 3.13 and 3.14 show an end-to-end system, including the gateway function (uplink/downlink) to connect the system to the Internet (or other services). The outbound link uses TDM techniques to download information for each spot beam. The RF maximum outbound link symbol rates for third-generation HTS systems are about 45 MSps, but will increase to 250 MSps with fourth-generation HTS systems. The return (inbound) channel protocols are mostly proprietary and use combinations of MF-TDMA and Aloha access protocols. The number of subscribers that can be served per beam is related to the reuse factor (between 25:1 and 50:1), the average bit rate desired per user, and the average EIRP per beam (EIRP is determined by the Traveling Wave Tube Amplifier (TWTA) size and the beams' sizes, which determine the beam gains).

3.5   FREQUENCY BANDS OF OPERATION

While HTSs can operate at any band and although Ku-band HTS systems are also being deployed, as noted, the Ka band is the current target of major commercialization of HTS services. There are some differences worldwide about the Ka-range frequencies, in particular, because the concept of the Ka band is not formally defined in the Radio Regulations (RR) of the International Telecommunications Union (ITU). The Ka band is generally considered to

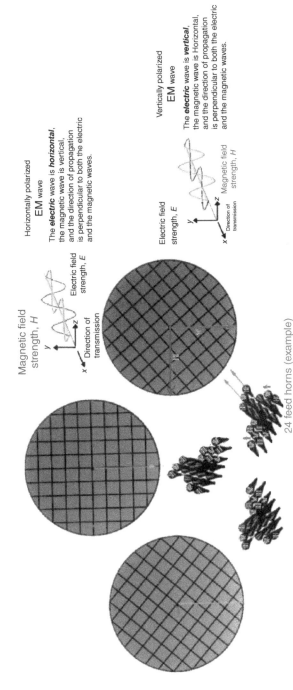

Figure 3.10 Dual polarization transmission from each feed horn assembly/reflector (three-antennas approach).

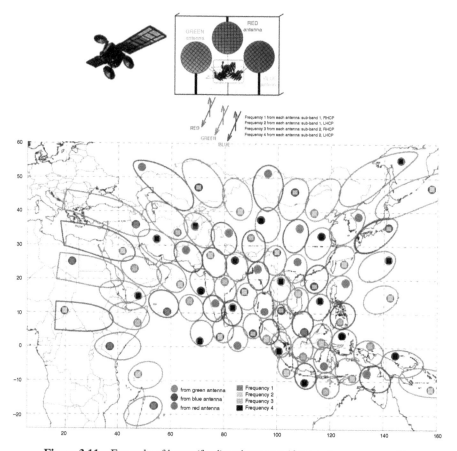

Figure 3.11   Example of beam (feed) assignments (three-antennas approach).

span the frequency range 17.3–31 GHz.[8] A "preferred" range for FSS Ka applications is the 1.1 GHz band, covering the range 27.5–28.6 GHz for the uplink and 17.7–18.8 GHz for the downlink; in the United States, 28.1–28.6 GHz for the uplink

[8]Note: in some of the bands, the concept of equivalent power flux-density (epdf) applies. The idea is to protect specified ground users (stations) from interference by certain classes of satellites: the limits define the maximum *permissible* interference that non-GSO FSS systems can cause to GSO FSS networks. The epfd limits were introduced by SkyBridge and adopted by the World Radio Conference 2000 (WRC-2000), and although the SkyBridge non-GSO satellite network was intended to operate in the Ku-band, WRC-2000 adopted epfd limits for portions of both the C-, Ku- and Ka-bands. At the current time there are no satellites operating using this concept. The motivation is based on desire of reusing GSO frequencies by a non-GSO constellation outside the GSO by avoiding the GSO by about ±10°. The epfd (for the uplink and for the downlink) values are calculated so that they would increase the unavailability by no more than 10% on the most sensitive links. Therefore, in practicality, a non-GSO FSS system will cause even less than a 10% increase in unavailability on a GSO FSS link, that is, the values are so low that they will have a negligible effect on the GSO FSS links [CHR201201].

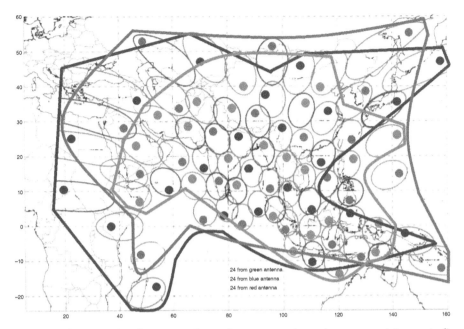

Figure 3.12 Example of antenna polygonal contours and spot beam support (conceptual) (three-antennas approach).

and 18.3–18.8 GHz for the downlink is prevalent. Figure 3.15 provides a simplified view of the available Ka bands. Note that almost the entirety of Ka band is also allocated to the terrestrial Fixed Service (FS), and many countries have licensed LMDS (Local Multipoint Distribution System) services in the Ka band; specifically, in all three ITU Regions (see Chapter 1), the terrestrial FS is coprimary in most of the FSS Ka band. Note, incidentally, that currently there are no satellites in orbit using the new 21.4–22 GHz BSS band but satellites using this band are being planned for future DTH applications. We focus here on the main traditional Ka bands, but other bands (as per Figure 3.15) may be utilized in the future. As discussed in the next section, high-frequency electromagnetic waves are fairly susceptible to rain fade: in heavy rain fall and in heavy rain fall regions, signal degradation can occur. For downlinks rain dissipates 3–10 times more energy at Ka band (20 GHz) than at Ku band (11 GHz); for uplinks rain dissipates 60–400 times more energy at Ka band (30 GHz) than at Ku band (14 GHz). To close the link and provide adequate margin at Ka, additional power is needed in the spacecraft (and/or more gain and/or higher efficiency in the antenna).

Antennas always need proper licensing. *Large gateway TX/RX antennas* are definitely required to comply with regulatory requirements. *Smaller end-user terminals* may receive blanket licensing, but the manufacturer still needs to go through technical filings and demonstrations that the equipment is generally compliant with pertinent overall regulation. For example, in the United States, the Federal Communications

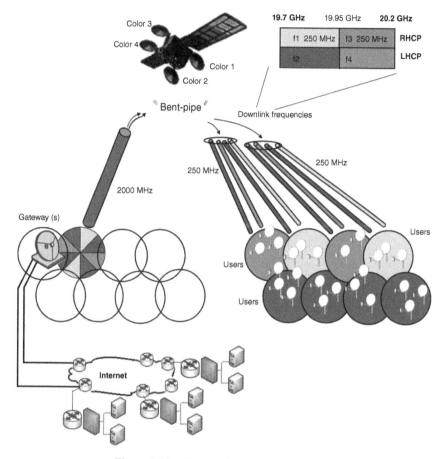

Figure 3.13 Gateway in a HTS environment.

Commission (FCC) has approved operation of blanket-licensed remote VSAT Ku/Ka terminals in specified bands. Related to Ka, those authorizations allow the spacecraft to operate (i) in the 28.6–29.1 GHz band on a secondary basis, and (ii) in the associated 18.8–19.3 GHz band on a nonconforming basis. For example, the ViaSat-1 FCC authorization allows the spacecraft to operate (i) in the 28.6–29.1 GHz band on a secondary basis, (ii) in the 28.10–28.35 GHz on a secondary basis, and (iii) in the 18.8–19.3 GHz band on a nonconforming basis. *Mobile antennas*, say on airplanes where the movement of the aircraft may (temporarily) point/position the antenna such that it could cause interference, require more rigorous validation and associated power management mechanisms (as discussed in Chapter 4).

In general, and in accordance with existing licensing procedures, Ka FSS operators must demonstrate that they are not creating GSO FSS interference, non-GSO FSS interference, and interference to terrestrial microwave users. These validations need to be made at a fundamental design level (is the equipment transmitting in a permitted

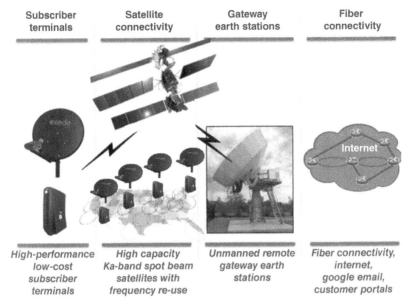

Figure 3.14   End-to-end system using HTS services.

band), as well as in consideration of other operators in the proximity (either on the ground or in space) via a frequency coordination process. Thus, Ka antenna operations (in the United States) must be compatible with the operation of adjacent GSO systems, non-GSO systems, and primary terrestrial users to use in the 18.3–19.3 GHz and 19.7–20.2 GHz downlink bands and the 28.1–29.1 GHz and 29.5–30.0 GHz uplink bands. The expectation is that the satellite network will cease operations in the 18.8–19.3 GHz downlink band (and the associated 28.6–29.1 GHz uplink band) in any spot beams, where the predicted physical alignment of such beams would fall within a specified minimum separation angle of a non-GSO operational link.

Fundamentally, in the United States, 47 CRF Section 25.132(a)(2), promulgated by the FCC, provides[9] that transmitting earth stations operating in the 20/30 GHz band must demonstrate compliance with Section 25.138. The antenna must meet the performance requirements in Section 25.138(a) in the direction of the GSO arc, as well as in all other directions, as illustrated by off-axis EIRP spectral density plots.

---

[9]The full reference is as follows: Code of Federal Regulations (CRF) – Title 47 – Telecommunication – Chapter I – Federal Communications Commission – Subchapter B – Common Carrier Services – Part 25 – Satellite Communications

Section 25.132 – Verification of earth station antenna performance standards.

Section 25.138 Blanket Licensing provisions of GSO FSS Earth Stations in the 18.3–18.8 GHz (space-to-earth), 19.7–20.2 GHz (space-to-earth), 28.35–28.6 GHz (earth-to-space), and 29.25–30.0 GHz (earth-to-space) bands.
Section 25.209 – Antenna performance standards.

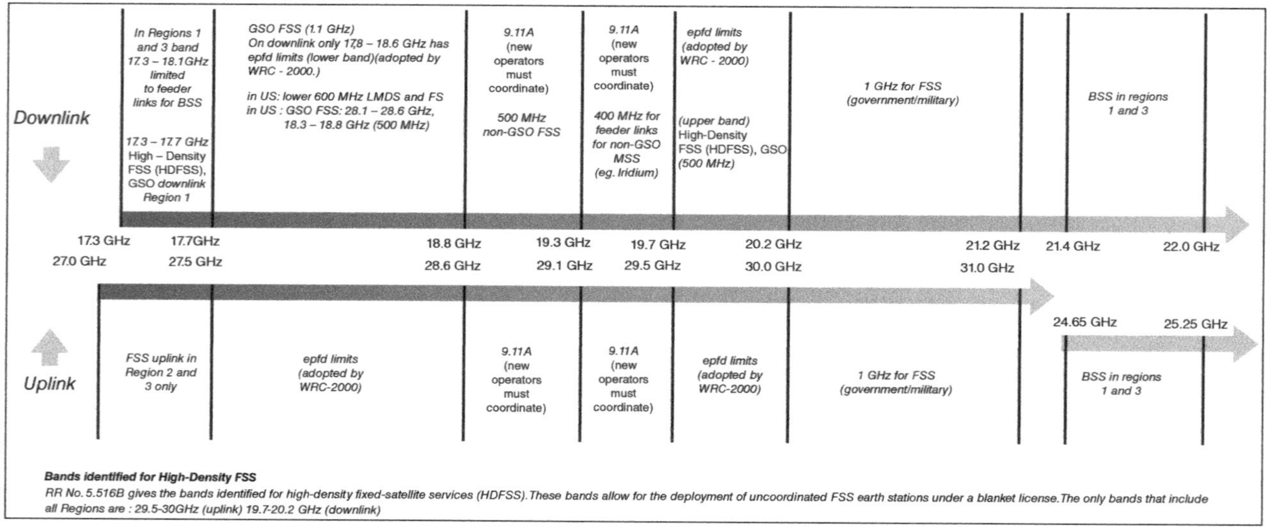
Downlink
In Regions 1 and 3 band 17.3 – 18.1GHz limited to feeder links for BSS
17.3 – 17.7 GHz High – Density FSS (HDFSS), GSO downlink Region 1
GSO FSS (1.1 GHz)
On downlink only 17.8 – 18.6 GHz has epfd limits (lower band)(adopted by WRC - 2000.)
in US: lower 600 MHz LMDS and FS
in US : GSO FSS: 28.1 – 28.6 GHz, 18.3 – 18.8 GHz (500 MHz)
9.11A (new operators must coordinate)
500 MHz non-GSO FSS
9.11A (new operators must coordinate)
400 MHz for feeder links for non-GSO MSS (eg. Iridium)
epfd limits (adopted by WRC - 2000)
(upper band) High-Density FSS (HDFSS), GSO (500 MHz)
1 GHz for FSS (government/military)
BSS in regions 1 and 3
17.3 GHz
17.7GHz
18.8 GHz
19.3 GHz
19.7 GHz
20.2 GHz
21.2 GHz
21.4 GHz
22.0 GHz
27.0 GHz
27.5 GHz
28.6 GHz
29.1 GHz
29.5 GHz
30.0 GHz
31.0 GHz
24.65 GHz
25.25 GHz
Uplink
FSS uplink in Region 2 and 3 only
epfd limits (adopted by WRC-2000)
9.11A (new operators must coordinate)
9.11A (new operators must coordinate)
epfd limits (adopted by WRC-2000)
1 GHz for FSS (government/military)
BSS in regions 1 and 3
Bands identified for High-Density FSS
RR No. 5.516B gives the bands identified for high-density fixed-satellite services (HDFSS). These bands allow for the deployment of uncoordinated FSS earth stations under a blanket license. The only bands that include all Regions are : 29.5-30GHz (uplink) 19.7-20.2 GHz (downlink)

Figure 3.15 Ka bands.

Further, as established by the FCC, the power flux density at the earth's surface produced by emissions from a terminal must be within the -118 dBW/m^2/MHz limit set forth in Section 25.138(a)(6). In addition, to the extent required for the protection of received satellite signals pursuant to Section 25.138(e), the earth station must conform to the antenna performance standards in Section 25.209, as demonstrated by the antenna gain patterns. These issues are very important in aeronautical applications, as discussed in Chapter 4.

In closing this brief regulatory discussion, it has to be mentioned that HTS operators in the United States (and similarly in other parts of the world in accordance with local regulation), Ka FSS operators must demonstrate that they are not creating GSO FSS interference, non-GSO FSS interference, and interference to LMDS users; they must prove that they will operate in the 28.10–28.35 GHz band in a manner that will protect terrestrial microwave users, which are designated as the primary use of the band, from harmful interference. Consistent with the secondary nature of the GSO FSS allocation in this band, the operator's use of the 28.10–28.35 GHz frequency band for gateway earth stations must be on a nonharmful interference basis relative to LMDS. As an illustrative example, ViaSat has implemented measures to ensure that their proposed gateway/TT&C earth station will operate in a manner that will protect LMDS stations from harmful interference. ViaSat has located the gateway antenna in a remote area where LMDS is unlikely to be deployed, and it conducted a technical analysis to determine a "worst case" potential required separation distance from an LMDS terminal, assuming no shielding were employed at the gateway earth station. They determined that the required separation distance between an LMDS terminal and the gateway for the worst case alignment is 9.5 km. This is the minimum distance between an LMDS terminal and the ViaSat gateway that may require ViaSat to take measures to mitigate interference into that LMDS terminal. The actual required separation distance may be smaller depending on the characteristics of the surrounding terrain and variations in the LMDS system from the assumptions used in this analysis [FCC201201].

Ka-band-based technologies have matured to the point where the performance, reliability, and availability of Ka-band networks are comparable to Ku-band networks. This extends from the gateway radio frequency transmission (RFT) equipment, where the industry is bringing to market Ka-traveling wave tubes (TWTs) supporting output power up to 750 W, to the high-volume VSAT production incorporating state-of-the-art gallium arsenide (GaAs) monolithic microwave integrated circuits (MMICs) to produce reliable, cost-effective, and high-performance VSAT systems operating in Ka band [HUG201301]. At press time, there were more than 50 active Ka-band communication projects (either in orbit or planned for launch by mid-decade); however, the majority of these satellites included (or were planning to include) Ka-band connectivity as a subset of their entire payload. HTS payloads may also include transponders for several different satellite bands: a number of satellite operators are taking the approach to incrementally add Ka-band capacity into service, while simultaneously focusing on main traditional mission on the basis of 36/54 MHz Ku- and C-band coverage optimized for TV/data broadcast (DTH or cable headends, respectively). As is typically the case for modern satellite systems, HTSs are often multipurpose designs. HTS systems may thus provide large regional and hemispherical beams as well as spot beams.

3.6 LOSSES AND RAIN CONSIDERATIONS

Both free space losses and rain considerations have to be taken into account in HTS designs (as is in fact the case for all satellite channels). As the transmitted signal traverses a distance of space, its power level decreases at a rate inversely proportional to the distance traveled and proportional to the wavelength of the signal. This effect is because of the spreading of the radio waves as they propagate. As radio waves propagate in free space, the power falls off as the square of range: for a doubling of range, the power reaching a receiver antenna is reduced by a factor of four. For line-of-sight free space propagation, the loss L can be calculated by:

$$L = 20\log_{10}(4\pi d/\lambda) = 21.98 + 20\log_{10}(d/\lambda)$$

where

  d the distance between receiver and transmitter.
  λ free space wavelength $= c/f$.
  c speed of light (3×10^8 m/s).
  f frequency (Hz).

In addition to free space losses (attenuation), there are a multitude of other factors that have to be taken into account. These include, but are not limited to, attenuation, distortion, dispersion, intermodulation, fade, multipath, dropouts, and external and adjacent satellite interference, and cross-pol(arization) interference. Losses and interference degrade the reception of radio waves. Losses are typically because of absorption, for example, rain fade, among other causes. Interference can be caused, among other possibilities, by ground-based sources (e.g., terrestrial microwave links operating at the same frequency), celestial sources (sun spots), elements within the satellite system (e.g., cross-pol interference), or by transmissions by other satellite systems that use the same frequency bands. Rain impacts the performance of an RF link. Generally, the higher the frequency, the higher the attenuation caused by rain fall. Moisture can degrade the link such that the overall signal to noise ratio drops. In heavy rain there could be portions of time when the link is unusable (outage); heavy rain can also cause depolarization, where signals from one polarization appear in the opposite polarization. The Glossary provides a longer (although not exhaustive) list of impairments and issues that have to be taken into account when designing an RF system in general and a satellite link in particular.

Electromagnetic waves are absorbed in the atmosphere; the absorption is a function of the wavelength. Oxygen (O_2) and water vapor (H_2O) are responsible for the majority of signal absorption. There are areas in the electromagnetic spectrum where there are local maxima; the first maximum occurs at 22 GHz because of water, and

TABLE 3.1   Rain Rates in Various Zones and Percentage of the Time That the Specified Rate is Exceeded

			Rain rates in mm/h													
			A	B	C	D	E	F	G	H	J	K	L	M	N	P
Percentage time that amount given is exceeded	1.0%	Time (h) 87.60	0	1	2	3	1	2	3	2	8	2	2	4	5	12
	0.3%	26.28	1	2	3	5	3	4	7	4	13	6	7	11	15	34
	0.1%	8.76	2	3	5	8	6	8	12	10	20	12	15	22	35	65
	0.03%	2.63	5	6	9	13	12	15	20	18	28	23	33	40	65	105
	0.01%	0.86	8	12	15	19	22	28	30	32	35	42	60	63	95	145

the second maximum occurs at 63 GHz because of oxygen. Note that the amount of water vapor and oxygen in the atmosphere decreases with an increase in altitude because of the decrease in pressure. Table 3.1 depicts the amount of rain (in mm/hr) in various rain zones – A through P – defined by the ITU[10] and the percentage of the time that the specified precipitation rate is exceeded. One would determine which rain zones the earth station antenna in question is located by consulting the table to determine the precipitation rate in mm/hr and the graph to determine the attenuation. When one considers designing a distribution network consisting of one hub and multiple remotes sites that are geographically distributed over the satellite downlink beam (e.g., for commercial digital video distribution), one needs to keep in mind that sites have typically a variation of the antenna type/size (impacting the gain), the noise (temperature – e.g., because of the look angle), as well as different rain fade margin requirements.

Taking into account the main contributors to attenuation, we can conclude that the (pragmatic) total attenuation through unobstructed atmosphere is the sum of free space path loss, the attenuation caused by oxygen absorption and attenuation caused by water vapor absorption, and the attenuation caused by rain when present; namely,

$$\text{Atten}_{\text{Total}} = \text{Atten}_{\text{FreeSpacePathLoss}} + \text{Atten}_{\text{Oxygen}} + \text{Atten}_{\text{WaterVapor}} + \text{Atten}_{\text{Rain}}$$

[10]Rain models are used to calculate the amount of attenuation seen by the RF link. There are a number of models available, but all models ultimately use the following equation to calculate the rain loss:

$$A = a * R^b * L \text{ (dB)}$$

where A is the attenuation in dB, R is the rain rate seen along the susceptible portion of the RF path, and L is the length of the transmission path which is susceptible to rain. The coefficients a and b depend on RF frequency and polarization. The rain rate is a constant that depends on the site location of the RF terminal. There are a number of rain models in existence in varying degrees of revision. Practically, there are two rain models that are used extensively – the Global Crane Model and the ITU Rain Model. The models differ in the techniques and data used in calculating the various parameters in the attenuation equation. The ITU model is gaining wider acceptance of late and is updated periodically. The Global Crane model is still in widespread use but has not been updated in a number of years.

The issue of amplifier power both on the ground and on the satellite is important for the proper operation of a satellite link. Sufficient power is needed to be able to "close the link" (i.e., have a net power margin > 0). The transmitted power path loss for the Ku-band satellites is higher than C band, because of the higher frequency at Ku band. The path loss for a C-band signal is −196.5 dB, while path loss for a Ku-band signal is −205.8 dB. Hence, under the severe weather conditions such as heavy rain and snow, the link loss for Ku band is even higher; Ku-band systems, therefore, need more conservative rain fade margins designed into the link budget. The Ka band is impacted even more.

Some of the areas in the world where HTSs services make sense because of the dearth of infrastructure, such as tropical zones in South East Asia, Central America, and West and Central Africa, have precipitation levels of over 3,000 mm annually; thus rain fade may be an issue for Ka-band services in these locations.

Larger margins need to be included in the Ka HTS designs because of rain fade considerations, compared with C- or Ku-band operation. The Adaptive Coding and Modulation (ACM) schemes discussed in Chapter 2 are applicable, in particular, for the outbound channels (to the user terminals) to minimize or eliminate the rain fade margin, which is high in the 18–20 GHz downlink Ka FSS band. ACM changes the modulation and coding rates at the gateways, on the fly (each packet), in response to the feedback on the received C/N at the user terminal (see Figure 3.16). All HTSs use the DVB-S2 standard, and DVB-S2X will likely be used in the future. As we saw elsewhere in this book, ACM reduces the information rate during rain fades and resumes at the clear sky level after the rain fades (the range of C/N margin, when comparing 16 PSK clear sky to 1/2 BPSK, is approximately 12 dB).

3.7   HTS APPLICATIONS

A number of applications can be supported with HTSs, as seen in Table 3.2. The use of HTS for mobility applications is discussed in Chapter 4.

Some proponents see DTH as a key driver for HTS, while others see Internet access and mobility as the major drivers (as noted in the introductory section of this chapter).

DTH services are currently delivered in both the BSS and FSS bands. For DTH broadcasters, the use of high-power spot beams makes it possible to achieve greater flexibility, reach, and reliability in serving consumers with relatively small antennas, to match the growth in flat-screen availability in households in parts of the world such as India, China, and Asian island nations. Small beams reusing frequencies also make it possible to provide more localized DTH services for regional linguistic markets. As discussed earlier, new Ka bands are being made available in some ITU regions, but at higher frequencies (thus with rain attenuation/higher-power implications) [ITU201201]. WARC-1992 allocated the 21.4–22 GHz band to BSS in ITU regions 1 and 3 (Europe, Africa, Asia, and Oceania), with an effective date of April 1, 2007. The new 21.4–22 GHz BSS band will typically be used to provide a national service; however,

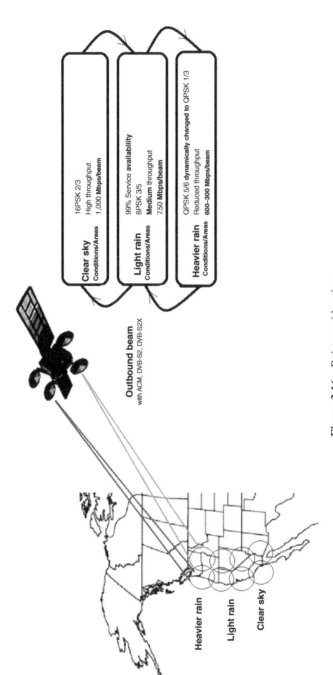

Figure 3.16 Rain considerations.

TABLE 3.2 HTS Applications, Partial List

Application	Description	Considerations
Internet access	Provide high-speed internet access for stationary users. Evolving Ka-band multi-spot-beam HTS can provide cost-effective broadband services to areas with an underdeveloped terrestrial infrastructure.	Appropriate beam design, namely, link availability considerations. In addition, the hub must be designed to be able to manage resources (e.g., capacity) effectively during high-contention intervals to maintain class of service metrics and fair access (preventing individual subscribers to monopolize the satellite capacity).
Commercial VSAT networking	Provide high-quality VSAT services.	Currently, Ku-based HTS/spot beams are better suited for this application, but as described in the text, Ka approaches can also be used in moderate-rain areas. For business applications, key features will include guaranteed bandwidth throughput, latency, and jitter specifications according to a published Service Level Agreement (SLA) spec, very high network availability, support of private IP addressing, VLAN tagging, prioritization of traffic-based service classes, and support for VPNs and other forms of encryption.
Mobility	Provide Internet access to people on the move, either terrestrially (especially for prosumers, scientific utilizers, etc. with mobile antennas), or for aeronautical/maritime environments.	As mobile terminals move across multiple beams, there is a need for multiple "hand-offs" from beam to beam, especially in aeronautical applications.

3G/4G backhauling	Provide trunking services to extend coverage to rural areas and emerging markets.	3G and 4G cellular systems support high channel rates (e.g., 10–20 Mbps for outbound links and 5–10 Mbps for inbound links for 3G; and 100 Mbps for outbound links and 50 Mbps for inbound links with LTE), which, in turn, require higher bandwidth backhaul channels. 3G and 4G/LTE are being rolled in urban areas with terrestrial backhaul on the basis of fiber or microwave connectivity. Providing coverage in rural areas and emerging markets is challenging in general but is an emerging niche for HTS-based satellite backhaul (initially, the opportunity is more focused on 3G's expansion to exurban areas in the developed world, but 4G services also provide an opportunity as 4G will graduate from urban areas to other less-dense markets). Link availability considerations.
Video distribution	DTH services in emerging countries, especially those with smaller geographies or distinct ethnic groups.	
M2M	Provide global coverage of M2M traffic, such as telemetry, sensor data, Internet of Things (IoT), and Unmanned Aerial Vehicles (UAVs).	The cost of the terminal devices must be kept in mind for large population wireless sensor networks (WSN) and/or IoT [MIN200701], [MIN201301]. Refer back to observations in Chapter 1 on M2M modem costs/targets.

if the service area is large enough (e.g., India, China, Russia), it could be covered with several spot beams with each spot beam catering to a specific language or cultural group. "Spot beam" DTH services that serve defined local areas are already utilized in the United States, where local broadcast services are focused on specific regions by different satellite spot beams. SPACEWAY®1 and -2 HTSs are being used in this way by DTH operator DirectTV; the SPACEWAY 1, -2, and -3 spacecraft were originally built to provide high-speed Internet connectivity using multiple Ka-band spot beams; however, at the time of construction there emerged a stronger-than-expected demand for DTH TV; thus, Hughes instructed Boeing to modify the SPACEWAY 1 and -2 satellites to provide simple turnaround "bent pipe" DTH broadcasts (and disable the regenerative on-board processing of the original system that was to be used for broadband satellite communications – the multiple spot beam configurations of these spacecraft meant that they could be used for local-into-local transmission; only the SPACEWAY 3 satellite has been used as originally intended [CAS201201]).

3.8   COMPARISON BETWEEN APPROACHES

It is important to realize that satellite service providers should have a menu of offerings as one size does not fit all (as some satellite operators sometimes appear to believe). Some applications require robustness (e.g., military, real-time telemetry), while other applications (e.g., Web browsing) are more forgiving. Some applications/users place a higher priority on network reliability, user throughput, and application performance; others are looking for basic, cost-effective connectivity. These considerations may lead to different designs.

The majority of commercial communications satellite networks, including HTSs, are GSO networks. Currently, there are some industry debates about HTS in terms of the advantages/disadvantages of Ka versus Ku operation and GSO versus non-GSO. Naturally, there are vested interests of the part of the vendor community, each proposing a particular architecture. Figure 3.17 provides a comparison between Ku- and Ka-based operation, as well as the GSO and non-GSO systems. In practical terms, HTS-based solutions, as an aggregate, will include a mix of frequency (and orbits), with each choice optimized for specific environments and applications. At lower frequencies (e.g., Ku band) the links require lower power/gain to overcome propagation impairments, and HTS systems have tended to use wider spot beams than is the case at the higher frequencies. Power allocation algorithms are needed for HTS operations; these schemes optimize the frequency bandwidth, the satellite transmission power, the modulation level, and the coding rate to each beam (MODCODs), to properly manage the user distributions and the interbeam interference conditions. One example of power optimization algorithm is discussed in reference [WAN201401].

At this juncture it is relatively difficult to obtain commercially Ku-band orbital slots from the ITU, in particular, over industrialized regions. These orbital slots are the life blood of a satellite operator, and without additional spectrum, the growth potential of an operator is severely restricted; furthermore, many of the orbital

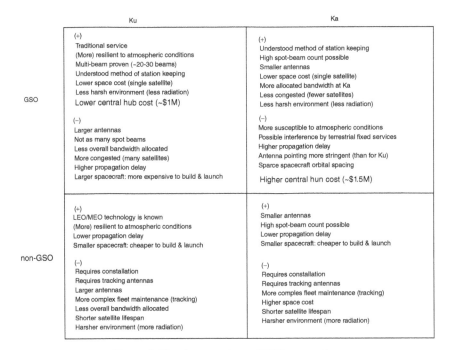

	Ku	Ka
GSO	(+) Traditional service (More) resilient to atmospheric conditions Multi-beam proven (~20-30 beams) Understood method of station keeping Lower space cost (single satellite) Less harsh environment (less radiation) Lower central hub cost (~$1M) (–) Larger antennas Not as many spot beams Less overall bandwidth allocated More congested (many satellites) Higher propagation delay Larger spacecraft: more expensive to build & launch	(+) Understood method of station keeping High spot-beam count possible Smaller antennas Lower space cost (single satellite) More allocated bandwidth at Ka Less congested (fewer satellites) Less harsh environment (less radiation) (–) More susceptible to atmospheric conditions Possible interference by terrestrial fixed services Higher propagation delay Antenna pointing more stringent (than for Ku) Sparce spacecraft orbital spacing Higher central hun cost (~$1.5M)
non-GSO	(+) LEO/MEO technology is known (More) resilient to atmospheric conditions Lower propagation delay Smaller spacecraft: cheaper to build & launch (–) Requires constellation Requires tracking antennas Larger antennas More complex fleet maintenance (tracking) Less overall bandwidth allocated Shorter satellite lifespan Harsher environment (more radiation)	(+) Smaller antennas High spot-beam count possible Lower propagation delay Smaller spacecraft: cheaper to build & launch (–) Requires constellation Requires tracking antennas More comples fleet maintenance (tracking) Higher space cost Shorter satellite lifespan Harsher environment (more radiation)

Figure 3.17   A comparison of HTS approaches.

slots are vigorously guarded by the nations over which the equatorial slots lie for national-only use.[11] On the contrary, Ka-band orbital slots are generally underused: although virtually every Ka-band slot has multiple filings, only a few of the slots have actually been used, which means that it is easier for an operator to obtain rights to a Ka-band orbital slot from the ITU [HUG201301]. Another benefit of Ka-band, as implied above, is the availability of greater amounts of spectrum compared to the Ku band. While a typical Ku-band satellite might operate across 500 MHz of spectrum, a Ka-band satellite might operate across 1,100 MHz or more of spectrum for the gateway beams.

A key disadvantage of a non-GSO HTS systems is that a constellation of (at least) eight spacecraft) are needed to provide global coverage, and all satellites in the constellation must be launched to provide continuous service (the number of spacecraft needed to provide continuous coverage depends on the altitude of the satellites – the lower the altitude the larger the number of satellites required). Tracking antennas are needed on the ground. The fact that a constellation of satellites is needed implies that capital costs of launching a non-GSO HTS system is typically high: generally,

---

[11]Some international operators strike deals with national entities to share and/or manage an orbital slot assigned (by the ITU) to that country. Typically, said operators establish a minimalistic *pied a terre* in that country to meet the minimal basic letter-of-the-law regulatory requirement of the local administration.

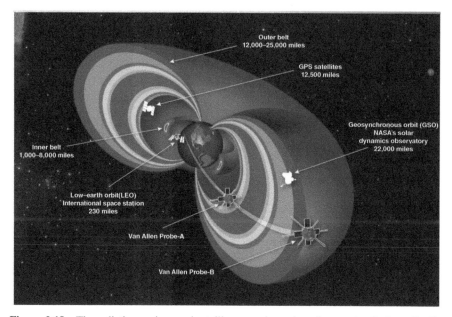

Figure 3.18   The radiation regions and satellites near the region of trapped radiation. *Credit:* NASA.   http://www.nasa.gov/mission_pages/sunearth/news/gallery/20130228-radiationbelts .html.

an initial budget of around $2B is needed,[12] while a GSO spacecraft in orbit may cost around $150 m range. Also, because it takes time (3–4 years or more) to build a constellation the operator's cash flow can be challenging during that period. Furthermore, the electromagnetic environment for Low Earth Orbit (LEO) and MEO satellites is relatively harsh, so that the lifespan of an LEO satellite is typically shorter, thus requiring a more frequent replacement (the lifespan of LEOs is in the range of 7 years as compared with around 15 years for a GSO satellite). The radiation belts are two donut-shaped regions encircling the earth, where high-energy particles, mostly electrons and ions, are trapped by earth's magnetic field. This radiation can affect the performance and reliability of satellites. The inner belt extends from about one to eight thousand miles above earth's equator; the outer belt extends from about 12 to 25 thousand miles. Figure 3.18 (from NASA) depicts the radiation regions and satellites near the region of trapped radiation.

[12] In the late 1990s, two non-GSO satellite networks were proposed to provide consumer Internet connectivity: Teledesic and SkyBridge. They were never launched mainly because of the initial large capital costs to implement the networks, absence of reasonably priced consumer terminals, and the difficulty of suitable installation sites as the earth station had to have a clear view of most of the sky to "see" the non-GSO satellites [CHR201201].

3.9   A VIEW OF KU-BASED HTS SYSTEMS

Most HTS systems deployed to date have been designed for mass markets and to operate in Ka band where small aperture antennas can provide link closure in the narrow spot beams. However, satellite operators are also now applying HTS technology and spot-beam antennas to new Ku-band spacecraft. As these HTS systems proliferate, operators of VSAT networks will have new technology choices when implementing solutions tailored to the specific application environment at hand [CAP201201]. Mass markets are not perceived as comprising VSAT users but as individual consumers; VSAT networks, on the contrary, are typically considered to encompass an enterprise (e.g., a large national chain of stores, gas stations, or fleets).

As discussed, different applications may benefit from different HTS solutions and price points (said price points being applicable to both of the initial equipment/installation cost and the monthly recurring costs). In some instances (e.g., residential service, wireless sensor/M2M services, in-field scientific/nature research), a "best effort" Internet service may be good enough; commercial and military applications may possibly require a more robust solution. Solutions that require more robustness tend to fall in the Ku VSAT class; terminals in this class employ relatively inexpensive antennas in the 1.2–1.8–2.4 m range. Ka-band VSAT systems are currently less common, and, therefore, are more expensive than Ku-band systems of similar performance, because of lack of economies of scale in the fabrication process (Ka-based VSAT terminals in the 1.2–1.8 m range are not as of yet produced in large quantities, and, therefore, remain more expensive than Ku-band terminals). By contrast, solutions for web browsing (whether at home or on the move), especially at reasonably acceptable price points of $50–150/month, will avail themselves of the "mass market" submeter systems designed for DTH reception; these systems are generally not directly suitable for industrial-grade environments in terms of both performance and hardware reliability.

A press time, discussion centered on the use of the legacy Ku band and the evolving Ka band. The use of spot beams with frequency reuse allows both the Ku-band and Ka-band systems to achieve high overall spectrum efficiencies and HTS systems in both bands use UWB transponders. Proponents of the Ku-band approach make the case that Ka-band system may at times sacrifice link performance (e.g., availability) in favor of coverage by using larger spot beams: spot beam systems are typically limited in geographic coverage; nonetheless, some HTS systems provide large fields of spot beams that, in aggregate, create continental and even global coverage; however, others offer only a relatively small number of fixed or steerable spots in targeted areas. Ku-band spot beams and Ka-band large spots beams are similar in beamwidth and, so, are generally comparable in system coverage. Ka-band small spot beams, however, generally cover only about 10% or 15% of the area covered by a large spot beam.

CapRock, a purveyor of VSAT services and an advocate of Ku-based services, recently undertook a cost comparison aimed at assessing high-grade commercial/military services in the context of HTS [CAP201201]. Figure 3.19 assembled from data from their study shows the relative cost of providing a constant bit rate (CBR[13]) service as a function of service availability, under different climatic conditions for the three classes of HTS systems cited earlier: Ku-band spot beams, Ka-band small spot beams, and Ka-band large spot beam. Relative performance characteristics were assessed for 1.2 m VSATs[14] located in temperate, tropical, and arid regions. The analysis isolated the effects of satellite technology and frequency band by using similar VSAT-to-satellite look-angles in the various regions, as well as common gateway locations. To the extent possible, given the different frequency bands, common earth terminal characteristics were used in the study according to the published White Paper [CAP201201]. Furthermore, teleport locations and gateway terminal performance parameters were selected such that the VSAT-to-satellite links dominated the availability. The results shown in the figure are for the outbound (i.e., GWSV – VSAT) link, which is the most difficult to manage in these VSAT systems, because of the need to manage satellite power resources as well as satellite bandwidth. The data in Figure 3.19 indicate that, for the assumptions made in the study, the cost of providing a high-availability service is always lowest with a Ku-band HTS system. The cost difference between a small spot Ka-band HTS and a Ku-band HTS narrows as the availability of the service is reduced, and in many locations the Ka-band small spot system is at near parity with the Ku-band system for availabilities less than around 99%. The Ku-band HTS has an advantage for customers who demand very high-service reliability. The large spot Ka-band HTS systems do not appear to be cost-competitive for providing CBR services in any of the regions considered; the costs were around 2× to 3× the costs for the other systems regardless of the service availability. As discussed earlier, Ka-band radio signals are more severely impacted by rain and other transient propagation conditions than lower frequency signals; consequently, Ka-band links require higher fade margins for a given service availability than lower frequency links, and Ka-band HTS spacecraft are designed to provide these margins. CapRock concludes that, for the assumptions made in their study, this can result in a cost penalty for the Ka-band systems when customers demand high service reliability. However, the situation can in some cases be turned to the advantage of Ka-band services, for customers whose service needs can tolerate lower availability – such as mass market or consumer clients.

Notwithstanding the discussion above, the practicality remains that Ka-band HTSs have a clear sky advantage, which is a consequence of the fact that, intrinsically, these HTSs are designed with a large fade margin; relatively seldom, however, do the links actually require the power engineered into the link margin, especially in the North America–Atlantic–Europe zone. In other words, the propagation impairments that

[13] Packet-oriented services may exhibit different economics.
[14] One often sees Ka-band VSATs designed with larger antennas in heavy rain regions, but that represents a drawback for many VSAT customers and if allowed could also be advantageous at Ku-band; larger VSAT antennas were not considered in the CapRock study.

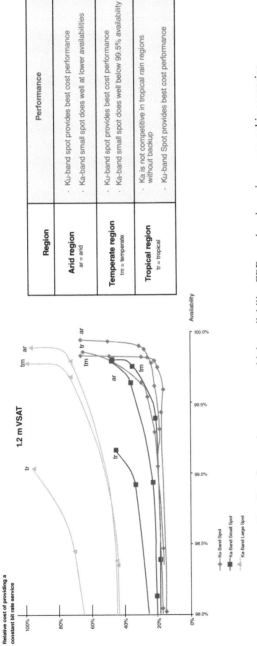

Figure 3.19 Cost Comparison to support high-availability CBR services in various geographic scenarios.

the link margins are designed to mitigate are reasonably rare occurrences and when not required to address transmission impairments, the power and bandwidth resources available for the link margin are exploited to operate the link at a higher data rate (e.g., with the DVB-S2X capabilities of adapting the power and/or modulation and/or forward error correction coding used on the link). Although link margins may be utilized in this manner at any frequency, the large link margins engineered by the satellite operators into Ka-based HTSs make this an intrinsic advantage for Ka-band HTSs. Downlink data rates can be increased by factors up to 10, depending on the regions and availabilities in question, and 100-fold increases in uplink data rates are also in scope. Although the high data rates that can be achieved by adapting the link parameters in real time to exploit unused link margin are obtainable only at lower availability than the link minimum data rate, typically only a small fraction of VSATs may be operating under impaired conditions at a given time; thus, the total VSAT network throughput can be significantly enhanced by this technique. This heuristic compromise is, in particular, effective for variable bit rate (VBR) and best effort services [CAP201201]. The technique of averaging the data rates of large quantities of VSAT sites across multiple regions, including arid climates, is used by Ka providers to calculate higher averaged network data rates. Industry analyses demonstrate that Ka-band availability in the range of 99.7% can be achieved even in high rain fade areas (e.g., Florida in the United States, and Southeast Asia). Forward channel and return channel mitigation techniques are typically used in Ka applications (and also at C band and Ku band). These are:

- Forward channel to the remote stations (gateway uplink):
  - Adaptive coding and modulation of the forward channel (e.g., with DVB-S2X).
  - Automatic level control by the spacecraft.
  - Gateway radio frequency transmission diversity.
  - Uplink power control at the Gateway stations.
  - Use of larger antennas to achieve higher EIRP.
- Return channel from the remote stations (terminals' uplink):
  - Adaptive coding of the return channel (e. g., with DVB-S2X).
  - Dynamic symbol shifting of the return channel.
  - Uplink power control at the remote stations.
  - Use of larger antennas to generate higher EIRP.

In conclusion, by availing themselves of the high rates available during clear sky conditions, Ka operators increase the overall average transmission data rates.

3.10   HTS DESIGN CONSIDERATIONS

In the mid-1980s, when VSATs were being introduced to the market, the cost of a terminal was of the order of $10,000, with data throughput of 9.6–64 kbps

[MIN198501], [MIN198601], [HUG201301]. At press time the cost of a Ku- or Ka-band VSAT was in the hundreds-of-dollars range, with throughputs of 10–20 Mbps. The fundamental architecture of the VSAT system has not changed significantly over the years: VSAT systems typically utilize a star topology with a hub station transmitting a forward channel to the remote stations, and the remote stations make use of an FDMA/TDMA channel for the return channel; this architecture is also being employed with HTSs spot beam systems. Nonetheless, deploying HTS systems and services requires thorough knowledge of the market to be addressed. In turn, this will drive key design factors including, although not limited to: (i) the operational bands to be utilized; (ii) the location and size of the user beams; (iii) the amount of spectrum per user beam; (iv) the need for contiguous or noncontiguous coverage (beam hand-off requirements, in particular, for aeronautical and maritime applications); and (v) the location and spectrum for gateway stations.

Highly efficient gateway stations are required with HTSs. A HTS with high bandwidth spot beam channels and multiple spot beams will result in hub/gateway stations that support from 1-to-10 Gbps of capacity. A classical VSAT hub station supporting a standard transponder generally requires hub station hardware that supports in the range of 100 Mbps of capacity; typically, this VSAT system is implemented in hardware fitting one-half rack of electronics. If the operator were to use conventional VSAT hardware, the HTS gateway could require 25–50 racks of equipment, which is somewhat impractical from a space and power consumption perspective. Instead, "high-density" hardware is needed in the HTS environment. New technology has appeared on the market that is able to achieve a gateway "density" of over 1 Gbps per rack; hence, a 10 Gbps gateway can be implemented in a compact 10-rack layout, with significant efficiencies in footprint, power consumption, and environmental conditioning (e.g., see [HUG201301]).

3.11 SPOT BEAM ANTENNA DESIGN BASICS (SATELLITE ANTENNA)

Earlier we mentioned two design approaches for the satellite antenna(s) used in HTS – the SFB and the MFB; in this section a more detailed view of the on-board antenna technology is provided.

Multiple beam antennas are essential enabling HTSs. Multibeam antennas operating at Ku- and Ka-band frequencies are designed to allow a single satellite to simultaneously serve hundreds of thousands of users throughout a continental region such as, although not limited to, Europe, United States, or South East Asia. Whenever a single antenna aperture is required to form a multiple set of spot-like or shaped beams, array-based solutions are used. Options include (i) directly radiating arrays and (ii) complex hybrid antennas composed of a number of radiating elements feeding optical systems [ANG201001]. With the exception of simple solutions working on a single feed per beam basis, complex BFNs are often used as the illuminating assembly; the function of the BFN is to provide proper phase and amplitude excitations to the elements constituting the array.

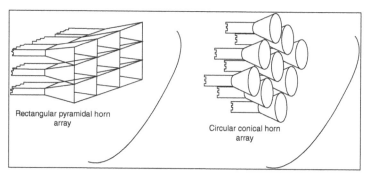

Rectangular pyramidal horn
array

Circular conical horn
array

Figure 3.20   Examples of array antennas.

An antenna array (also called a *phased array*) is a set of two or more antennas arranged in a grid. The number of antennas in an array can be as small as 2, or as large as several thousands (see Figure 3.20). The signals from the antennas are combined and/or processed to achieve improved performance over that of a single antenna. A phased array antenna is composed of a group of individual radiators that are distributed and oriented in a linear or two-dimensional spatial configuration. The amplitude and phase excitations of each radiator can be individually controlled to form a radiated beam of any desired shape (directive pencil-beam or a fan-beam shape) in space. The position of the beam in space is controlled electronically by adjusting the phase of the excitation signals at the individual signals. Therefore, beam scanning, for example, for radar application, is achieved with the antenna aperture remaining fixed in space without the involvement of mechanical motion in the scanning process. These antennas are also used for communication applications, directing high-gain beams toward distant locations either from a ground antenna or from a satellite antenna (to define a spot beam on the ground) [JOH198401]. The phased array can be used to (i) increase the overall gain, (ii) provide diversity reception or transmission, (iii) cancel interference from a particular set of directions, (iv) determine the direction of arrival of the incoming signals (e.g., for radar applications); (v) "steer" the array so that it is the most sensitive in a particular direction, and (vi) to maximize the Signal to Interference Plus Noise Ratio (SINR) [BEV200801]. Typically, the performance of an antenna array increases with the number of antennas (elements) in the array; the drawback is the increased cost, size, weight, and complexity, which is a key consideration in satellite applications. Dynamic phased array antennas tend to be expensive and are more often found in military applications.

Traditionally, most satellite antennas were not designed to cover a single point on the ground or even a regular shape, but rather to cover a certain geographically/politically dictated landmass area from the geostationary orbit. There are two techniques to design antenna that produce an irregular beam cross section (see Figure 3.21). The first technique is to employ an offset reflector but to distort the reflector from the normal paraboloidal shape; the design is relatively simple, but it lacks flexibility and precision, especially for complex geographies. The second approach is to use an array of small circular component beams packed together to

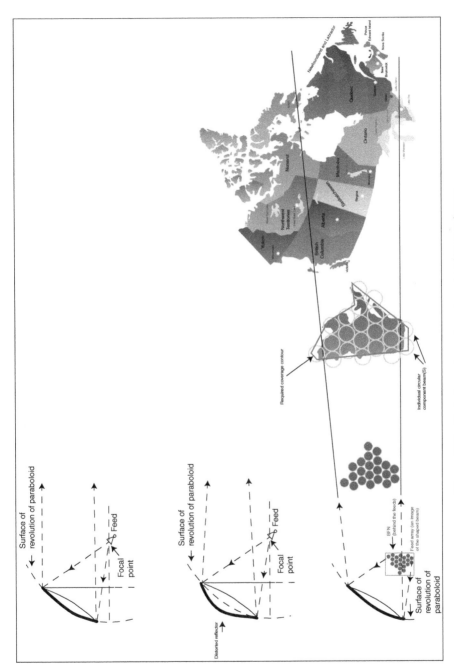

Figure 3.21 Techniques to produce shaped and/or spot beams.

form the required beam shape and use the standard paraboloidal reflector. As it can be seen in Figure 3.21, a group of circular feed horns packed into an array, all radiating simultaneously, can form a shaped beam of any planar topological shape.

Note in Figure 3.21 that the antenna feed array is arranged in an inverted fashion (with respect to the desired topology on the surface of the earth) and positioned at the focal plane of the undistorted offset feed reflector. Each feed horn in the array generates one circular component beam, and the feed array is fed via a BFN. The BFN takes the signal to be transmitted and divides it in such a way that each feed horn in the array is excited by an appropriate amount of the original signal. Each individual feed horn illuminates the entire reflector, not just a portion. Because each of the feed horns in the field array forms one single-component beam in the far field, there will be a need for as many feed horns as there are component beams. The reflector and the feed horn are sized to produce the component beam efficiently [AGR198601]. The amount of excitation for each feed horn is adjusted in both the amplitude and the phase to maximize the power within the coverage area while at the same time minimizing the sidelobes (radiation) outside the coverage area. The beam efficiency increases as the number of component beams within a coverage area increases. However, for a specific coverage area, the size of each component beam decreases (implying, e.g., more hand-offs for a mobile user); also, the reflector size must increase as the number of feed horns increases. In turn, the BFN becomes more complex (which could add weight); finally, the focal length will increase (requiring more spacecraft structure materials and adding to the flight mechanics considerations).

The use of phased array antennas in the sky and on the ground will provide greatly increased flexibility. While this technology is not being broadly used at the current time, the expectation is that it will be incorporated in the next generation of HTSs. Initially, there will be cost hurdles, but there are operational ("run-the-engine") cost reduction possibilities; by being able to shape coverage from the sky, one is not beholden to the antenna pattern for a given geography. For example, the US market for satellites services is seen as shrinking, but if one could reshape the beam and coverage to go where the demand is (e.g., South America) one could get to market quickly and at sunk-cost economics. On the ground one can instantly switch from one satellite to another. In addition, building electronics into the antenna will make them more efficient. Systems using metamaterials technology are being built by Kymeta as cited in Chapter 1, and the interested reader should research further. Using metamaterials has the potential to dramatically cut the costs of phased array antennas.

3.11.1   Single Feed per Beam Antennas

With an SFB approach, each beam is created by a single horn. To avoid gaps in the illumination of the coverage on the ground, overlapping spot beams are needed; this overlap can be achieved using a single oversize shaped reflector, or with passive and/or active lenses. A commonly used design is to create a virtual overlapping using four standard unshaped reflector antennas, one for each color; each reflector aperture creates nonoverlapping spots of a single color as depicted in Figure 3.22. The ensemble of four antennas is oriented in such a manner that a four-color environment with

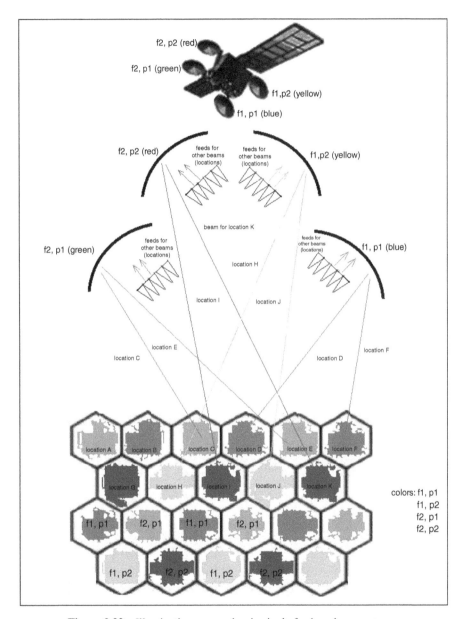

Figure 3.22   Illumination approaches in single feed per beam antenna.

overlapping spots is achieved with the four antennas [SCH201101]. In theory, eight reflectors would be required, four for transmit (Tx) and four for receive (Rx), but one can settle with only four reflectors if complex Tx/Rx feed chains are used; specifically, because the reflector diameter is chosen for the transmit frequency, it is effectively too large for the signals at the receive frequencies (the spots would become too small and not overlap any longer); therefore, an advanced feed horn design can provide proper illumination of the reflector at Tx frequencies and a specific underillumination in the receive band frequencies.

3.11.2 Multiple Feeds per Beam Antennas

Figure 3.23 depicts a conceptual multiple feeds per beam feed system. MFB antennas create the beams using arrays of small horns; fairly sophisticated (and complicated) BFNs are needed. Adjacent beams share horns and because of this physical overlap of the feed apertures, the overlapping on-the-ground spots can be produced using only one reflector aperture. This results in a significant reduction in mass and cost and simplifies the accommodation of the antennas on the spacecraft. For larger areas, several hundreds of couplers and phase shifters have to be designed and accommodated in the different layers of the BFN. Because separate antennas for transmit and receive are used, the reflector diameter can be sized optimally for frequency and spot diameter [SCH201101].

There is no single "better" choice for HTS with regard to SFB and MFB. SFB antennas have a slightly better gain performance than MFB antennas. Therefore, for mid-scope compact geographies (e.g., CONUS or Pan European applications) and large spacecraft, SFB antennas may prove to be advantageous. However, for very large geographies (e.g., Far East), the scan losses can become high; in these cases it may be advantageous to design a spacecraft for use in these orbital slot locations that replace the four SFB antennas with four MFB antennas (say two for Tx and two for Rx, where each MFB antenna provides only a half of the coverage, implying that the scan losses can be significantly be reduced; note that the cost and mass for both concepts are comparable). As hinted above, the advantage of the MFB design is the fact that only two reflectors (one for Tx and one for Rx) are needed for a compact geography. On large spacecraft both antennas can be accommodated on the same side panel (the second side panel could be used for C- or Ku-band antennas). Smaller satellite buses are often employed, and, in general, it is difficult to accommodate four large reflectors on these buses. Industry studies show that for smaller spacecraft and for smaller geographies (e.g., Western European) MFB antennas may be the optimal choice.

Multifeed antennas can also be used for contour beams [SCH201101]. Contour beams have a number of applications; for example, there may be a desire on the part of an operator to focus on the Caribbean area (only); or, in Europe, there may be interest in "linguistic beams" for DTH broadcast services (Figure 3.24 depicts a typical linguistic beam application). The level of frequency reuse is lower than for spot beams, but it is possible to increase the satellite capacity. A high isolation between areas with the same color is required; if the distance between contour beams using the same color is low, a large reflector is required to achieve a high isolation. An advantage of multifeed antennas is the high degree of freedom for the excitation coefficients; this makes it easier to suppress sidelobes for a high isolation between the different coverages.

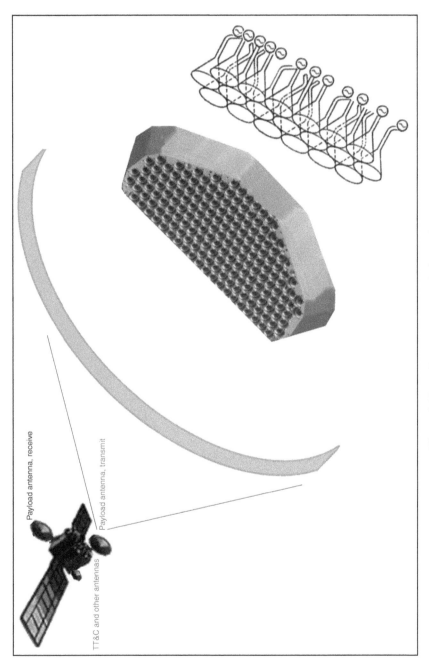

Figure 3.23 MFB antenna example for HTSs.

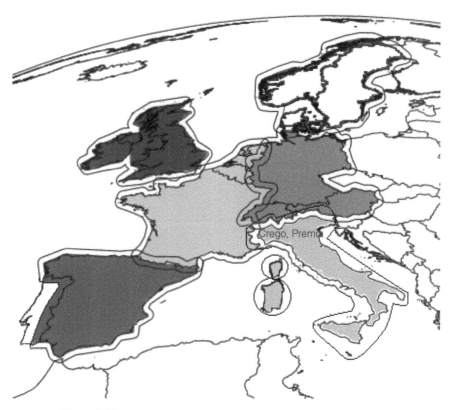

Figure 3.24   Shaped beams to cover specific areas/regions (example).

3.12   EXAMPLES OF HTS

This section highlights some of the key HTS systems; the list mentioned in this section is not intended to be exhaustive, but illustrative of the approach and of the early adopters. The material is based on published information by the providers, which is cited herewith to provide exposure of these providers to the reader.

While there have been notable all Ka-band launches of late, as of press time, many satellite operators were operating smaller Ka-band payloads consisting typically of 5–10 Ka-band spot beams on a satellite that also carries a C- and/or Ku-band payload; these operators may or may not want to eventually launch a full-scale, dedicated Ka-band HTS. These smaller Ka-band payloads are an effective way to test the market and to build up a client base for an eventual full-scale, dedicated HTS if the market warrants it [CAS201201]. The following is a list of HTS launched in recent years:

- Anik F2 (July 2004).
- Americom (SES) AMC-15/16 (October 2004).
- Thaicom 4 (August 2005).

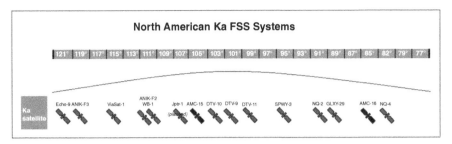

Figure 3.25   Ka deployments in North America.

- SPACEWAY 3 (August 2007).
- WINDS (February 2008).
- KA-SAT (December 2010).
- Yahsat Y1A (April 2011).
- ViaSat-1 (October 2011).
- Yahsat Y1B (April 2012).
- EchoStar® XVII (July 2012).
- HYLAS 2 (July 2012).
- Amazones 3 (February 2013).
- First four O3b constellation spacecraft (June 2013).
- Inmarsat-5 F1 (December 2013).
- Next eight O3b constellation spacecraft (4 in June 2014 and 4 additional planned for December 2014).

The O3b satellite constellation as well as Inmarsat's Global Xpress constellation do not target the consumer market but they are still considered HTS.

The United States has more than 80 Ka-band geostationary orbital slots assigned to satellite operators by the FCC. Figure 3.25 depicts the North American Ka deployment, both for spot beams and for wideband Ka. Also, refer to Figure 3.3. Globally, the overall HTS market share in 2012 was as follows [GLO201301]: ViaSat (with ViaSat-1) 20%, Thaicom 19%, Echostar 14% (Echostar XVII), Arabsat 7%, Ciel 7%, Eutelsat 5%, Telsat 4%, others 24%. ViaSat and EchoStar mainly targeted at the North American broadband access market. Thaicom and Eutelsat will dominate the next tier as they build up their client bases and services on their spacecraft, respectively, Thaicom 4 and KA-SAT HTS.

3.12.1   ViaSat-1 and -2

With[15] a measured throughput of 134 Gbps, ViaSat-1 was the highest capacity communications satellite in the world at press time. Using a total of 72 service spot-beams

---

[15]The author wishes to thank ViaSat Corporation and Mr. Bruce Rowe for this material.

TABLE 3.3   Exede Services and Pricing (Illustrative) (As of Press Time)

Exede by ViaSat
Retail Broadband Service Plans

EVOLUTION	CLASSIC 10	CLASSIC 15	CLASSIC 25
★ UNLIMITED ACCESS Email & Web Pages EVEN MORE DATA: 5 GB for everything else	▥ **10 GB** each month Data Allowance for all uses	▥ **15 GB** each month Data Allowance for all uses	▥ **25 GB** each month Data Allowance for all uses
◆ **Early Bird Free Zone** Unmetered access to everything 3 a.m. - 8 a.m.	◟ **Late Night Free Zone** Unmetered access to everything 12 midnight - 5 a.m.	◟ **Late Night Free Zone** Unmetered access to everything 12 midnight - 5 a.m.	◟ **Late Night Free Zone** Unmetered access to everything 12 midnight - 5 a.m.
⬇ 12 Mbps Download Speed	⬇ 12 Mbps Download Speed	⬇ 12 Mbps Download Speed	⬇ 12 Mbps Download Speed
⬆ 3 Mbps Upload Speed	⬆ 3 Mbps Upload Speed	⬆ 3 Mbps Upload Speed	⬆ 3 Mbps Upload Speed
$64.99/mo	$49.99/mo	$79.99/mo	$129.99/mo

(61 over the continental United States, 1 over Alaska, 1 over Hawaii, and 9 over Canada), this single satellite has more capacity than *the sum of all* the other FSS and BSS communications satellites (C, Ku, and Ka band) that covered North America at the time of its launch in October of 2011. This HTS technology allowed ViaSat to launch its consumer Exede® service in 2013, offering 12 Mbps downstream speed (and 3 Mbps upstream) for $50/month. This price is comparable on a dollars/Mbps basis to most urban terrestrial systems (see Table 3.3).

The broadband offered through ViaSat-1 is enabled through a ground system based on the ViaSat SurfBeam® 2 architecture; other innovations are as follows:

- New high-density fiber gateway servers and infrastructure and next-generation customer premises equipment capable of handling the speeds required for broadband service over ViaSat-1;
- Next generation protocol acceleration technology that minimizes the impact of latency on the user experience, including providing web browsing performance that is typically superior to 4/1 Mbps DSL-like terrestrial Internet service; and,
- Incorporation of the WiMax standard enhanced for use over satellite, combined with dynamic physical layer (dynamic modulation, forward error correction coding, and power) to more effectively combat rain fade and provide service availability comparable to that of satellite TV.

This enhanced speed, capacity, and acceleration enables subscribers to access standard web content as well as SD and HD video; any third party VoIP application or service such as Skype can be used over Exede Internet, and, in addition, ViaSat introduced its own Exede VoIP service in 2013.

ViaSat-1 has been recognized as a disruptive force in satellite broadband, earning a Guinness World Records title as the world's highest capacity communications satellite. A February 2013 FCC report showed ViaSat's Exede satellite Internet service was outperforming all other ISPs in the United States in delivering promised speeds to subscribers. The report revealed that 90% of Exede subscribers were receiving 140% or better of the advertised 12 Mbps speed during peak periods – a major shift from the previous generations of satellite Internet service, which were considered to be a "last resort" for those with no other Internet alternative [SAT201401].

A companion spacecraft, ViaSat-2, was near completion as of press time (to be launched mid-2016), and has the following characteristics:

- Advances broadband satellite state of the art.
- Builds on Boeing 702HP platform.
- Integrating ViaSat payload and system designs.
- Provides a ~2x gain in bandwidth economics.
- Provides a ~7x increase in geographic coverage.
- Provides a novel combination of:
  – throughput capacity,
  – geographic coverage,
  – operational flexibility.
- Enables
  – Fiber To The Node (FTTN)-class consumer broadband speeds,
  – total CONUS service,
  – aero and maritime global mobile expansion,
  – international expansion,
  – key operational efficiencies.

Figure 3.26 depicts the ViaSat-2 coverage achieved with a multitude of beams; the coverage area includes North America, Central America, the Caribbean, a small portion of northern South America as well as the aeronautical and maritime routes across the Atlantic Ocean between North America and Europe.

3.12.2   EchoStar

3.12.2.1   EchoStar XVII satellite with JUPITER™   The EchoStar XVII (EchoStar 17, also known as Jupiter 1) satellite with JUPITER high throughput technology utilizes 60 Ka spot beams, each of which is separated by a combination of frequency and polarization. EchoStar XVII is a HTS operated by Hughes Network Systems, a subsidiary of EchoStar. It is positioned for North American services in the GSO at a longitude of 107.1° W. EchoStar XVII was built by Space Systems/Loral and is based on the LS-1300 satellite bus. Its beam layout has been combined with the earlier SPACEWAY 3 satellite to optimize coverage over North America. The Hughes SPACEWAY 3 satellite (launched in August 2007) is an early example of many of the advanced technologies used in HTS systems today, including small spot beams and frequency reuse. The SPACEWAY 3 is a Ka-band system that is

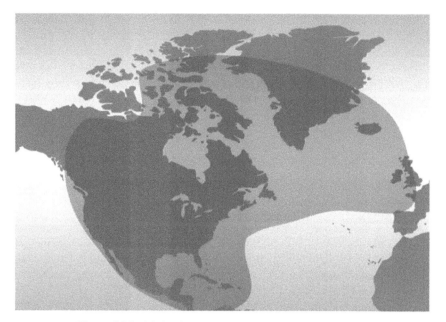

Figure 3.26   ViaSat-2 coverage (from a multitude of beams).

able to deliver high throughput where it is needed, owing in part to the phased array antenna on the spacecraft, which enables dynamic allocation of power, as illustrated in Figure 3.27. SPACEWAY 3 has the ability to support the reconfiguration of the beam sizes and shapes, and to dynamically allocate capacity per beam as the traffic needs change.

Hughes Network Systems has stated that they have achieved carrier data rates exceeding 1 Gbps on their system. JUPITER high throughput technology is the basis for HughesNet® Gen4 service introduced to the market in October of 2012. HughesNet Gen4 customers enjoy high speeds, from 10 to 15 Mbps, and a richer Internet experience, including a VoIP voice calling option. The services offered at press time are shown in Table 3.4 and included as an example of the typical rates and fees available with HTSs.

TABLE 3.4   HughesNet Gen4 Service Plans

EchoStar XVII	Power	Power PRO	Power MAX
Downstream (Mbps)	10	10	15
Upstream (Mbps)	1	2	2
Anytime (GB)	10	15	20
Bonus bytes (GB)	10	15	20

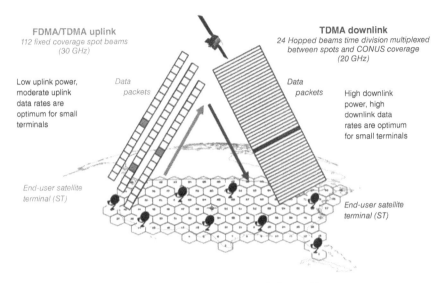

FDMA/TDMA uplink
112 fixed coverage spot beams
(30 GHz)

Low uplink power,
moderate uplink
data rates are
optimum for small
terminals

Data
packets

TDMA downlink
24 Hopped beams time division multiplexed
between spots and CONUS coverage
(20 GHz)

Data
packets

High downlink
power, high
downlink data
rates are optimum
for small terminals

End-user satellite
terminal (ST)

End-user satellite
terminal (ST)

Figure 3.27   SPACEWAY 3 transmission architecture.

3.12.2.2 SES (GE Americom) AMC 15/16   SES deployed one of the first commercial Ka payloads in North America in 2004 with AMC-15 and AMC-16. The spacecraft are leased to EchoStar. SES launched these hybrid Ku/Ka FSS satellites to enter the market (but to limit the risk of an all-Ka system). AMC-15 and AMC-16 are clone satellites built by Lockheed Martin. They have 12 spot beams of 125 MHz, 10 covering CONUS, and 2 over Hawaii and Alaska. The transponders provide 3 × 40 MHz carriers per beam. The spacecraft include an IF switch to route traffic between beams/carriers. Figure 3.28 depicts a beam layout and a possible application. The spacecraft have been used by EchoStar Corp. In late 2014 EchoStar announced it would replace AMC-15 (also known as EchoStar 105) with a new satellite, providing EchoStar with 24 × 36 MHz Ku-band transponders; the spacecraft will be positioned at the 105 degrees West orbital position, offering coverage of the Americas, including Alaska, Hawaii, Mexico and the Caribbean. EchoStar 105 will be based on the Eurostar E3000 platform. EchoStar 105 will also be equipped with a payload of 24 C-band transponders, known as SES-11 and will be operated ("flown") by SES.

3.12.3   Eutelsat KA-SAT

KA-SAT, launched[16] and flown by Eutelsat, was the first all Ka-band HTS in Europe. KA-SAT has been developed to provide cost-effective broadband access and IP services. Launched end-2010 and positioned at 9° East, KA-SAT is an integral part of

---

[16]This information is based on Eutelsat materials.

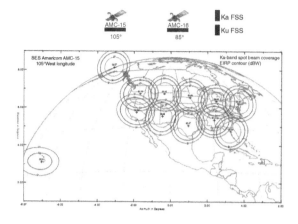

AMC-15/16 Ka-band Frequency Plan

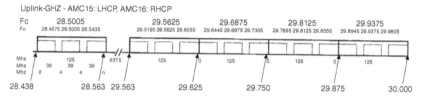

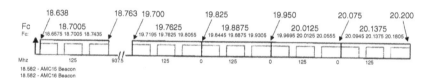

Figure 3.28   AMC-15/16 payload.

a network of 10 ground stations interconnected by an IP MPLS fiber network linked to four major European POPs. KA-SAT provides coverage through 82 spot beams using four different frequencies for efficient frequency reuse. This process multiplies the available capacity by a factor of 20, to attain a broadband capacity of 90 Gbps (see Figure 3.29).

Tooway™ on KA-SAT is an advanced, high-speed Internet service via satellite for households and businesses. Tooway is available and performs equally in all KA-SAT coverage areas across Europe, the Mediterranean Basin, and some areas of the Middle East. It is also a cost-effective solution for local administrations to provide imme-diate access to broadband where the limited number of connections does not jus-tify investment for terrestrial infrastructures, such as copper or fiber cables. Each Tooway-enabled home is individually equipped with a satellite kit that transmits and receives directly from the satellite. Tooway flat rate packages offer up to 18 Mbps download speeds and up to 6 Mbps upload speeds.

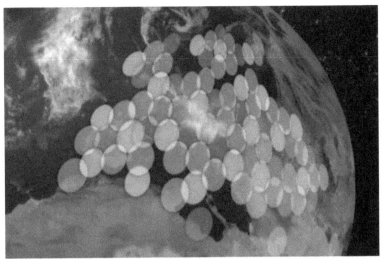

KA-SAT coverage over Europe and the Mediterranean Basin

Figure 3.29   EUTELSAT KA-SAT.

KA-SAT enables bandwidth-intensive, professional applications to be delivered at new, low threshold prices. Professional services on KA-SAT include broadband access and corporate network solutions for businesses, video transmission services (Satellite News Gathering - SNG) for the news and broadcasting industries, as well as data transmission for government applications. With the same advantageous bandwidth and pricing ratios, KA-SAT can also deliver carrier services, including high throughput IP trunking and mobile network backhauling. For professional services, KA-SAT provides secure broadband access of up to 50 Mbps downloads and 20 Mbps uploads to businesses in "notspots," as well as corporate networks solutions, including network extensions, end-to-end VPN networks and backup services. KA-SAT corporate network solutions can extend and complete existing networks, seamlessly integrating new sites to existing terrestrial intranets and carrier-supplied networks. These services are available across KA-SAT's European-wide footprint and ground network. With NewsSpotter, the SNG service on KA-SAT, broadcasters can send SD or HD live video from the field to the studio using lightweight compact equipment and a simple automatic process of bandwidth allocation. Video and Data transmission services are also available for government applications such as emergencies, border control, e-medicine. For carrier services, KA-SAT provides competitively priced and reliable IP transport services for fixed and mobile network operators and carriers supporting, in addition to GSM backhaul connectivity to complete 2G/3G networks in underserved areas, point-to-point IP trunking and backup of network links.

3.12.4   Intelsat EPIC

Intelsat is planning a global service HTS system called $Epic^{TM}$. The initial deployment envisioned two satellites to be deployed in different orbital locations, expected to be operational in 2015–2016; planned coverage spans nearly all global land masses. Epic is designed to provide high throughput and efficiency in an open architecture platform. Intelsat Epic is fully integrated with Intelsat's existing infrastructure; it incorporates C, Ku, and Ka spot beams in a high-performance platform that delivers significantly more capacity and more throughput per unit of spectrum. Epic supports:

- Fixed and wireless telecommunications operators.
- Telecommunications service providers for the oil and gas industry.
- Government and military communications.
- Private data network service providers.
- Maritime and aeronautical data service providers.
- Global organizations, including corporations and intergovernmental organizations.
- DTH and other television distribution and broadcast service providers.
- Multiple spot beam of 2-degree (or less) beam width.
- 160 Mbps spot beam capacity.
- 40 Mbps widebeam capacity.
- 25–60 Gbps of capacity.

As seen in this chapter, there are several HTS either in operation today or nearing deployment. While each of them has distinct features, a common design element among most of these systems (typically Ka band) is a network topology that, according to Intelsat, limits connectivity and has lower isolation of cochannel spots; as a result, most of these systems are designed with an architecture that is proprietary and closed. This topology is a limitation for many operators. Intelsat Epic allows connectivity among multiple spot beams, including star and mesh, as well as loopback within the same user beam; this guarantees backward compatibility with existing networks and forward compatibility with the flexibility to evolve the network design and technology as, and when, customers want. Intelsat Epic takes advantage of satellite antenna technology that enables multiple smaller beams to be deployed. This is similar to how consumer-focused Ka-band platforms have been deployed over small regions, but in this case the implementation is expanded to the frequency band and beam configuration that is most appropriate for each region, application, and customer set.

Intelsat Epic applies multispot technology to Ku band and C band, as well as Ka band, providing increased throughput on an efficient basis. The physics of satellite communications are well understood: for the same satellite power, the same spot beam size, and the same terminal size, all frequency bands provide the same throughput in clear sky conditions. The selection of the best frequency for a given application is, therefore, driven by many other considerations. From a satellite antenna design perspective, smaller spot sizes are easier to achieve at higher frequencies. Some examples of the minimum size of spot-beam coverages across the different frequency bands, with a standard satellite antenna size, are:

- *C band:* ≈1,750 km (1,100 miles).
- *Ku band:* ≈1,000 km (620 miles).
- *Ka band:* ≈350 km (220 miles).

The wider beam width associated with C band translates to larger spot sizes; however, it is possible to generate small spots even in C band with larger satellite antennas. The use of multi-spot C band and Ku band in a bent pipe architecture will allow for an open network architecture that is backward-compatible in most instances with operators' current network infrastructure.

The Intelsat Epic satellites will provide four to five times more bandwidth capacity than traditional satellites from frequency reuse through spot beams. In addition, because of the higher power available per spot beam and a design that minimizes RF interference between spots, the efficiency (number of bits per MHz) will be multiplied by two to three times. In total, the expected aggregate throughput on an Intelsat Epic satellite will vary according to the application served and satellite but is anticipated to be in the range of 25–60 Gbps. Whereas most HTS offer the best effort, contended service using proprietary ground equipment, Intelsat Epic offers Committed Information Rate (CIR) bandwidth using operator-selected ground platforms. The use of spot beams on Intelsat Epic provides two additional benefits that translate into improved spectral efficiency offered to customers; that is, the Mbps that can be achieved in a MHz of satellite bandwidth (both relate to the use of smaller beam size): first is the increase in receiver performance (higher G/T), and the second is the higher downlink power (EIRP) that can be provided. The combination of increased MHz on Intelsat Epic and improved spectral efficiency enables service providers to achieve significantly higher throughputs (overall Mbps).

3.12.5   Global Xpress

We discuss this service in detail in Chapter 4.

3.12.6   Other Traditional HTS

A short list of other HTS are as follows:

- IPSTAR (Thaicom 4) was the first commercial HTS operating in the Ku band. It is positioned at 119.5 E. It has 84 Ku-Spot Beams (2-way); 8 Ku-Spot Beams (Augment) (2-way); 3 Ku-Shaped Beams (2-way); and 7 Ku-Broadcast Beams (1-way) (see Figure 3.30).
- The Russian Communications Satellite Company (RSCC) had two HTS satellites at press time, as seen in Figure 3.31.
- *The Hispasat Amazonas 3.* This satellite was launched in 2013. The Amazonas 3 satellite supports the following payload: 33 Ku-band transponders; 19 C-band transponders; and 9 Ka-band spot beams. Assuming that these spot beams are 500 MHz each, this satellite will enable 9 GHz of capacity for the Ka-band data services alone.

Figure 3.30   IPSTAR.

- *Hylas-1.* In December 2010, the two first European Ka-band multi-spot-beam satellites, Avanti's Hylas-1 and Eutelsat's KA-SAT (discussed earlier) were launched. Both satellites were built by Astrium (Airbus Military, Astrium, and Cassidian are now Airbus Defense and Space) and are now fully operational in orbit. Hylas-1 is operated commercially by Avanti Communications of London and is the first satellite dedicated to delivering broadband Internet services to Europe. Its payload can be reconfigured in orbit, so that satellites can adapt to market demands, essentially future-proofing them during their typically 15-year lifetime.
- Canada's Anik F2, in addition to C-band and Ku-band capacity, carries 45 Ka-band circular spot beams with 30 of these licensed to Wildblue Communications, and 15 dedicated to Telesat's Canadian customers. Total throughput will range between 3 and 4 Gbs. Anik F2, a Boeing 702 spacecraft, uses a half-dozen wideband 492-MHz transponders to beam traffic to the six planned gateways, three each in the United States and Canada, relaying multiple MF-TDMA return carriers from users grouped into six to eight beams in the process. WildBlue delivers 2-way Internet satellite in the United States starting at $50/month.

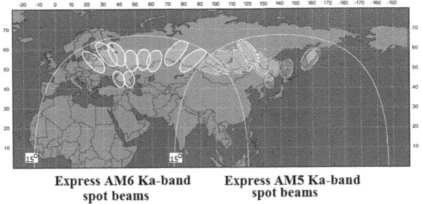

Express AM6 Ka-band
spot beams

**Express AM6 Ka-band
spot beams**

**Express AM5 Ka-band
spot beams**

Figure 3.31   Russian communications satellite company (RSCC) HTS.

3.12.7   O3b

O3b is a satellite constellation built with IP and mobile networks in mind. The O3B non-GSO network is planned to operate in the 28.6–29.1 GHz (uplink), 18.8–19.3 GHz (downlink); in this band, GSO and non-GSO satellite networks have equal frequency rights. The OB3 network does not suffer from the usual GSO-intrinsic latency disadvantages because of the fact that the service area, which is the area ±45° North and South latitude, can be covered continuously by a relatively small constellation of initially 8 satellites in equatorial orbit. O3b's Ka MEO constellation reduces latency and is in particular suitable for interactive applications. From any one place on earth (within the northern and southern 45-degree latitude mark) the satellites will "fly by" and provide access; but because of the fly by environment, a remote station requires a tracking antenna to track the satellite as it flies across the earth. In fact, the O3b remote stations require two tracking antennas so that the remote can perform "make-before-break" decisions and ensure that connectivity can be established with the next satellite to fly over prior to the current satellite disappearing from view (as it continues its flight around the earth) [HUG201301]. The MEO-HTS reduces delay by 75% compared with GEO (150 ms round trip instead of 500 ms for 8,062 km instead of 36,000 km). O3b's low latency improves the quality of voice and data service. On a per-spacecraft basis, it has a lower cost to build and launch compared with GEO. The features are as follows [BUR201301]:

- Each satellite has 12 steerable antennas
  - 10 antennas used for customer beams.
  - 2 antennas used for Gateway connectivity.
  - Each beam is independently steerable to any location within ±45 latitude.
  - Beams can be stacked at the same location to provide additional capacity.

- Circular polarization is used for RHCP and LHCP.
- Each Gateway beam is connected to five customer beams.
  - A single satellite supports two groups of these 1:5 configurations.
  - Loopback beams can also be configured to provide localized connectivity up and down in the same beam.
- Each customer beam is configured with
  - A 216 MHz Ka-band transponder in the forward direction.
  - A 216 MHz Ka-band transponder in the return direction.

See Figures 3.32 and 3.33 for beam information. The satellites will operate in the GSO and non-GSO Ka band as follows (also see Figure 3.34).

- *Downlink:* 17.8–18.6 GHz/18.8–19.3 GHz.
- *Uplink:* 27.6–28.4 GHz/28.6–29.1 GHz.

The service is not intended for direct delivery to consumers but for telecom operators who can afford the more expensive tracking antennas needed; in turn, he telecom operators will resell the service to consumers using terrestrial distribution systems as WiMax or a collocated set of antennas at a cellular site (for cellular backhaul). Note that the telecom operators must use two steerable antennas to track the spacecraft and retain signal connectivity by moving the path from one satellite in the constellation to another. Typical PLL LNBs of the O3b antennas operate as follows:

- 17.852–18.588 GHz with a local frequency of 16.8 GHz (IF frequency on 1,052-1,788 MHz – local oscillator (LO) Stability ±1.5 ppm) (Noise temperature 101 K typical).

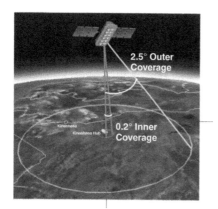

Outer coverage is defined as 2.5° off-boresight. This antenna coverage provides coverage area for customer services, while also providing sufficient antenna gain approximately 700 km in diameter

Inner coverage is defined as 0.2° off-boresight. This inner antenna coverage provides optimal antenna gain for premier Tier 1 services approximately 50 km in diameter

Figure 3.32   O3b beam configuration. Source: O3b

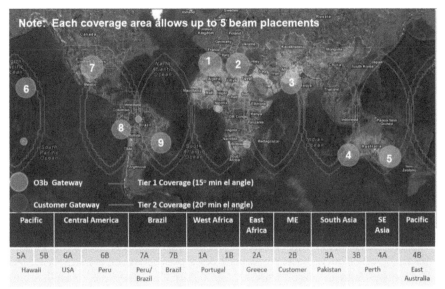

Figure 3.33 Eight satellites' regional coverage in O3b (example). Source: O3b

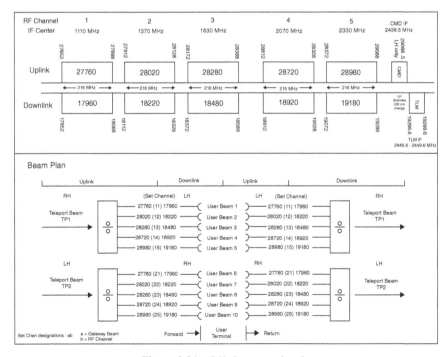

Figure 3.34 O3b frequency bands.

- 18.372–19.271 GHz with a local frequency of 17.4 GHz (IF frequency on 972–1,871 MHz – LO Stability ±1.5 ppm) (Noise temperature 101 K typical).

The BUC is typically 5 W RF (+37 dBm; 88 W input power) or 10 W RF (+40 dBm; 170 W input power) (SSPA and upconverter), to upconvert L-band IF signal to Ka-band RF signal (27.652–29.071 GHz), more specifically, operation at 27.652–28.388 GHz with 26.600 GHz LO and 28.172–29.071 GHz with 27.200 GHz LO.

The need for mechanically tracking the satellite constellation elements horizon to horizon was noted above. Mechanical systems are generally available. However, some manufacturers (e.g., Kymeta) are now developing flat panel metamaterial antennas that are capable of tracking and instantaneously switching between satellites. As of press time the vendor demonstrated both the receive and transmit capability. This antenna is currently Ka band, but the vendor was reportedly planning a Ku-band version.

Kymeta's Metamaterials Surface Antenna Technology (MSA-T) enables wide-angle, all-electronic beam steering from a proprietary, PCB-like surface that can be manufactured using a mature and affordable lithography manufacturing infrastructure. The reconfigurability is achieved through the use of a standard PCB-like circuit board composed of several thousand subwavelength resonators that can be individually tuned [KYM201401]. This PCB-like board is attached to a conventional feed structure. Thus, as the RF energy propagates through the system, individual tunable elements can be activated (i.e., turned "on") to scatter a portion of this RF energy out of the guided mode. It is the pattern of activated tunable elements that determines the shape and direction of the radiated wave through the formation of a reconfigurable grating. Changing the pattern of activated elements changes the shape and direction of the beam. The net result is an antenna with the dynamic performance of a phased array but without the need for phase shifters, related amplifiers, and other components.

3.12.8   Wideband Global Satcom (WGS)

The *Wideband Global Satcom (WGS)* system is a constellation of highly capable military communications satellites that leverage cost-effective methods and technological advances in the communications satellite industry. With launches in October 2007, April 2009, December 2009, January 2012, May 2013, and August 2013, WGS Space Vehicles are the US Department of Defense's highest capacity communications satellites. Each WGS satellite provides high-capacity service in both the X and Ka frequency bands, with the unprecedented ability to cross-band between the two frequencies on board the satellite. WGS supplements X-band communications, provided by the Defense Satellite Communications System, and augments the one-way Global Broadcast Service through new two-way Ka-band service. Each WGS satellite is digitally channelized and transponded. These characteristics provide a an improvement leap in communications capacity, connectivity, and flexibility for US military forces and international partners while seamlessly integrating with current and future

X- and Ka-band terminals. Just one WGS satellite provides more SATCOM capacity than the entire Defense Satellite Communications System constellation. International partners participating on the program are Australia, Canada, Denmark, Luxembourg, The Netherlands, and New Zealand. WGS provides worldwide flexible, high data rate, and long haul communications to soldiers, sailors, airmen, marine corps, the White House communication agency, the US State Department, international partners, and other special users [AIR201401].

Part of the Wideband SATCOM Division of the Space and Missile Systems Center's MILSATCOM Directorate, the WGS system is composed of three principal segments: Space Segment (satellites), Control Segment (operators), and Terminal Segment (users). MILSATCOM is responsible for development, acquisition, fielding, and sustainment of the WGS program. Block II follow-on satellites 7, 8, 9, and 10 are anticipated for launch in FY15, FY16, FY17, and FY18, respectively. Satellites are launched either via the Delta IV or the Atlas V Evolved Expendable Launch Vehicles. *General characteristics of WGS are as follows:*

- *Primary Function:* High-capacity military communications satellite.
- *Primary Contractor:* Boeing Defense, Space, and Security.
- *Payload:* Transponded, cross-banded-X, and Ka-band communications suite.
- *Antennas:* Electrically steerable, phased arrays X-band transmit and receive; mechanically steered Ka band; and fixed earth-coverage X band.
- *Capability:*
  - WGS 1–7 39 125-MHz Channels via digital channelizer/router; 2.1 Gbps capacity.
  - WGS 8–10 19 500 MHz Channels via digital channelizer/router; 2.1 Gbps capacity.

REFERENCES

[AGI200101] Agilent Technologies, "Digital Modulation in Communications Systems – n Introduction", Application Note 1298, March 14, 2001, Doc. 5965-7160E, 5301 Stevens Creek Blvd, Santa Clara, CA, 95051, United States.

[AGR198601] B. N. Agrawal, *Design of Geosynchronous Spacecraft*, Prentice-Hall, Englewood Cliffs, NJ, 1986.

[AIR201401] Air Force Materials, 2014, Retrieved at http://www.losangeles.af.mil/library /factsheets/factsheet.asp?id=5333.

[ANG201001] P. Angeletti, M. Lisi, "Multimode Beamforming Networks", European Space Agency White Paper, 2010, Keplerlaan 1, Noordwijk, 2200 AG, The Netherlands.

[ANS200001] ANS T1.523-2001, Telecom Glossary 2000, American National Standard (ANS), an outgrowth of the Federal Standard 1037 series, *Glossary of Telecommunication Terms*, 1996.

[BEV200801] P. J. Bevelacqua, *Antenna Arrays: Performance Limits and Geometry Optimization*, Dissertation Presented in Partial Fulfillment of the Requirements for the Degree Doctor of Philosophy, Arizona State University, May 2008.

[BUR201301] D. Burr, MEO Satellite Applications to Support Mobility, Colloquium on Satellite Services for Global Mobility 14 October, 2013.

[CAP201201] CapRock Staff, "Not All Bands Are Created Equal: A Closer Look at Ka & Ku High Throughput Satellites", Harris CapRock White Paper, 2012.

[CAS201201] High Throughput Satellite Networks, White Paper, June 2012. CASBAA Executive Office 802 Wilson House 19–27 Wyndham Street Central, Hong Kong.

[CHR201201] J. Christensen, "ITU Regulations for Ka-band Satellite Networks", Asiasat White Paper, July 2012. Asiasat (Asia Satellite Telecommunications Company Limited), 12/F, Harbour Centre, 25 Harbour Road, Wanchai, Hong Kong.

[FCC201201] Applications to the FCC by ViaSat for its large aperture TT&C antenna (7.3 meter Model VA-73-KA gateway-type antenna in the Ka-band) in the vicinity of Rapid City, South Dakota to communicate with the ViaSat-1 satellite. Also, Application to the FCC by ViaSat for its mobile Ka antenna.

[GAM199901] T. W. Gamelin, R. E. Greene, *Introduction to Topology*, Second Edition, Dover Publications, Mineola, NY, 1999. ISBN 0-486-40680-6.

[GLO201301] Global Assessment of Satellite Supply & Demand, 10th Edition. A Region-Specific Supply and Demand Analysis of the Commercial Geostationary Satellite Transponder Market for 2012–2022, September 2013, Northern Sky Research, www.nsr.com.

[HUG201301] Staff, "The View from Jupiter: High-Throughput Satellite Systems", White Paper by Hughes, July 2013. Retrieved at www.hughes.com/resources/the-view-from-jupiter-high-throughput-satellite-systems.

[ITU201201] ITU, *Regulation of Global Broadband*, Satellite Communications, April 2012, International Telecommunications Union Report, Geneva, Switzerland.

[KYM201401] Kymeta Corporation, 12034 134th Court NE, Suite 102, Redmond WA 98052.

[LAN201301] J. Landovskis, "Avantech Wireless Products", High Throughput Satellites Engineering Roundtable Conference, London, UK. December 5th and 6th, 2013, Avantech, www.advantechwireless.com.

[MAC199401] A. J. Macula, "Covers of a Finite Set." Mathematics Magazine 67, 141–144, 1994.

[MIN197501] D. Minoli, "Use of Matrices In The Four Color Problem", PME Journal, VOL. 5, NO. 10, Spring 1975, pp. 503–511.

[MIN197502] D. Minoli, "A Note On The Relation Between The Incidence And The Adjacency Matrix Of A Graph", The Matrix and Tensor Quarterly, VOL. 1975, NO. 4, pp. 145–150.

[MIN197701] D. Minoli, K. Schneider, "A Technique For Establishing The Minimum Number Of Frequencies Required For Urban Mobile Radio Communication", IEEE Transactions on Communication, September 1977, pp. 1054–1056.

[MIN198501] D. Minoli, "Bypass Strategies At Prudential-Bache", Interview With Computer World, April 29, 1985, pp. 1, 4.

[MIN198601] D. Minoli, "Aloha Channels Throughput Degradation", 1986 Computer Networking Symposium Conference Record, pp. 151–159.

[MIN200701] D. Minoli, K. Sohraby and T. Znati, *Wireless Sensor Networks*, Wiley, 2007.

[MIN201301] D. Minoli, *Building the Internet of Things with IPv6 and MIPv6, The Evolving World of M2M Communications*, Wiley, 2013).

[PAW201301] B. Pawling, "Separating Fact from Fiction: HTS Ka- and Ku-Band for Mission Critical SATCOM", Microwave Journal, August 14, 2013, http://www.microwavejournal .com.

[ROM198901] R. Rom, M. Sidi, *Multiple Access Protocols, Performance and analysis*, Springer-Verlag, New York, June 1989.

[RUI201301] N. De Ruiter, B. Prokosh, *"High Throughput Satellites: The Quest For Market Fit – A Vertical Market Analysis of Major Drivers, Strategic Issues and Demand Take Up"*, An Euroconsult Executive Report. November 2013. 86 Blvd. Sebastopol, 75003 Paris, France. www.euroconsult-ec.com

[SAT201401] ViaSat + Xplornet – Gaining Canadian Coverage (SATCOM – Capacity), Satnews Daily, May 2nd, 2014.

[SCH201101] M. Schneider, C. Hartwanger, H. Wolf, "Antennas for multiple spot beam satellites", Presentation at the German Aerospace Congress, September 27–29, 2011, Bremen, Germany. CEAS Space J (2011) 2:59–66 DOI 10.1007/s12567-011-0012-z, Springer.

[WAN201401] H. Wang, A. Liu, X. Pan, J. Li, "Optimization of Power Allocation for a Multibeam Satellite Communication System with Interbeam Interference", Journal of Applied Mathematics, VOL. 2014, January 2014, Article ID 469437, 10.1155/2014/469437.

4

AERONAUTICAL MOBILITY SERVICES

This chapter[1] assesses emerging mobility applications and services that can be provided with satellite-based systems. Aeronautical and maritime applications are becoming important both in support of human communication (e.g., Internet access) and for machine-to-machine (M2M) communication (e.g., engine telemetry and goods tracking). Polls show that over 80% of passengers traveling on planes now have a smartphone or laptop and would, thus, benefit from connectivity, if the service is priced reasonably. Business travelers may need to connect to their intranet, access information on the Internet, and communicate via email or Voice over Internet Protocol (VoIP) while in-flight; the casual traveler may want to connect to the Internet or access social media platforms – also there appears to be an emerging trend for the mobile consumption of entertainment streaming video. This chapter looks at satellite-based connectivity for people on the move in airplanes, in ships, and in terrestrial vehicles, but with most of the focus on aeronautical applications. In the context of airplanes, these services are being referred to as in-flight communication (IFC), which some also call in-flight connectivity. Maritime services are covered in the next chapter.

[1] Services, service options, and service providers change over time (new ones are added and existing ones may drop out as time goes by); as such, any service, service option, or service provider referred to in this chapter is mentioned strictly as illustrative examples of possible examples of emerging technologies, trends, or approaches. As such, the mention is intended to provide pedagogical value. No recommendations are implicitly or explicitly implied by the mention of any vendor or any product (or lack thereof).

Innovations in Satellite Communications and Satellite Technology: The Industry Implications of DVB-S2X, High Throughput Satellites, Ultra HD, M2M, and IP, First Edition. Daniel Minoli.
© 2015 John Wiley & Sons, Inc. Published 2015 by John Wiley & Sons, Inc.

4.1 OVERVIEW OF THE MOBILITY ENVIRONMENT

Until recently, in-flight connectivity by satellite has been largely provided over traditional L-band and Ku-band, and has generally supported only low-bandwidth sessions
(another common way that IFC has been supported over the years is via air-to-ground
(ATG) mechanisms – but still at low bandwidth). Major airlines are now rapidly
upgrading their fleets to support live in-flight entertainment (IFE) and broadband
Internet connectivity, and some airlines are planning to install integrated in-*flight*
entertainment and *communications* (IFECs) devices in the planes' seatbacks. At press
time major airlines were equipping their aircraft with small Ku-band tracking antennas to provide broadband Internet and live video streaming service, but Ka-based
services were also emerging, as covered in Chapter 3. The footprint of the supporting
satellite can be one large oceanic beam, or a handful of beams (e.g., an East North
Atlantic–Europe – and a West North Atlantic – United States – as is the case for
the SES-6 spacecraft Ku beams of the same name), or can be a tessellation of smaller
Ka beams. Several satellites, each serving a portion of a route (especially a very long
route), can also be used to provide broader coverage with automatic handoff between
systems. Figure 4.1 provides a pictorial overview of the environment, while Figure 4.2
depicts the potential utilization of spot beams to provide such a service (e.g., over
the Atlantic), which is the emerging model. Clearly, it is easier to provide services
to ships because antenna size (and spacecraft tracking) is less of an issue on ships.

Proponents indicate that the Ka High-Throughput Satellite (HTS) technology is
expected to fundamentally alter the economics of aeronautical broadband. Lower
priced HTS capacity will animate demand in the aviation sector: the expanding service availability and the expanding geographic coverage of 3/4G HTS systems will
drive the utilization of such HTS-based solutions in this vertical market. Ka-band
HTS are now beginning to be utilized for mobility services because of the following
factors, among others:

- Lower per-bit cost due to improved spectrum efficiency achieved with
  frequency reuse and narrow spot beams, characteristic of HTS systems, as
  discussed in Chapter 3;
- Greater amount of spectrum available in Ka-band (up to 1.1 GHz), which results
  in higher channel speed as experienced by each mobile customer;
- Higher frequency and higher received power of spot beams facilitate the use of
  smaller plane/ship terminals, while supporting better overall performance;
- Being smaller, Ka-based terminals are simpler and quicker to install than
  existing Ku-band systems, minimizing downtime for aircraft and naval vessels.
  These terminals are more compact and lighter (this being especially important
  in aeronautical applications), and the terminals have a lower cost than those
  currently available for Ku-band.

The large majority of the airlines charge the passengers extra for in-flight connectivity services, but the business model and economics of IFC services have yet to be
proven, with no clear normative model established yet. In fact, a key supplier states
that

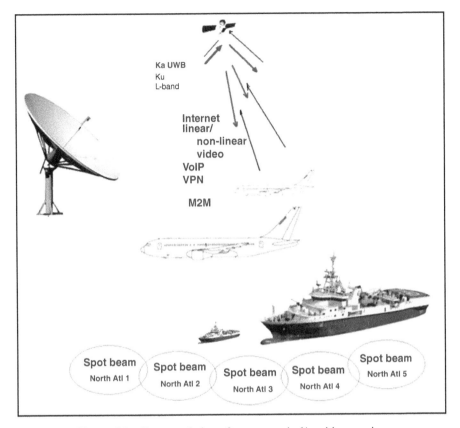

Figure 4.1   Conceptual view of an aeronautical/maritime service.

… Passenger connectivity is an on-going competitive battleground. Solutions have been around for some time, but do not yet enable a viable commercial offering on speed, coverage, and price …

[INM201401].

The IFC market segment is very cost-sensitive; hence, the unproven willingness of passengers to pay for connectivity services limits the market growth in the immediate short term [RUI201301], [WLI201301]; press time estimates were that only a small minority of passengers use the service (about 5%). Ka-based systems attempt to change these factors. In-flight phone services that became available in planes in the 1990s proved not to be economically viable because of the high prices charged and, hence, the ensuing low consumer demand. [2] As a heuristic, one could state that since the average cost of a movie theater ticket in the United States in 2013 was around $8

---

[2]The ATG telephone service was priced at around $3/min, or about 6–10X what domestic long distance service was priced at. It should be noted that home broadband costs around $2/day while some hotels charge $10–16 per night (while others give it for free), it would be interesting to see how many people personally actually sign up and pay for such a hotel service when not on a corporate trip

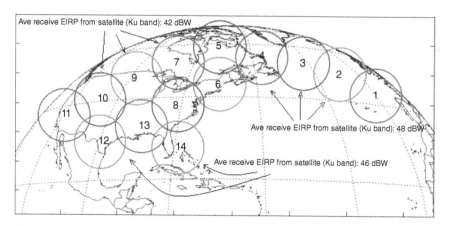

Figure 4.2 Example of possible use of HTSs (Ku and/or Ka) for aeronautical services.

for a compelling experience of 2 h of entertainment [BOW201301], and further state that since an Internet connection is only "half-as-compelling" as a visit to a movie theater, then the typical consumer may be willing to pay around $2/h for connectivity, or around $10–12 for a transcontinental flight; the same might apply to an easterly flight from the United States to Europe (considering that many passengers may only use the system for a couple of hours and then may want to sleep for the rest of the flight)

There are other barriers to deployment of the technology. In-flight connectivity systems represent a nontrivial investment for airlines, with Very Small Aperture Terminals (VSATs) equipment costs at approximately a-quarter-of-a-million dollars at this time and a comparable expenditure for installation. A consideration is whether the aircraft simply provides a Wi-Fi hotspot service where users utilize their own equipment, or if the goal is to provide an integrated solution also encompassing the seatback IFE, or, as yet another option, utilize tablets-handout-based IFEs. Including the cost of IFE devices, equipping a wide-body aircraft with seatback screens can cost up to $3 million; in addition, the aircraft are typically grounded for 1–3 days to install IFEC systems, impacting the aircraft's ability to generate revenue.[3] The certification process for terminals can also be a market entry barrier: a Supplemental Type Certification (STC) is required for any IFE retrofit operation (and validations are even more intensive if Wi-Fi-based systems are being deployed), which results in a regulatory approval process that can take up to a year for each aircraft type. Fortunately for the airlines, the terminal manufacturers generally undertake the blanket product certification process, at least for the discrete subsystems. Nonetheless, some take an altogether positive view of the market opportunities as follows:

---

[3]The IFE market is a highly consolidated market. At present, the two market leaders are Panasonic Avionics and Thales, together representing more than 90% of the IFE market [WLI201301]. On average, the cost of equipping a single seat with traditional IFE is approximately $8,000. Handing out tablets (at $250 each) during the flight would be much less expensive, but it increases labor of distribution, collection, storage, damage, and possible theft.

The multitude of commercial, government and private aircraft that traverse the skies everyday represent a tremendous opportunity for broadband service providers. With the proliferation of tablets, smartphones, and other mobile devices, passengers' desire to be online during their flight is ever increasing. Those customers demand the same quality of connection that they experience at home or in the office. High-speed IP connectivity onboard aircrafts enables a host of customer-driven applications, including in-flight consumer Wi-Fi and entertainment/video on demand services, as well as increasing operational efficiencies for the [airline] carrier.

[INT201401].

Implementing mobile platforms using recently deployed Ka-band geosynchronous orbit (GSO) satellites is, in the view of many industry observers, the logical evolution of the deployment of mobile applications over GSO Ku-band satellites that has taken place over the past decade, also because of the advantageous overall economics. Indeed, it is now well established in the industry and in regulatory bodies [such as the Federal Communications Commission (FCC) in the United States] that use of GSO fixed satellite services (FSSs) spectrum resources for mobile platforms can be accomplished without causing incremental interference compared with what would be the case with a traditional fixed antenna. Specifically, the FCC has permitted aeronautical applications of Ku-band FSS spectrum where no service-specific rules exist, by waiving the US Table of Frequency Allocations; in addition, the International Telecommunications Union (ITU) has recognized the increased use of GSO FSS networks to provide services to earth stations mounted on mobile platforms, including the Ka-band [VIA201201]. These arguments support the establishment of aeronautical application (or for mobile applications in general, for that matter), without being constrained to only using Mobile Satellite Services (MSSs) bands or approaches.

In fact, we noted in passing in Chapter 1 that the distinction between traditional FSS and MSS is blurring. Formally "fixed" satellite services entail transmitting signals to ground antennas that remain in a fixed location, whereas "mobile" satellite services transmit signals to a moving or transportable terrestrial antenna (e.g., for shipboard services). Traditionally, MSS applications have used lower frequencies such as L-band and S-band; however, in order to support higher link capacities, MSS applications are now moving to higher frequencies, where additional spectrum bandwidth is available. Thus, the convergence of the two, formally distinct, services is being accelerated by the availability of relatively inexpensive Ka-band capacity that can serve both communities of users. This is not a completely new development: note, for example, that WRC-03 adopted a regulation allowing earth stations onboard vessels (ESVs) to operate on a primary basis using frequencies in the FSS C-band and Ku-band; WRC-03 allowed aeronautical mobile satellite service (AMSS) applications to operate in the band 14.0–14.5 GHz on a secondary basis. These ESV and AMSS applications are implemented within frequencies assigned to FSS services, but specific technical parameters are used for the antenna platforms, to ensure that the interference levels reaching satellites in geostationary orbit never exceed the levels agreed to in the frequency coordination process [CAS201201]. A number of national and international regulatory bodies, notably the ITU, are now proceeding with work to accommodate "formally" mobility services within the FSS in the Ka-band.

Observers state that as this convergence is being accommodated by regulators, the deployment of commercial Ka-band mobile applications becomes increasingly more economical. Based on this discussion, we use the term "mobility services" to imply satellite-oriented services directed to moving entities in the proximity of the ground (i.e., on the ground, in the air, or on the seas), regardless of the legacy terminology.

4.2 AERONAUTICAL SYSTEMS

4.2.1 Market Opportunities

Traditionally, satellite terminals on aircraft have been mostly installed to support safety communication requirements, as a backup for cockpit operations; at this juncture, however, interest in satellite communication is being driven by in-flight passenger entertainment and for connectivity services (business intranet and/or Internet). Table 4.1 depicts the routes with major concentration of airline traffic. Airplane-based connectivity has been provided either as ATG communication or as satellite communication; with satellite-based communications the L-band (for MSS-based solutions) or the Ku-band have been used in the recent past, but increasingly providers are shifting to HTS-based Ka solutions. HTS systems have been designed to provide as much as 100 Mbps to each aircraft in flight.

The commercial aviation market is currently supported by a number of niche service providers. These service providers, who make use of the satellite assets (usually provided by others), are clearly focusing initially on the largest markets, specifically CONUS, Transatlantic, and Europe. Key service providers at press time included Gogo®, Yonder (ViaSat), Panasonic, and AT&T.

Euroconsult, a market research firm, estimated in 2013 that approximately 12,200 commercial aircraft and approximately 16,800 business aircraft will provide in-flight connectivity services by 2022 (total 29,000) [WLI201301]; this compares with

TABLE 4.1 Routes with Major Concentration of Air Traffic

Market ID	From	To (and Vice Versa)	Air Traffic	Average Flight Length
1	North America	North America	H	2.5
2	North America	South America	M	8
3	North America	Europe	H	8
4	Europe	Europe	H	2
5	Europe	Africa	M	8
6	Europe	East Asia	H	6
7	East Asia	East Asia	H	4
8	East Asia	Far East Asia	H	5
9	Far East Asia	Far East Asia	H	4
10	Far East Asia	Australia & Islands	M	6
11	Far East Asia	North America	M-H	15

M = Medium; H = High

approximately 5,800 and 11,500, respectively (total 17,300), at press time (2015). Figure 4.3 shows that between 2015 and 2022, the total number of aircraft with IFC services will increase substantially: this represents about 11% Compound Annual Growth Rate (CAGR) for commercial aircraft and about 5.5% CAGR for business aircraft – about 7.5% CAGR for the combined total. Euroconsult also forecasts that by 2022 there will be 33,700 commercial aircraft in operation, up from 23,800 in 2012, and it estimates that 25,900 business jets will be in operation by 2022, up from 17,900 in 2012. Thus, in 2015, 22% of the commercial aircraft had IFC services; the percentage is expected to be 36% for 2022; in 2015, 57% of the business aircraft had IFC services; the percentage is expected to be 65% for 2022 (in total, 37% of all aircraft had IFC in 2015 and 49% in 2022). Geographically, North America will dominate the market through 2022 (contributing about two-thirds the total market by various counts). Table 4.2 provides some summary market data from various industry sources (including but not limited to [WLI201301]) with emphasis on VSAT/HTS aspects of the service, and Figure 4.3b provides a revenue forecast for the commercial airline IFC services, synthetized from data published by Euroconsult.

As just noted, up to the present, a higher percentage of business aircraft compared to its total population have or have had IFC than is the case for commercial airlines; this may well follow the technology trends of the past four decades where many (telecom and computing) technologies are first enjoyed by the business community and eventually migrate to the general consumer community (notable in those categories: computers, cellular phones, and overall Internet access). The large majority of passengers now carry portable electronic devices; wireless technologies inside the cabin may be used to support IFC to these user-provided devices; otherwise, airlines may opt to funnel connectivity over their IFEC devices. At the end of 2012,

TABLE 4.2   Some Market Data from Various Industry Sources

	Commercial Aircrafts	Business Aircraft
Ku VSAT	1,800 aircraft (2015) → 3,100 (2022) 1.8 Gbps capacity (2015) → 11 Gbps (2022)	1 Gbps capacity (2015) → 2 Gbps (2022)
Ka VSAT	~200 aircraft (2015) → 2,700 (2022) Negligible capacity → 20 Gbps (2022)	Negligible capacity (2015) → 5 Gbps (2022)
Aggregate	$400 M airline revenue (*all technologies*) (2015) → $1,300 M (2022) Negligible Ka revenue (2015) → $300 M (2022) $100 M sat operator revenue (2015) → $500M (2022) Equipment revenues, tracks deployment timetables, several hundred million per year	~300 VSAT aircraft (2015) → 1,500 (2022) $90 M MSS revenue (2015) → $140 M (2022) $30 M FSS sat operator revenue (2015) → $100 M (2022) Equipment revenues, tracks deployment timetables

TABLE 4.3 Comparison of Technologies

	Ku-Band	Ka-Band
ADVANTAGES	Established technology and terminal products Currently widely deployed Lower costs of deployment Large number of satellite operators and spacecraft	Higher bandwidth Smaller antenna Services optimized for IP connectivity Global coverage (e.g., with Inmarsat GlobalXpress)
DISADVANTAGES	Regional coverage Lower bandwidth More concern with creating FSS inference to other spacecraft due to large deployment of Ku	Commercial unproven technology High costs of deployment Small beam-to-beam handoff complexity

56 commercial airlines were operating connectivity services or had announced plans to offer the service, and that number has increased since then.

4.2.2 Technology Approaches to Aeronautical Connectivity

As noted, four different types of technology have been used in recent years to provide IFC: ATG, satellite L-band (MSS), satellite Ku-band, and satellite Ka-band technologies.

- ATG approaches make use of ground towers to maintain a path to the aircraft; clearly, coverage is limited to the land masses. ATG systems are less expensive than satellite-based systems and, thus, services can be provided to passengers at lower prices. Currently, ATG technology is primarily used in North America with Gogo as a key provider (China and Europe began testing ATG systems recently but deployment plans are unknown). In 2012, ATG technology was in operation on 1,800 commercial aircraft and 1,300 business jets.
- MSS-based approaches use L-band satellite technologies. MSS service providers were the first to enter the satellite-based IFC market. MSS provides global coverage. Initially, the focus was on cockpit communication services, primarily on business and military aircraft; then it expanded to passenger communication. Currently, MSS dominates the business aviation market while (the commercial aircraft market is more fragmented in terms of service options/deployments). Inmarsat and Iridium are the two satellite infrastructure providers (they also provide direct end-user services); as for key service operators, OnAir provides connectivity services to both the business and

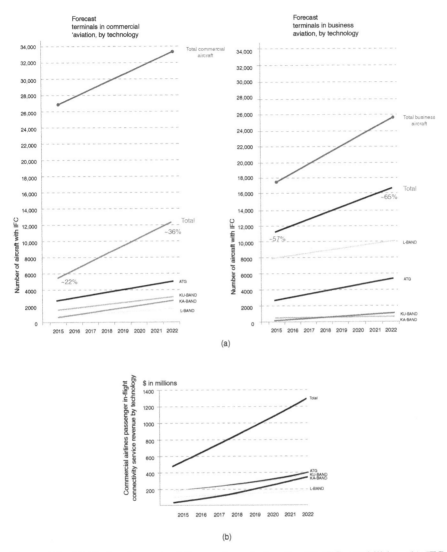

Figure 4.3 (a) Market forecast of the number of aircraft with IFC capabilities. (b) IFC revenue forecast.

commercial aircraft market and LiveTV, with emphasis on live television, provides some services in the L-band.

- Ku-band-based approaches (Ku VSATs) represent a third wave of IFC capabilities. The services typically have regional coverage; as a consequence, there are many more satellite operators in the Ku-band market. The coverage is prevalent in the northern part of the globe, covering the North Atlantic corridor, the United States, Europe, part of Asia, and the Middle East. Two examples include ViaSat's Yonder service with Ku-band connectivity to the business jet market on a number of satellites with a large area of coverage over the Atlantic and Pacific oceans, North America, Europe, Asia-Pacific, and the Middle East; and, Row 44 that operates over the Hughes Ku-band network and is currently focused primarily on narrow body aircraft serving the North American and European routes.

- Ka-band-based approaches (Ka VSATs) are now evolving, particularly employing HTSs (and narrow beams). As discussed in Chapter 3, Ka-band services offer higher bandwidth and smaller antennas; while the technology is more impacted by rainfade than Ku-band, this issue is of marginal interest in aeronautical applications since, in general, flights will be above the clouds. Several satellite infrastructure providers have launched or announced the launch of Ka-band satellites (as discussed in Chapter 3), notably Eutelsat with KA-SAT, ViaSat with ViaSat-1 -2, and Inmarsat with GlobalXpress (GX). ViaSat currently operates Ka-band aeronautical services in North America through its *Exede in the Air* service. A number of satellite operators (including ViaSat, Intelsat, Inmarsat) have designed their in-construction and/or just-launched HTS systems to support the busiest airline routes (specifically North Atlantic, CONUS, Europe).

A brief note about L-band frequencies follows next. As we saw in Chapter 1, the IEEE Standard 521–1984 positions the L-band at 1–2 GHz. This frequency range includes the Global Positioning (Service/) System (GPS) and other GNSSs (Global Navigation Satellite Systems) such as the Russian Glonass, the European Galileo, and the Chinese Beidou. It also supports SARSAT/COSPAS [Search And Rescue Satellite Aided Tracking/*Co*smicheskaya *S*istema *P*oiska *A*variynyh *S*udov (Space System for the Search of Vessels in Distress)] search and rescue payloads that are carried on board US and Russian meteorological satellites. In addition, the ITU Radio Regulations allocate the band 1,452–1,492 MHz to the Fixed, Mobile, Broadcasting, and Broadcasting Satellite Service on a coprimary basis in all ITU regions with the exception of the mobile aeronautical service in Region 1; in Region 2 (the Americas), the use of the band 1,435–1,535 MHz by the aeronautical mobile service for telemetry has priority over other uses by the mobile service; in Europe, the L-band (1,452–1,492 MHz) is currently allocated for use by terrestrial and satellite digital audio broadcasting (DAB) services in most European countries, where the terrestrial segment is 1,452–1,479.5 MHz (27.5 MHz), and the satellite segment is 1,479.5–1,492 MHz (12.5 MHz). Of particular interest for the discussion in this chapter, the band also includes the MSS communication band as follows: 1,525–1,559 MHz (downlink) and 1,626.5–1,660.5 MHz (uplink). The Aeronautical Mobile-Satellite (Route) Service (AMS(R)S) allocation was identified since 1972 for the safety communication services. At WRC-97, the AMS(R)S allocation was made generic to MSS and a regulatory footnote was introduced to preserve priority

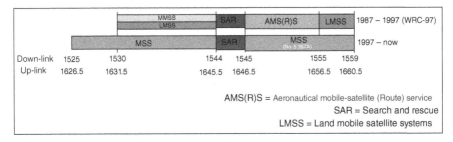

Figure 4.4 MSS L-band frequency allocation.

to aeronautical safety communications [AZZ201001] (Figure 4.4). In the United States, other bands have been allowed for ATG.[4]

As seen in Figure 4.3, at the current time, ATG technology has the largest deployment in commercial aircraft. This is followed by Ku-band-based systems and some L-band systems; Ka-band systems are just being introduced as discussed so far. For business jets, L-band systems had the largest deployment followed by ATG systems. By the start of the next decade, Ka-band system should rival Ku-based systems in terms of deployed terminals; Ka will experience the highest growth rates, but the other technologies (including ATG) will also see steady growth. For the business aircraft, L-band MSS and ATG will continue to grow steadily, with some deployment of Ku- and Ka-based VSATs (with about 8,000 aircraft, currently MSS represents about 70% of the installed base; while this percentage will go down to 60% by the beginning of the new decade – giving way to other technologies – the absolute number of aircraft with MSS systems will continue to grow to about 10,000 aircraft). Table 4.3 shows a comparison between Ku-band and Ka-band approaches in terms of relative advantages and disadvantages.

Table 4.4 depicts some of the key aeronautical initiatives pegged to airlines, as of press time. As illustrative examples of the pricing approaches, Gogo offers (as of press time) *Gogo Unlimited* for $59.95 per month, targeted at frequent fliers, and a *Gogo All-Day Pass* for $16.00 (24 h of continuous access).

---

[4]For example, in 2004, the FCC adopted an Order revising its rules pertaining to the 4 MHz of spectrum in the 800 MHz air-to-ground services in 849–851 MHz and 894–896 MHz. In adopting these rules, it sought to "promote key spectrum policy objectives that would lead to greater technical, economic, and marketplace efficiency" while responding to evolving market demands. The Commission weighed the possible band plans and technical considerations at that time, and concluded that no more than 3 MHz of spectrum was required for a licensee to deliver high-speed air–ground services using then-existing broadband technologies, and that a service offered over 1 MHz of spectrum could provide a meaningful competitive alternative to other air–ground services, including satellite services. It also was concerned that 1 MHz of spectrum might end up "lying fallow" if one party controlled the air–ground band's entire 4 MHz of spectrum. As a result, the Commission adopted Section 22.853, which provides that "[n]o individual or entity may hold, directly or indirectly, a controlling interest in licenses authorizing the use of more than 3 MHz of spectrum (either shared or exclusive) in the 800 MHz commercial aviation Air-Ground Radiotelephone Service frequency bands." The Commission expressly noted, however, that it would "consider a waiver of the eligibility rule based on a showing that market conditions and other factors would favor common control of more than 3 MHz without resulting in a significant likelihood of substantial competitive harm." [FCC201301]

TABLE 4.4   Aeronautical Services on Major Airlines (examples, as of press time) (subject to change)

Airline	Frequency Band	Service Provider	Notes
Aer Lingus	HTS Ka	Live TV/Eutelsat	Wi-Fi is available on all A330 aircraft on all transatlantic routes. Pricing: Aer Lingus Wi-Fi Pass 1 h: €10.95/$14.95 Aer Lingus Wi-Fi Pass 24 h: €19.95/$24.95
Air Canada	Ku	Gogo (select flights)	Gogo Inflight Internet service is available on select Air Canada flights traveling between Toronto or Montreal and Los Angeles over the United States Internet access for only $9.95
Air France-KLM	Ku	Panasonic	Trial underway in 2013. HTS Ku-band after signing a long-term agreement with Intelsat for over 1 Gbps of capacity on IS-29e and IS-33e
AirTran Airways	Ku	Gogo	Web, email, online shopping, and social networking on every AirTran Airways flight through the partnership with Gogo $11.00–$49.00 for computer devices $4.95–$19.95 for mobile devices
Alaska Airlines	Ku	Gogo (select flights)	Alaska Airlines offers Gogo Inflight Internet on almost all of their aircraft operating within the Lower 48 United States and specific areas in Alaska Gogo has flexible pricing options ranging from $1.95–$39.95 including per-flight, day passes and several different subscriptions
Allegiant Air	Ku	Row 44	
American Airlines	Ku	Gogo (select flights)	In-flight Wi-Fi is available on American Airlines' 767–200, selected 737–800, and selected MD-80 flights
Azul Brazilian Airlines	Ku	Live TV	
Cathay Pacific Airways	Ku	Panasonic	HTS Ku-band after signing a long-term agreement with Intelsat for over 1 Gbps of capacity on IS-29e and IS-33e

TABLE 4.4 *(Continued)*

Airline	Frequency Band	Service Provider	Notes
Delta Air Lines	Ku	Gogo (select flights)	In-flight Wi-Fi is available on Delta Airlines' 500 aircrafts and on selected 757–200 and selected MD9-50 flights
El Al Israel Airlines	HTS Ka	ViaSat's Exede in the Air/Eutelsat's KA-SAT	Service in 2015
Frontier Airlines	Ku	Gogo (select flights)	Available on ERG170 and ERJ190 flights
Icelandair	Ku	Row 44	Icelandair was planning to offering Wi-Fi on its Boeing 757 flights between North America and Europe in the mid-2014, now that a deal is in place with Ku-band connectivity provider Row 44
Japan Airlines	Ku	Panasonic, Gogo	A fee-based, in-flight Wi-Fi connection service that supports passengers' own smart phones, notebook computers, and other wireless LAN devices. Upon connection through Wi-Fi, passenger will be able to enjoy reading web pages, checking your email, and updating social media networks. Available routes: Tokyo (Narita) – New York (JL005/006)/Chicago/Los Angeles/Frankfurt/Jakarta (JL725/726) and Tokyo (Haneda) – London/Paris(JL045/046) 1 h plan: Usage time: 1 h; usage fee: $11.95. Flight plan: usage time: 24 h; usage fee: $21.95
JetBlue	HTS Ka	LiveTV/Viasat	Agreement with ViaSat (2010) to deploy the first Ka-band commercial aviation broadband network on >170 aircraft using ViaSat-1 in North America

(continued)

TABLE 4.4 *(Continued)*

Airline	Frequency Band	Service Provider	Notes
			In-flight Wi-Fi is available on JetBlue's A320 flights; free Wi-Fi-enabled devices (Wi-Fi limited to email, instant messaging, DirectTV, and shopping websites)
Lufthansa Airlines	Ku	Panasonic	HTS Ku-band agreement with Intelsat for over 1 Gbps of capacity on IS-29e and IS-33e
Mango Airlines (South Africa)	Ku	Row 44	
Norwegian Air Shuttle	Ku	Row 44	
SAS	Ku	Panasonic	HTS Ku-band after signing a long-term agreement with Intelsat for over 1 Gbps of capacity on IS-29e and IS-33e
Southwest Airlines	Ku	Row 44	In-flight Wi-Fi via satellite
Turkish Airlines	Ku	Panasonic	
United Airlines	HTS Ka and Ku	Ka LiveTV/ ViaSat's Exede in the Air/Gogo (select flights)	United Airlines started rolling out services, with installations ramping up. ViaSat *Exede In The Air* service In-flight Wi-Fi is available selected 757–200 flights $11.00–$49.00 for computer devices $4.95–$19.95 for mobile devices
US Airways	Ku	Gogo (select flights)	In-flight Wi-Fi is available on US Airways' selected A321 flights. Gogo $11.00–$49.00 for computer devices $4.95–$19.95 for mobile devices
Virgin America	Ku	Gogo	In-flight Wi-Fi is available on Virgin America's A319 and A320 flights. Gogo $11.00–$49.00 for computer devices $4.95–$19.95 for mobile devices
WestJet Airlines	Ku	Live TV	
Business aircraft	HTS Ka and Ku	Depends on geography of operation	

4.2.3 Aeronautical Antenna Technology and Regulatory Matters

It has been noted by the International Telecommunications Union – Radio (ITU-R) that "GSO fixed-satellite service networks are being used at an increasing rate to provide services to earth stations mounted on mobile platforms. GSO FSS networks are currently providing valuable broadband telecommunications services to aircraft, ships, trains and other vehicles in the 14.0–14.5 GHz (earth-to-space) and in the 10.7–12.7 GHz (space-to-earth) bands, for example Resolution 902 (WRC-03). The growing demand for service to these mobile platforms has caused service providers to turn to the 17.3–30.0 GHz FSS band to meet the need for increased broadband speed, capacity, and efficiency. Advances in satellite antenna technology, particularly the development of 3-axis stabilized antennas capable of maintaining a high degree of pointing accuracy even on rapidly moving platforms, have allowed the development of mobile earth stations with very stable pointing characteristics. Similarly, the application of low power density waveforms has likewise enabled the use of smaller antennas and lower performance pointing systems while still maintaining off-axis Effective Isotropically Radiated Power (EIRP) density within prescribed limits. When properly managed and controlled, the technical characteristics of these mobile earth stations are indistinguishable from fixed earth stations when viewed from an interference perspective to FSS networks" [ITU201101].

There are three factors related to antenna technology in the aeronautical context:

1. The need to design antennas that provide the required technical TX/RX performance (G/T, cross-pol isolation, etc.), while minimizing interference to other GSO satellites, and also being light weight and relatively inexpensive;

2. The need to have the Federal Aviation Administration (FAA) (or other national regulatory entities) approve the system (for safety, aircraft interference, etc.); and

3. The fact that airlines look for a complete, integrated terminal (not just an antenna) to be installed in the aircraft.

Regulatory matters (of focus here) cover the antenna design and the frequency band of operation. Key considerations for aeronautical application do relate to the antenna construction. Most antennas currently deployed on aircrafts are a hybrid of mechanical and phased array tracking. These antennas are typically designed to meet the ITU-R requirement of generating less than 6% increase in interference to the existing noise levels; off-axis transmit antenna gain characteristics usually do not meet ITU-R standards over the entire arc, but transmit power is lowered so that they do meet the power density in the GSO arc. Satellite service providers and terminal manufacturers also have to comply with certification procedures (e.g., with the FAA[5]); these certifications are country dependent and product dependent, which typically entail considerable technical and financial effort (the equipment certification issue is not further discussed herewith).

[5]For example, an STC is required by the FAA for any IFE retrofit operation. Wi-Fi installations are subject to further assessments.

As discussed in Chapter 3, for the US 47 CRF, Section 25.132 provides that transmitting earth stations operating in the 20/30 GHz band must demonstrate compliance with Section 25.138. The antenna must meet the performance requirements in Section 25.138(a) in the direction of the GSO arc as well as in all other directions, as illustrated by off-axis EIRP spectral density plots. Furthermore, as established by the FCC, the power flux density (PFD) at the earth's surface produced by emissions from a terminal must be within the $-118\,dBW/m^2/MHz$ limit set forth in Section 25.138. In addition, to the extent required for protection of received satellite signals pursuant to Section 25.138, the earth station must conform to the antenna performance standards in Section 25.209, as demonstrated by the antenna gain patterns.

Specifically in the context of Ka, aeronautical antennas must operate consistent with the existing regulatory framework for the Ka-band. Under that framework, the key element for ensuring compatibility with the FCCs 2° spacing policy (in the United States) is (i) compliance with the off-axis-EIRP density levels specified in Section 25.138 in the uplink direction, (ii) compliance with the PFD levels referenced in Section 25.138 in the downlink direction, and (iii) in the case of any exceedance of those levels, coordination with potentially affected satellite systems. Moreover, the longstanding, coprimary MSS allocation in the upper 500 MHz of the Ka-band (19.7–20.2 GHz, 29.5–30.0 GHz) contemplates mobile applications; thus, the FCC has acknowledged the likelihood of future licensing of mobile applications in the Ka-band once technology emerged that ensures compatibility with existing FSS applications of the Ka-band. That technology, in fact, exists today, and its efficacy has been proven over the past several years.

More specifically for (both Ku and) Ka applications, the antenna design must be such that during actual operational conditions (i) these terminals remain pointed at the intended satellite with a maximum pointing error of ±0.5° in the azimuth direction and ±1.35° in the elevation direction, and (ii) the transmit output of the terminal will be inhibited in less than 100 ms should these tolerances be exceeded (whether by the motion of the aircraft or otherwise), and will not resume until the pointing of the terminal is again within these tolerances. Within these tolerances, the off-axis EIRP density limits of Section 25.138 must be met in the GSO plane. With a good design (e.g., with a mix of quality mechanical tracking and/or phased array operation), the 3σ (σ = standard deviation) pointing error might be only ±0.27° in azimuth, and, as a practical matter, the system should never require the cessation of transmissions due to azimuth pointing errors. Elevation pointing errors should thus only cause the terminal to cease transmissions less than a fraction of 1% of the time.

Pragmatically, aeronautical antenna developers may find their antennas exceed the Section 25.138 levels in certain parts of the elevation plane; perhaps, the off-axis EIRP density of the main lobe exceeds the Section 25.138(a)(2) mask in the elevation plane, and/or the off-axis EIRP density exceeds the mask at some discrete "grating" lobes in the elevation plane, far removed from the main lobe. These grating lobes could intersect the GSO arc when aircraft are operated in a limited number of geographic areas such that the antenna is oriented at a skewed angle, relative to the satellite, of approximately 25°. If these emissions cannot be rectified by technical adjustments to the antenna design, then an operator utilizing such antennas would need to coordinate the potential exceedances of the Section 25.138 off-axis EIRP density levels with all potentially affected GSO and non-geosynchronous (non-GSO) systems.

At the international level, the ITU-R document *Technical and operational requirements for GSO FSS earth stations on mobile platforms in bands from 17.3 to 30.0 GHz, Rep. ITU-R S.2223*, [ITU201101], addresses FSS earth station operation on mobile platforms in the frequency range from 17.3 to 30 GHz. The crux of the recommendations by the ITU-R regarding the matter is as follows (cited directly from the reference):

1. It is clear that implementation of FSS earth stations on mobile platforms would be simplified in bands that are not shared with terrestrial services, as this reduces the sharing situation to one of sharing between satellite networks. In such cases, in order to address potential interference with other cofrequency GSO FSS networks, it is essential that FSS earth stations on mobile platforms comply with the off-axis EIRP limits contained in Recommendation ITU-R S.524-9, or with any other limits coordinated with neighboring satellite networks. In addition, any network of such earth stations should be operated such that the aggregate off-axis EIRP levels produced in the earth-to-space direction by all cofrequency earth stations within such networks, in the direction of neighboring satellite networks, are no greater than the off-axis EIRP levels produced by other specific and/or typical FSS earth station(s) operated in conformance with Recommendation ITU-R S.524-9, or with any other limits coordinated with neighboring satellite networks. These requirements will ensure that such earth stations are essentially equivalent to stationary FSS earth stations from the perspective of static uplink interference potential.

2. Realizing that earth stations on mobile platforms operate in a dynamic environment (i.e., the position and orientation of the platform can change with time), it is important to address this aspect in specifying an essential set of technical and operational requirements. The design, coordination, and operation of earth stations on mobile platforms should be such that, in addition to the static requirements discussed earlier, the interference levels generated by such earth stations account for the following factors:

   • Mispointing of the earth station antenna. Where applicable, this includes, at least, motion-induced antenna pointing errors, effects caused by bias and latency of their pointing systems, tracking error of open- or closed-loop tracking systems, misalignment between transmit and receive apertures for systems that use separate apertures, and misalignment between transmit and receive feeds for systems that use combined apertures.

   • Variations in the antenna pattern of the earth station antenna. Where applicable, this includes, at least, effects caused by manufacturing tolerances, aging of the antenna, and environmental effects. Networks using certain types of antennas, such as phased arrays, should account for variation in antenna pattern with scan angles (elevation and azimuth). Networks using phased arrays should also account for element phase error, amplitude error, and failure rate.

- Variations in the transmit EIRP from the earth station. Where applicable, this includes, at least, effects caused by measurement error, control error and latency for closed-loop power control systems, and motion-induced antenna pointing errors.

FSS earth stations on mobile platforms that use closed-loop tracking of the satellite signal need to employ an algorithm that is resistant to capturing and tracking adjacent satellite signals. Such earth stations must be designed and operated such that they immediately inhibit transmission when they detect that unintended satellite tracking has occurred or is about to occur. Such earth stations must also immediately inhibit transmission when their mispointing would result in off-axis EIRP levels in the direction of neighboring satellite networks above those of other specific and/or typical FSS earth stations operating in compliance with Recommendation ITU-R S.524-9 or with any other limits coordinated with neighboring satellite networks. These earth stations also need to be self-monitoring and, should a fault be detected which can cause harmful interference to FSS networks, must automatically mute any transmissions.

In addition to these autonomous capabilities, FSS earth stations on mobile platforms should be subject to the monitoring and control by a Network Control and Monitoring Center (NCMC) or equivalent facility, and these earth stations should be able to receive at least "enable transmission" and "disable transmission" commands from the NCMC. It should be possible for the NCMC to monitor the operation of the earth station to determine if it is malfunctioning.

4.2.4   Terminal Technology

We noted earlier that airlines look for a complete, integrated terminal (not just an antenna) to be installed in the aircraft. This would include the modem and other packet processing such as Transport Control Protocol (TCP) Acceleration and Hypertext Transfer Protocol (HTTP) Prefetch; compression (software) modules to optimize use of bandwidth; spacecraft telephony VoIP or pico-cell-based mobile phone connections support; hotspot support with Quality of Service (QoS) management to/from the passenger; security firewalling; and other packet classification and filtering. Figure 4.5 depicts a simplified such terminal system (other systems are discussed later in the chapter.)

4.2.5   A Specific Example of Antenna Engineering (ViaSat)

In this section, the various design desiderata discussed in the previous sections are pulled together using a specific example to expose the reader to the various technical aspects involved, particularly with regard to the aeronautical antenna. This section uses ViaSat Mantarry M40 antenna as an example, and it includes *nearly verbatim*, for useful pedagogical purposes, portions of a public 2012 domain filing to the

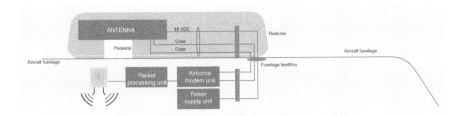

Figure 4.5   Possible external fuselage approach.

FCC by ViaSat for such system*. Beam-to-beam handoff at 500 miles can be a challenge. ViaSat systems are designed to address this requirement. The airborne terminal is an integral part of the ViaSat mobile system delivering truly broadband Internet connection speeds to passengers. The aero mobile terminal delivers typical connection speeds of 70–100 Mbps to the aircraft and 2.5–20 Mbps from the aircraft. The technical and regulatory challenges are highlighted by the discussion that follows later.

4.2.5.1 Bands of Operation   ViaSat's proposed Ka mobile terminal operations (covered by this filing) are consistent with the Ka-band allocations in the US Table of Frequency Allocations (the "US Table"). As an initial matter, operation of the proposed mobile terminals in the 19.7–20.2 and 29.5–30.0 GHz bands is consistent with the coprimary MSS allocation in the US Table. Although there are no service rules for MSS in these bands, the requirements of Section 25.138 could be applied by analogy. By demonstrating compliance with the requirements of Section 25.138, the proposed terminals thus could be deemed compatible with adjacent Ka-band FSS satellite operations. Moreover, as the Commission has recognized in the context of Ku-band, ViaSat's proposed mobile operations essentially are an application of the FSS, which is allocated on a primary basis in the 18.3–19.3, 19.7–20.2z, 28.35–29.1, and 29.5–30.0 GHz bands. The mobile terminals will operate within ViaSat's existing Ka-band FSS network. Using a highly accurate pointing mechanism, the emissions from the terminals effectively will be fixed toward the satellite points of communication and would be no more interfering than any FSS application. However, to the extent necessary, ViaSat requests a waiver of the US Table in these frequency bands to permit the proposed mobile terminals to operate as described in this application. In addition, ViaSat requests a waiver to allow these terminals to operate in the 18.8–19.3 GHz portion of the band in the absence of a GSO allocation in that band segment.

As a general matter, operation of the proposed terminals in the 18.3–19.3 and 19.7–20.2 GHz downlink bands and the 28.35–29.1 and 29.5–30.0 GHz uplink bands is compatible with the operation of GSO systems and non-GSO systems in these band segments, as well as coprimary terrestrial allocations in segments of the downlink bands. Notably, ViaSat has either coordinated the proposed antenna with,

---

*The author wisher to thank ViaSat Corporation and Mr. Bruce Rowe for this material.

or has received confirmation that the proposed antenna can be coordinated with, (i) all operating Ka-band GSO satellite networks within 6° of ViaSat-1 at 115.1° W.L. and WildBlue-1 and ANIK-F2 at 111.1° W.L., (ii) all potentially affected Ka-band GSO satellite networks outside of the 6° range, and (iii) the one potentially affected Ka-band non-GSO network. ViaSat has completed coordination with O3b, which is the only relevant commercial non-GSO system in the 18.8–19.3 and 28.6–29.1 GHz band segments.

GSO FSS Operations Section 25.132(a)(2) provides that transmitting earth stations operating in the 20/30 GHz band must demonstrate compliance with Section 25.138.15. While there are no rules for mobile operations in the Ka-band, operating the proposed terminals consistent with the technical parameters of Section 25.138 would ensure compatibility with satellite systems operating in the Ka-band. This approach is consistent with the ITU's recommendation in Report ITU-R S.2223 that GSO FSS earth stations on mobile platforms in bands from 17.3 to 30.0 GHz comply with the off-axis EIRP limits coordinated with neighboring satellite networks.

The FCC has acknowledged the potential for MSSs in the Ka-band to be able to coexist with FSS. When the Commission designated the 19.7–20.2 and 29.5–30.0 GHz bands for GSO FSS, it maintained the MSS coprimary allocation in the US Table of Frequency Allocations because it believed "that the development of technology may enable these two different types of systems to coexist in the same frequencies in the future." As in the case of the Ku-band, mobile systems can operate on FSS platforms in the Ka-band without causing harmful interference to FSS operations. Moreover, as discussed earlier, ViaSat has coordinated the proposed antenna, or will soon complete such coordination, with all potentially impacted satellite operators. The FCC has found coordination to be adequate in the context of Ku-band mobile aeronautical operations, and that such satellite operators are capable of assessing the potential interference impact of such mobile operations.

Thus, the proposed terminal operations are compatible with and will not cause harmful interference into FSS systems. As described later, the proposed antenna complies with the Section 25.138 EIRP spectral density limits in the GSO plane. Furthermore, the antenna control unit (ACU) and closed-loop tracking system allow the terminal to be pointed accurately at the satellite while in motion, thereby protecting adjacent satellite operations. Namely, ViaSat will ensure that (i) these terminals remain pointed at the intended satellite with a maximum pointing error of ±0.5° in the azimuth direction and ±1.35° in the elevation direction, and (ii) the transmit output of the terminal will be inhibited in less than 100 ms should these tolerances be exceeded (whether by the motion of the aircraft or otherwise), and will not resume until the pointing of the terminal is again within these tolerances. Within these tolerances, the off-axis EIRP density limits of Section 25.138 will be met in the GSO plane. Notably, because the 3σ pointing error is only ±0.27° in azimuth, as a practical matter, the system should never require the cessation of transmissions due to azimuth pointing errors. Elevation pointing errors should only cause the terminal to cease transmissions less than 0.27% of the time.

The antenna does not comply with the Section 25.138(a)(2) EIRP spectral density limits in certain areas of the elevation plane. However, ViaSat satisfies the requirements of Section 25.138(b) to ensure that adjacent GSO systems are adequately protected from any higher power operations. The antenna pattern shows off-axis exceedances for the main lobe and four grating lobes along the elevation axis and well outside of the GSO. GSO FSS networks will never be impacted by the exceedance of the main lobe along the elevation axis, and the grating lobes would intersect the GSO arc only when the aircraft is traveling within certain geographic locations in which the GSO arc appears skewed with respect to the local horizon of the antenna, or when the aircraft is banking at certain angles while in flight. Due to the high speeds at which aircraft travel, any intersection of a grating lobe with the GSO arc likely would be fleeting. Moreover, due to the large off-axis angles where these grating lobes occur, the actual level of interference to any GSO satellite is well below the 6% delta T/T threshold that triggers satellite coordination.

Non-GSO FSS Operations in the 18.8–19.3 and 28.6–29.1 GHz bands need to be addressed. Pursuant to the terms of the Commission's authorization of ViaSat-1, operation of the GSO FSS system in the 28.6–29.1 GHz band is on a secondary allocation, and in the 18.8–19.3 GHz band is on a non-conforming basis. The Commission has approved operation of the ViaSat-1 satellite in these bands, and has acknowledged that ViaSat can operate in these bands while protecting the primary non-GSO FSS operations. The same previously approved capability of ViaSat-1 to cease operations in these bands in the event of an inline event between ViaSat's communications and the non-GSO system's communications will also avoid interference from communications with proposed terminals into non-GSO systems. Each of the proposed terminals will be dynamically controlled and can shut down operations in the bands in which non-GSO systems have priority when a non-GSO satellite is within the minimum line-of-sight separation angle established through coordination.

As summarized earlier, while the sidelobes of the proposed antenna exceed the Section 25.138(a)(2) limits in the elevation plane at the main lobe and at the four discrete points identified, ViaSat has coordinated the operation of the proposed antenna with O3b, which currently is the only potentially impacted non-GSO FSS system. ViaSat will coordinate its aeronautical terminal operations with any future potentially affected non-GSO applicants.

Terrestrial coordination also needs to be addressed. When the FCC adopted allocations for the Ka-band, it established sunset provisions for the coprimary status of certain terrestrial users in the FSS downlink bands in order to protect and facilitate deployment of FSS operations. Terrestrial microwave users maintain coprimary status in the 18.3–18.58 GHz band until November 18, 2012. In accordance with the blanket licensing rules, no coordination with terrestrial or other users is required on the GSO frequencies. The mobile nature of the proposed terminals does not change the satellite downlinks from ViaSat-1, WildBlue-1, and ANIK-F2. The PFD at the earth's surface produced by emissions from each of the satellite points of communication are within the $-118\,dBW/m^2/MHz$ limit set forth in Section 25.138(a)(6). Therefore, the RF environment in which the grandfathered terrestrial users operate will not change as a result of the proposed terminal operations. Moreover, ViaSat may either accept any potential for interference from the coprimary terrestrial users until the sunset date or relocate such users. ViaSat will accept the potential for interference from such users until the relevant sunset date.

Regarding blanket licensing of Ka terminals, because ViaSat's proposed operations are consistent with the policies underlying Section 25.138, which establishes the requirements for routine processing of blanket-licensed Ka-band terminals, a waiver of Section 25.138 is unnecessary for blanket licensing of the proposed terminals. Moreover, blanket licensing as proposed here is fully consistent with the FCC's precedent. The Commission has implemented blanket licensing procedures on a case-by-case basis (and in the absence of service rules) where circumstances have warranted such an approach. The Commission's policy justifications underlying its adoption of blanket earth station licensing procedures in rulemaking proceedings and declaratory rulings are equally applicable to the subject application. Allowing processing flexibility in this case will promote the expanded use of spectrum and the rapid development and deployment of new technologies. Such an approach serves the public interest by reducing administrative costs and delays and by accelerating system deployment, which facilitates the delivery of service to end-users. Blanket licensing of the proposed terminals will speed the delivery of mobile broadband services to consumers on commercial aircraft. Therefore, flexibility in processing this application is warranted and is consistent with recent precedent.

4.2.5.2 Network The proposed Ka terminals will operate in the same SurfBeam 2 Ka-band network as residential customers using the fixed VSAT equipment. Building upon its experience with Ku-band-based AMSS and ESV mobile broadband, ViaSat has incorporated the functions necessary to support mobility into the management functions of the SurfBeam 2 network. The network allows the aircraft to fly across the service area and seamlessly switch from spot beam to spot beam within the current operational satellite and to switch between satellites as coverage dictates.

Generally, when within the coverage footprint of ViaSat-1, the terminals will operate using ViaSat-1 spot beams to take advantage of its higher power and G/T and thereby enjoy improved throughput. As the aircraft flies across areas not supported by ViaSat-1, the AES will switch to capacity on the WildBlue-1 or Anik-F2 spacecraft.

The link budgets in Table 4.5 are included to illustrate typical links on each of the satellites. Because the SurfBeam 2 architecture employs adaptive coding and modulation, the terminals could transmit at any code and modulation point within the library of available choices. The available symbol rates are 625,000 symbols per second, or kilobaud (kBd), 1.25, 2.5, 5, and 10 MBd. Service traffic typically uses the 1.25 MBd or higher symbol rates. While the service may be operational when the aircraft is on the ground, in general operation will be while the aircraft is in flight and above most rain attenuation. The link budgets in Table 4.5 demonstrate that no margin shortfalls are present for clear sky operation (margins in the 0.7–2.4 dB range.)

The SurfBeam 2 architecture is designed to always operate at the lowest power density modulation and code point that allows the link to close. The network employs active power control and reduces power when conditions permit, keeping the E_s/N_0 margin at 1 dB or less. When the modem has sufficient excess transmit capability, it will automatically switch to the next symbol rate and increase data rate, keeping the

TABLE 4.5   Link Budgets

Forward Link Budgets

General:		Units	ViaSat-1	WB-1 22 MBd	AF-2 22 MBd
	Symbol rate	MBd	416.7	22.0	22.0
	Data rate	Mbps	550.9	43.6	43.6
	Modulation and coding rate		QPSK, $r=2/3$	8PSK, $r=2/3$	8PSK, $r=2/3$
	Carrier bandwidth	MHz	500.0	26.4	26.4
	Uplink frequency	GHz	28.60	29.75	29.75
	Downlink frequency	GHz	18.80	19.95	19.95
Uplink:					
	Tx E/S EIRP per carrier	dBW	65.2	74.0	68.0
	Atmospheric and rain losses	dB	0.50	0.50	0.50
	Free space loss	dB	213.3	213.3	214.0
	G/T toward Tx E/S	dB/K	24.2	13.0	15.0
	C/1 - intra system	dB	22.9	19.2	19.1
Downlink:					
	EIRP per carrier toward Rx E/S	dBW	60.7	53.6	52.0
	Rx E/S pointing loss	dB	0.3	0.5	0.5
	Atmospheric and rain losses	dB	0.5	0.5	0.5
	Free space loss	dB	209.3	209.8	209.8
	Rx E/S G/T incl. radome	dB/K	12.0	12.5	12.5
	C/1 - intra system	dB	16.0	15.9	17.9
End-to-End:					
	C/N - thermal uplink	dB	18.0	28.4	23.7
	C/1 Up - ASI	dB	29.1	43.4	38.7
	C/N - thermal downlink	dB	5.0	10.5	8.9
	C/1 Down - ASI	dB	20.8	32.5	30.9
	C/(N+1) - total actual	dB	4.3	8.9	7.8
	C/N - required	dB	3.6	7.2	7.2
	Excess margin	dB	0.7	1.7	0.6

(continued)

TABLE 4.5   (*Continued*)

Return Link Budgets

General:		Units	VS-1 10 MBd beam interior	WB-1 5 Msps beam interior	AF2 5 MBd beam interior
	Satellite location	°E	−115.1	−111.1	−111.1
	Symbol rate	MBd	10.00	5.00	5.00
	Data rate	Mbps	12.50	5.00	5.00
	Modulation/ coding rate		QPSK, $r=5/8$	QPSK, $r=1/2$	QPSK, $r=1/2$
	Carrier bandwidth	MHz	12.5	6.25	6.25
	Uplink frequency	GHz	28.60	29.75	29.75
	Downlink frequency	GHz	18.80	19.95	19.95
Uplink:					
	Tx E/S EIRP per carrier	dBW	44.8	43.7	43.7
	Radome Loss	dB	1.2	1.2	1.2
	Tx E/S EIRP density	dBW/40 kHz	19.6	23.9	23.9
	Tx E/S ant pointing loss	dB	0.7	0.7	0.7
	Atmospheric/ rain loss	dB	0.5	0.5	0.5
	Free space loss	dB	212.9	213.3	213.3
	G/T toward Tx E/S	dB/K	22.7	16.0	16.0
	C/1 - intra system	dB	18.2	15.5	17.5
Downlink:					
	EIRP/cxr toward Rx E/S	dBW	39.2	35.1	35.9
	Atmospheric/ rain loss	dB	0.5	0.5	0.5
	Free space loss	dB	209.4	209.3	209.2
	Rx E/S antenna diameter	m	7.3	8.1	8.1
	Rx E/S G/T	dB/K	37.4	39.0	38.3
	C/1 - intra system	dB	15.0	16.4	15.1

TABLE 4.5   (*Continued*)

End-to-End:					
	C/N - Thermal uplink	dB	9.7	8.0	8.0
	C/1 Up - ASI	dB	11.8	9.8	10.3
	C/N - Thermal downlink	dB	25.3	25.9	26.2
	C/1 Down - ASI	dB	35.4	22.6	21.2
	C/(N+1) - Total actual	dB	6.5	4.9	5.1
	C/N - Required	dB	4.2	2.6	2.6
	excess margin	dB	2.3	2.3	2.4

Figure 4.6   Mantarry M40 front view. Courtesy of ViaSat.

EIRP density at the minimum. This further reduces the likelihood that the system will impact traffic on other satellites.

4.2.5.3   Antenna and Pointing Accuracy   The antenna used in this application is a low profile waveguide horn array. The Mantarry M40 is a mechanically steered waveguide horn array antenna. The M40 designation reflects the number of feed horns across the width of the aperture. The M40 shown in Figure 4.6 is 40 horns wide with several horns at the upper corners of the array deleted to provide radome clearance. The height of the array is 15.75 cm and the width is 78.75 cm.

These antennas have two transmit receive interface adapters (TRIAs), one for each polarization. The TRIAs are similar in design to the outdoor units used on ViaSat's current blanket licensed earth station, but modified for airborne use and with slightly higher output power. The TRIA feeds a passive feed network which divides and routes the power to each of the feed horns in the array.

The Mantarry antenna will be fuselage-mounted typically as depicted in Figure 4.7 and will be covered by a radome (the same radome may also house a receive-only antenna for DBS satellite TV services; the DBS satellite receive-only antenna and service are not associated with, or part of, the IFC application under discussion.)

Figure 4.7   Typical antenna mounting location.

The terminal is directed toward the intended satellite by the ACU, which receives input data from the inertial reference unit (IRU) that is part of the avionics navigation system of the aircraft. This input includes information, such as the current latitude, longitude, altitude, pitch, roll, and yaw. The ACU uses this information to calculate the initial pointing angles and polarization for the antenna to the desired satellite (ViaSat-1, WildBlue-1, or Anik-F2). Once the required pointing angles have been determined, the ACU will drive the antenna to the desired position and the modem will attempt to acquire the receive signal from the satellite. When the signal is received and the modem is able to properly identify and demodulate the carrier, the antenna will enter a closed-loop tracking mode.

By performing closed-loop tracking, the ACU is able to properly account for any installation alignment differences between the IRU/airframe and antenna, as well as bending of the aircraft body on the ground or in flight. The antenna system also incorporates local rate gyros to mitigate latency between the IRU and the Mantarry ACU and further improve pointing accuracy.

The mean pointing error is 0° in both the azimuth and elevation directions, and the standard deviation (σ) for each axis is given in Table 4.6 along with the peak pointing error (3σ or 99.73%). The pointing error values are different in the azimuth and elevation directions because the arrays are wider than they are tall. The M40 has a 5:1 width-to-height aspect ratio, and accordingly the elevation beamwidth is wider than the azimuth beamwidth by the same factor. Likewise, the target standard deviation for pointing accuracy follows the same ratio.

TABLE 4.6   Pointing Error

	1σ Azimuth	Elevation	3σ Azimuth	Elevation	Limit Azimuth	Elevation
M40	±0.09°	±0.45°	±0.27°	±1.35°	±0.5°	±1.35°

The ACU monitors the current and target pointing directions, and if the error limit in either the azimuth or elevation axis is exceeded, the transmit output from the modem is inhibited in less than 100 ms (20 ms typical). The pointing error threshold is programmable for each axis, and ViaSat proposes to inhibit transmissions should the pointing error exceed 0.5° in the azimuth direction, or 1.35° in the elevation direction. As noted earlier, because the 3σ pointing error is only ±0.27° in azimuth, the system should not inhibit due to azimuth pointing errors. Elevation pointing error should only cause the antenna to inhibit transmit less than 0.27% of the time (Table 4.6).

4.2.5.4 Antenna Patterns The antenna patterns generated by the M40 antenna differ from those typically encountered when considering circular or mildly elliptical reflector type antennas. The patterns are characterized by a narrow main beam and a line of sidelobes in the azimuth axis, a wide main beam, and line of sidelobes in the elevation axis, and relatively low-amplitude sidelobes elsewhere. Figure 4.8 depicts an $X-Y$ view of the azimuth and elevation patterns when looking directly into the boresight of the antenna. The figure illustrates the lobes that exceed the Section 25.138 limit.

Notably, there are four grating lobes in the transmit antenna patterns that are well removed from the main lobe. These grating lobes are only present for a limited range of skew angles centered around approximately 25° of skew. The location and amplitude of the grating lobes are a function of transmit frequency and typically are between 25° and 35° off-axis from the main lobe. While the amplitude of these grating lobes when operating at the highest clear sky EIRP is as much as 22 dB above the 25.138 off-axis EIRP density mask, the location of these lobes with respect to the GSO arc is such that the lobes do not intersect the GSO arc except when the aircraft

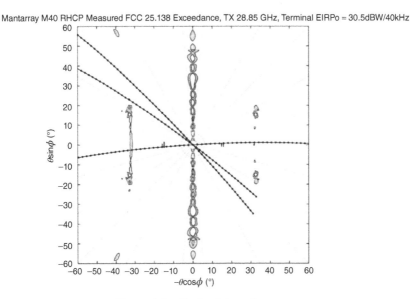

Figure 4.8   Grating lobes of antenna.

is located in a limited number of geographic areas. ViaSat has analyzed the potential impact of the spacecraft at the affected locations and found the actual level of interference to be minimal – less than 2% delta T/T at the lowest symbol rate of 625 kBd and only 0.2% at the 5 MBd symbol rate.

Figure 4.8 depicts the grating lobes as viewed looking into the boresight of the antenna. The three black lines represent the GSO arc from the perspective of the terminal at three different geographic locations: Carlsbad, CA; Melbourne, FL; and Germantown, MD.

The potentially affected satellite is SES AMC-16 at 85° WL and is 26° and 30° away for WildBlue-1/Anik-F2 and ViaSat-1, respectively. Even though the likelihood that the geographic alignment will occur is small, and the worst-case delta T/T is less than 2%, ViaSat has coordinated the operation of this antenna with the satellite operator.

Because width of the main lobe of the antenna increases between the azimuth and elevation axes as the antenna is rotated around the boresight, the alignment of the major axis of the antenna with the GSO must be considered. As the geographic location of the aircraft moves away in longitude from the longitude of the satellite (115.1° WL for ViaSat-1 and 111.1° WL for WildBlue-1 and Anik-F2), the GSO appears skewed with respect to the local horizon of the AES antenna. This skew angle is also affected by the banking of the aircraft while in flight. ViaSat has evaluated the worst-case skew angle within the operational service area of the AES antenna and determined it to be less than 50°. The M40 antenna is fully compliant in the main lobe with the 25.138 mask up to a skew angle of 60°. Accordingly, the M40 ACU monitors the skew and bank angle, and will inhibit transmissions if the combination of bank angle and geographic skew is equal to or greater than 60°.

In Figure 4.8, it can also be seen that in the elevation axis there is a narrow line of sidelobes that extends for a few degrees to either side of the elevation axis. The EIRP density of these sidelobes exceeds the Section 25.138 limit for elevation angles. While the sidelobes do not intersect with the GSO, they do however extend into the region where non-GSO satellites may operate. The only currently identified non-GSO satellite system in the Ka-band is the O3b network. ViaSat performed extensive simulations to determine the potential for impact to the O3b network and, following discussions with O3b, has coordinated the operation of the M40 antenna with O3b.

The FCC public domain application material included in this section for one operator/equipment developer is intended to clearly illustrate pedagogically to the reader the issues involved with aeronautical applications of HTS systems.

4.2.6 Beamforming and Ground-Based Beam Forming (GBBF) Systems

As hinted in Chapter 3, beamforming and ground-based beam forming (GBBF) systems may be used in next-generation (4G) HTSs. GBBF of various levels of sophistication are used in MSS. Hence, we provide here a short description of (some aspects) of these technologies. These technologies allow altering the coverage and the beam shape, enabling operators to dynamically allocate the capacity of beams – however, there are cost implications in deploying these technologies.

Numerous[6] narrow beams may be formed from a relatively few elementary feeds by a process known as beamforming (described, e.g., in US Patent numbers 5,115,248 and 5,784,030). Adaptive beamforming permits electrical reconfiguration of the direction of each spot beam, or the formation of beams with different sizes and shapes, each accomplished without the need to change any hardware element. A beamforming capability provides important benefits. For example, it permits a given satellite to operate from a number of different orbital locations. Thus, a satellite fleet operator licensed to operate GSO spacecraft at multiple orbital locations may use a common hardware design for all locations and electrically configure the beam as required to tailor the spot beam pattern based on the satellite's location. Moreover, beamforming allows a satellite, which typically has a 15-year life span, to be adapted on orbit to changing traffic patterns or new applications on the ground [USP201001]. Beamforming, however, is technically challenging to perform on a satellite, inasmuch as the amplitude and phase relationship of each feed element within an array must be precisely set and provide for both the forward (gateway to satellite to user) signal path and the return (user to satellite to gateway) signal path. Conventional space-based beamforming techniques include analog and digital beamforming networks (BFNs). Analog BFN's are generally colocated with the feed array, because it is otherwise difficult to compensate for losses or electrical path length variations between the feed apertures and the points of application of the beamforming coefficients. Volume and thermal constraints limit the number of analog BFNs that can be colocated with the feed array. Digital BFNs have a better ability to compensate for losses or electrical path length variations between the feed apertures and the points of application of the beamforming coefficients. Accordingly, they can be employed in the middle of the payload at a considerable electrical path distance from the feed array, provided that strict attention is paid to design practices minimizing amplitude and phase variations and calibration processes that accurately track the variations. The burdens associated with space-borne BFNs can be substantial and include system reliability degradation and added hardware mass, cost, power consumption, and thermal control requirements. Moreover, if the BFN is on the satellite, the ability to introduce improved technologies and react flexibly to changing market demand is limited during the life of the satellite. Moving BFN functions to the ground is, therefore, desirable, but ground-based beamforming systems must overcome several additional problems not inherent in space-based beamforming. Among these is the need to compensate for gateway and satellite component performance changes over temperature and life, satellite and ground station pointing errors, and signal propagation amplitude and phase dispersion effects, including Doppler shifts. These difficulties have limited the use of ground-based beamforming techniques. Some designs apply beamforming in only the return direction, or are limited to systems in which the feeder link signals are code division or time division multiplexed (however, frequency division multiplexing is more commonly used in space and offers significant cost and reliability advantages over code division and time division multiplexing) [USP201001]. Ideally one

---

[6]Major portions of this section are based on information found in US Patent US 7787819 B2 [USP201001].

would want to provide ground-based beamforming for both the forward and return communications path, usable for systems employing frequency division multiplexed signals.

Thus, a GBBF system is a large signal processor that is designed to coordinate and process up to several hundred beams at once (up to 500). The GBBF system creates hundreds of small, flexible, adaptive "spot" beams on the earth that allow mobile handsets to communicate directly with the satellite using smaller antennas, and achieving higher speeds than possible before. Although the beams are projected to the earth by the satellite, the ground system performs the beam-shaping signal processing. GBBF technology provides the following benefits to satellite systems [COM201401]:

- Faster and lower cost satellite deployment because the processing is on the ground, rather than part of the satellite bus;
- Ability to coordinate frequency use and remove interference for mass numbers of subscribers;
- Refocusing of satellite capacity to the areas of greatest need.

GBBF is most advantageous for missions that require spatial reutilization of communication spectrum (bandwidth) over the satellite field of view (FOV), as exemplified by, but not limited to MSS providing communications services to personal, often handheld, terminals. Figure 4.9 (from the cited patent) illustrates a simplified diagram of an exemplary MSS system to which a GBBF system is advantageously

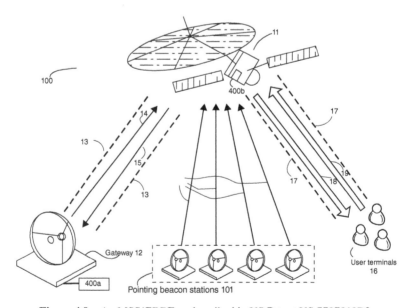

Figure 4.9 An MSS/GBBF as described in US Patent US 7787819B2.

applied. The MSS system includes a satellite, typically though not necessarily located at a GSO location defined by a longitude. The satellite is communicatively coupled to at least one gateway and to a population of user terminals. The user terminals comprise satellite terminals that may be handheld mobile smartphones or car phones, or may be embedded in laptop, tablets, or desktop personal computers, Internet coffee kiosks, or phone booths. At least one gateway is coupled to the public-switched telephone network and/or Internet. Each gateway and the satellite communicate over a feeder link, which has both a forward uplink and a return downlink. Each user terminal and the satellite communicate over a user link that has both a forward downlink and a return uplink. Pointing beacon stations are optionally employed to provide precise pointing feedback information to the GBBF system as described next.

The GBBF system is a distributed control system having substantial elements (see items labeled as **400***a* in figure 4.9) on the ground, preferably colocated with one gateway. These ground-based elements **400***a* communicate with pointing beacon stations, the colocated gateway, and, via the corresponding feeder link, the satellite. Certain space-based elements (see items labeled as **400***b* in figure 4.9) of GBBF are necessarily deployed on satellite.

The communications payload system has a satellite forward path connecting the forward uplink to the calibration network by way of a receiver and a transmitter. The satellite forward path also typically includes frequency converters, multiplexers, demultiplexers, amplifiers, filters, and other components. The communications payload system also has a satellite return path connecting the calibration network to return downlink by way of a receiver and a transmitter. The satellite forward path and the satellite return path are communicatively coupled through the calibration network to a forward downlink and to a return uplink, respectively, by way of a feed array that has multiple feed elements.

Four elements of GBBF are integrated into satellite communications payload system: a calibration network, a payload beacon, a payload pilot, and a tracking master reference oscillator (MRO). The calibration network includes low loss couplers that (i) permit user communications traffic to pass transparently in the forward and return directions and (ii) simultaneously generate signals having the same amplitude and phase characteristics as the user traffic signals at each feed element. These signals are passed to satellite return path for transmission back to at least one gateway. The payload beacon provides an encoded signal of known phase and amplitude to the calibration network for use in providing forward path signal amplitude and phase error correction. The payload pilot is a signal generator for use in providing Doppler frequency shift correction and forward uplink power control. The tracking MRO is a tracking MRO for use in providing Doppler frequency shift correction.

Some space-based elements **400***b* of GBBF system are necessarily deployed on the satellite, others are necessarily disposed on the ground, and still others are preferably placed on the ground but may be deployed on the satellite. The ground-based elements **400***a* of GBBF system are preferably colocated with any one gateway. Some elements of the GBBF system are represented as computation or signal generating modules, which may be implemented in any combination of hardware, software, and firmware. When implemented in software, the modules may be implemented in computer readable medium.

The GBBF system works cooperatively with certain standard conventional elements of the satellite communications network, for example, with a satellite return path and a satellite forward path of the communications payload system. The space-borne elements **400***b* unique to GBBF system are the calibration network, the payload beacon, the payload pilot, and the tracking MRO.

The GBBF system constitutes a beamforming network that controls the overall shape of beam pattern while in addition computing and applying beamforming coefficients that compensate for certain errors. Specifically, the GBBF system measures and corrects signal amplitude and phase errors associated with the satellite return path, the return downlink, the forward uplink, and the satellite forward path. Furthermore, the GBBF system controls the power of the forward uplink, minimizes errors associated with Doppler frequency shifts, and corrects for satellite pointing errors. The signals carried over each feeder link are frequency domain multiplexed. The fundamental functions supported are as follows:

- Return Path Signal Amplitude and Phase Error Correction;
- Forward Path Signal Amplitude and Phase Error Correction;
- Forward Uplink Power Control;
- Doppler Frequency Shift Error Minimization, and,
- Satellite Pointing Error Correction.

4.3   TECHNOLOGY PLAYERS AND APPROACHES

This section provides a press time view of some of the industry players. This section is not intended to be exhaustive, but illustrative. Table 4.7 depicts at a high level some of the key technology category and technology providers at press time. The total satellite services market for mobility (in all categories) at $6B in 2014 and about $10B in 2022 (from a variety of industry sources including but not limited to [NSR201201]; also see Figure 4.10).

4.3.1   Satellite Infrastructure Providers

Major satellite operators, including but not limited to Inmarsat, Intelsat, and ViaSat, are in the process of bringing HTS capacity online over the most heavily traveled flight routes.

4.3.1.1   Inmarsat   Inmarsat[7] has, since its creation in 1979, been providing maritime, terrestrial, and aeronautical connectivity solutions over its fleet of geostationary satellites. The Aeronautical Services introduced on the second- and third-generation

---

[7]This material is based on company information.

TABLE 4.7   Key IFC Technology Providers (Partial List)

Category	Player	Involvement
Satellite infrastructure providers (Satellite operators)	Inmarsat	GlobalXpress deployment
	Eutelsat	Launch of air access on KA-SAT
	Intelsat	Epic HTS; contract with Panasonic
	Iridium	Iridium Next
	SES	Provides satellite bandwidth to some operators
Service providers	Connexion by Boeing	Exited the market
	AeroMobile	Commercial provider
	Gogo	Commercial provider
	OnAir	Commercial provider
	Row 44	Commercial provider
	ViaSat	Commercial provider
IFE vendors	Thales	
	LiveTV	
	Panasonic Avionics	Rebranded of Matsushita products
	Rockwell Collins	

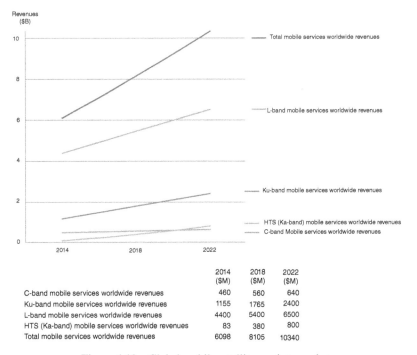

	2014 ($M)	2018 ($M)	2022 ($M)
C-band mobile services worldwide revenues	460	560	640
Ku-band mobile services worldwide revenues	1155	1765	2400
L-band mobile services worldwide revenues	4400	5400	6500
HTS (Ka-band) mobile services worldwide revenues	83	380	800
Total mobile services worldwide revenues	6098	8105	10340

Figure 4.10   Global mobile satellite services market.

satellites (Aero H, H+, I, L, Mini M Aero, and Aero C) are referred to as "Classic Services." The first aeronautical service, Aero H, was introduced in 1991 and has become the *de facto* standard of cockpit data communication for many of the world's airlines and private aircraft, giving Inmarsat a reputation of reliability and longevity. With the addition of Swift 64, in 2002, the first high-speed data service was added to the portfolio and has since proven to be a popular solution for business and military users who need ISDN (Integrated Services Digital Network) data connectivity. With SwiftBroadband, Inmarsat introduces a high-speed IP-based data service allowing even higher data throughput at a lower cost than the traditional circuit-switched services. SwiftBroadband is a Universal Mobile Telecommunications System (UMTS)-based service that is being provided over the fourth generation of Inmarsat satellites. The main difference between SwiftBroadband and earlier circuit-switched services such as Swift 64 is that it provides an "always on" background service and enables higher bandwidth than its predecessors. SwiftBroadband is being provided over a constellation of three Inmarsat 4 (I-4) satellites. The first two were launched in 2005 and the third was launched in 2008. The I-4 satellites are designed to provide higher bandwidth and to use the L-band spectrum more efficiently than previous generations of satellites. This is achieved by using very narrow spot beams that allow efficient reuse of the available spectrum and by dynamically allocating resources to the areas where they are most needed, that is, the network operates in an analogous manner to a terrestrial Global System for Mobile Communications (GSM) network. Each of the I-4 satellites has 16 times the traffic-bearing capacity compared to their predecessors. SwiftBroadband is an IP-based packet-switched service that provides an always-on data connection of up to 432 kbps throughput per channel on a contended basis. This is the same model that terrestrial broadband services use, and for the end-user this means that the performance experienced will vary depending on the number of users transferring data at any given point in time. Inmarsat allows up to four channels to be used per aircraft. If an application needs guaranteed bandwidth it can request a streaming class, which will provide a predetermined quality of service to the application. If the application support it, the system is also able to provide dynamic streaming, allowing a dynamic change between streaming classes without change of Packet Data Protocol (PDP) context. The streaming classes available are 8, 16, 32, 64, 128 kbps and SwiftBroadband X-stream, which offer full channel streaming. The predetermined streaming classes can be combined with a total maximum of 224 kbps per channel. This theoretical maximum is dependent on a number of factors such as the class of terminal being used and the angle of elevation to the satellite.

The FleetBroadband service on the Inmarsat-4s will be complemented by Inmarsat Global Xpress™ (GX); the aggregate satellite network combines both Ka-band and L-band technologies to provide full global coverage. In 2010, Inmarsat awarded Boeing a contract to build a constellation of three Inmarsat-5 Ka-band satellites under the GX rubric. The Inmarsat-5 satellites are being built by Boeing and are based on the firm's 702HP spacecraft platform. These satellites are part of a US$1.2 billion global satellite broadband network. Each Inmarsat-5 satellite antenna will provide 89 smaller Ka-band beams covering the portion of the Earth visible from its orbital location. The three Inmarsat-5 satellites needed for global coverage are being launched on

Proton rockets by International Launch Services from the Baikonur Cosmodrome in Kazakhstan. The Inmarsat-5 F1 was launched in December 2013; this spacecraft has now been placed in its final GSO and has successfully completed testing of the platform and communications payload. The second and third spacecraft were scheduled for launch in 2014, providing full global coverage and service by the end of 2014. A fourth I-5 is on order from Boeing for delivery in late 2016. It will provide an early available spare in the unlikely event of a launch failure of any of the first three, but in the event the satellite is not required as a launch spare, Inmarsat is developing an incremental business case to support the launch of the fourth satellite to increase capacity and strategically enhance network coverage.

Some of the spacecraft features include the following:

- The I-5 body – at 6.98 m (22.9ft), the height of a double-decker bus
- User beams – 89 Ka-band beams generated by two transmit and two receive apertures
- Spot beams – six steerable spot beams to direct additional capacity where it is needed
- Solar arrays – a wingspan of 33.8 m (111ft)
- Solar panels – five panels of ultra triple-junction gallium arsenide solar cells generate 15 kW of power at start of service and 13.8 kW by end of life
- Station-keeping thrusters – a xenon ion propulsion system (XIPS) handles in-orbit maneuvering
- Launch mass −6,100 kg
- Mission life span – 15 years

The spacecraft operate with a combination of fixed narrow spot beams that enable them to deliver higher speeds through more compact terminals plus steerable beams, so additional capacity can be directed in real time to where it is needed. GX aviation terminals will be provided exclusively by Honeywell, and Inmarsat is working with aircraft manufacturers to have both line-fit and retrofit solutions available.

The services offered by GX include in-flight broadband connectivity. The Ka-band service will be supported by the integration of an L-band FleetBroadband (FBB) backup service; with this approach, reliability will not be compromised in case of heavy rain, causing signal attenuation in the Ka-band, especially for maritime applications (aeronautical IFC services are less subject to rain-related issues, since flight below 10,000 ft is usually restricted to a few minutes after takeoff and a few minutes before landing). The goal of GX is to replace maritime, land, aviation, and government C- and Ku-band VSAT services offered by the company across the globe. The global network of Ka-band satellites will provide data rates up to 50 Mbps (20 cm antennas will support data rates of up to 10 Mbps; 60 cm antennas will support data rates up to 50 Mbps.) GX uses smaller terminals – as noted – at a lower operational cost compared to the Ku-band offering. The uplink data rate can be as high as 5 Mbps. Passengers can connect to the GX service with smartphones, tablets and laptops, or any other Wi-Fi-enabled device.

GX is also targeted to maritime communications, with the availability of higher bandwidth on a consistent, end-to-end global basis. It enables advanced applications such as, but not limited to, video distribution, telemedicine, and video surveillance. For extra resilience, Inmarsat GX will be complemented by the FleetBroadband service on the Inmarsat-4s, which deliver 99.9% overall network availability. As planes fly across the time zones, passengers will have a continuous, consistent service as traffic is handed seamlessly across each spot beam, and from one satellite to another. In 2014, Inmarsat announced that it would launch a Pan-European in-flight broadband service; British Airways was in talks to Become Launch Customer [BAL201401] (Figure 4.11).

In a somewhat related matter in the Spring of 2014, Inmarsat proposed to International Civil Aviation Organization (ICAO) a free global airline tracking service over the Inmarsat network, as part of the anticipated adoption of further aviation safety service measures by the world's airlines following the loss of flight MH370. This service is being offered to all 11,000 commercial passenger aircraft, which are already equipped with an Inmarsat satellite connection (virtually 100% of the world's long-haul commercial fleet). In addition to this free global airline tracking service, Inmarsat also offered both an enhanced position reporting facility to support reduced in-flight aircraft separation and a "black box in the cloud" service, under which – on the back of certain defined trigger events (such as an unapproved course deviation) – historic and real-time flight data recorder and cockpit voice recorder information can be streamed off an aircraft to defined aviation safety recipients. Inmarsat has been providing global aviation safety services for over 20 years and the ICAO, and International Air Transport Association (IATA) proposals represent a major contribution to enhancing aviation safety services on a global basis. Because of the universal nature of existing Inmarsat aviation services, the proposal from Inmarsat can be implemented right away on all ocean-going commercial aircraft using equipment that is already installed.

4.3.1.2 Intelsat Intelsat's Global Mobility Network[8] aims at providing reliable, high-performance aeronautical connectivity. The Ku-band mobility network delivers persistent, high-speed IP access for converged voice, data, and Internet applications – all on one contiguous global platform. Specially designed mobility beams, positioned across multiple Intelsat satellites, provide coverage across the world's major flight routes. Automatic Beam Switching (ABS) technology ensures a seamless, always-on broadband connection as the aircraft travels between coverage areas. Reliable and secure, Intelsat's mobility solutions are the clear choice for broadband and content providers operating in the aeronautical space.

Intelsat collaborated with Panasonic for the design of its first two HTS EPIC satellites (IS-29e and IS-33e) optimizing them for mobility applications. In order to meet specific traffic profiles (e.g., as shown in Table 4.1), Intelsat planned to provide spot beam coverage for high-traffic areas (Northern Atlantic and Pacific), along with wide beams for low-traffic areas. This hybrid HTS/wide-beam architecture improves

---

[8]This material is based on company information.

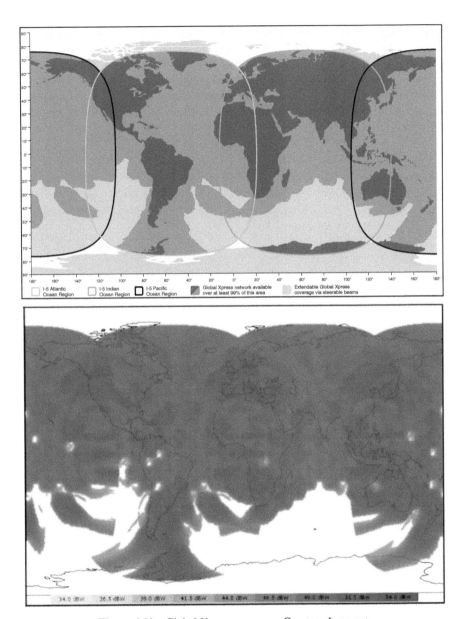

Figure 4.11   Global Xpress coverage. Courtesy Inmarsat.

end-to-end performance by reducing the number and effects of hand-offs between beams.

IntelsatOne Managed Mobility services offer access to mobile broadband through shared hardware and networking platforms and the global IntelsatOne terrestrial network. The service uses an infrastructure of 50+ satellites and 36,000 miles of terrestrial fiber. IntelsatOne Mobility allows customers to accelerate their broadband access through a hybrid service that combines space, teleport, shared networking platforms, and terrestrial data delivery. Features of the service include the following:

- Reliable, always-on broadband access for voice, data, and video services;
- ABS allows for seamless transfer from one satellite to another while maintaining connectivity with continuous provisioning of the service;
- Pre-engineered solution with data rates from 128 to 4,096 kbps in C-band, and up to 50 Mbps in Ku-band;
- Global broadband solution from a single iDirect-based IntelsatOne platform; and
- Defined IP rate for downlink and uplink channels for each customer.

Intelsat's mobility services include the Mobility Transponder Services Space segment-only services based on Intelsat's global fleet of satellites with coverage optimized for mobility users on the land, at sea, and in the air.

4.3.1.3 ViaSat   ViaSat as a core HTS infrastructure provider with its ViaSat-1 and −2 spacecraft was discussed in Chapter 3 and earlier in this chapter, where the reader can refer back to, at this juncture. Service features are discussed in Section 4.3.3.3.

4.3.1.4 Iridium   We discuss iridium in Chapter 5.

4.3.2 Vertical Service Providers to Airlines

4.3.2.1 Gogo   Gogo a key provider of IFC, particularly ATG services. In 1997, Aircel started installing in-flight phone systems; in 2011 the company was renamed Gogo Inc. Using Gogo's network and services, passengers with laptops and other Wi-Fi-enabled devices can get online on all domestic AirTran Airways and Virgin America flights and on select Air Canada, Alaska Airlines, American Airlines, Delta Air Lines, Frontier Airlines, United Airlines, and US Airways flights – as well as on thousands of business aircraft – bringing the total to more than 6,000 Gogo-equipped aircraft to date.

Gogo has three operating segments: commercial aviation, business aviation, and international aviation. In the United States, Gogo uses a terrestrial-based ATG network relying on its 3 MHz license in the ATG band. On the commercial side, Gogo has partnerships with North American airlines including Delta Air Lines, American Airlines, and US Airways. As of press time Gogo made broadband Internet available on more than 1,600 commercial aircraft; it also provided services to about 1,200 business jets. Gogo has filed an application to operate 1000 Ku-band transmit–receive earth stations onboard aircraft in order to expand its in-flight broadband service offerings from the contiguous United States and portions of Alaska served by the current ATG network to provide continuous in-flight connectivity on transoceanic and other international flights. Gogo received Blanket Authority for Operation of 1000 Technically Identical Ku-Band Transmit/Receive Earth Stations in the Aeronautical MSS, File No. SES-LIC-20120619-00574, Exhibit at 2 (filed June 19, 2012) [FCC201301].

Gogo's Ku-band-delivered in-flight service relies on three SES spacecraft. The SES-1 satellite enables airline passengers flying over the United States to access Gogo's high-speed Internet service, while the SES-6 satellite with its mobility beams provides seamless coverage over the North Atlantic, and the SES-4 spacecraft serves Europe. The FAA recently certified Gogo to install Ku-band equipment on Airbus A330, Bombadier Challenger 300, and Boeing 747–400.

Gogo is expanding its connectivity business and is also providing MSS solutions to the business aviation market [WLI201301], [RUI201301]. In addition, Gogo is the largest reseller of Iridium services to the business aviation market. Reportedly Gogo also has plans to provide Ka-band services as a distribution partner of Inmarsat's GX service.

The minimum requirements to access Gogo are a laptop with 802.11 a/b/g wireless capability (Wi-Fi) and a Gogo-supported web browser. Supported laptop browsers include Google Chrome; Microsoft® Internet Explorer® 6, 7, and 8 (Windows® XP or Vista™); Mozilla® Firefox®; and Safari™ on Mac OS® X. Gogo Inflight Internet also works with any Wi-Fi-enabled smartphone or tablet device that runs Android®, Apple®, Windows®, and Blackberry® platforms. Supported browsers include Apple Safari, Google Chrome, Microsoft Internet Explorer Mobile (version 8 or above), and Mozilla Firefox.

4.3.2.2   ViaSat   Earlier we made use of public filing information related to ViaSat[9] to discuss important technical aspects of aeronautical services. Here, we provide some illustrative market information, with respect to ViaSat providing an integrated service to airlines.

In recent years, ViaSat has deployed an AMSS network using Ku-band FSS spectrum and has provided technology to support ESV applications using Ku-band FSS spectrum; as noted, Ku-band mobile network currently has a coverage area that spans the globe to support satellite communications on private jets and government aircraft, as well as ships and other ocean-going vessels. JetBlue, the current licensee of the 1 MHz of ATG spectrum through LiveTV, asserts that it has partnered with ViaSat Inc. for the provision of in-flight broadband services via satellite after determining that the 1 MHz of air–ground spectrum was insufficient to provide a quality broadband service. ViaSat also has contracted to install its air-to-ground broadband service onto aircraft operated by United Airlines, with 370 total aircraft scheduled to be online by the end of 2015 [FCC201301].

---

[9]This material is based on company information.

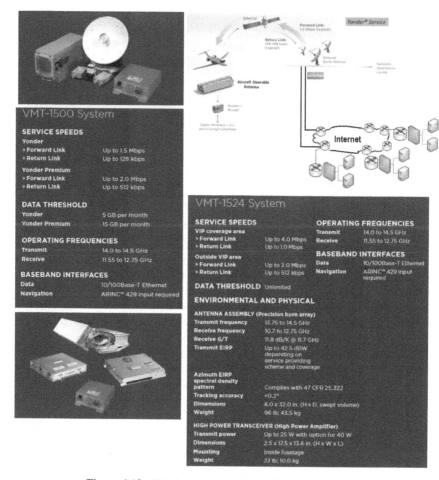

Figure 4.12 Yonder service elements. Courtesy of ViaSat.

ViaSat, thus, offers two paths to the IFC market, as follows:

- *Exede® In the Air* for Commercial Airlines. Exede Internet service by ViaSat is a highly differentiated service capable of delivering 12 Mbps to the connected passenger giving them the best in-flight experience available. Commercial airlines can deliver a high-quality in-flight Wi-Fi experience at an affordable price. Connected passengers can work, surf, shop, and play up to 8x faster than competitive services giving airlines an advantage.

- Yonder® In-Flight Internet for General Aviation. In addition to HTS commercial airline services, ViaSat offers Yonder Service, which is an in-flight Internet connectivity service for business and VIP aircraft. Yonder offers a variety of service plans; it offer global high-speed coverage over continents and bodies of water (see Figure 4.12 for some design specifics and Figure 4.13 for the

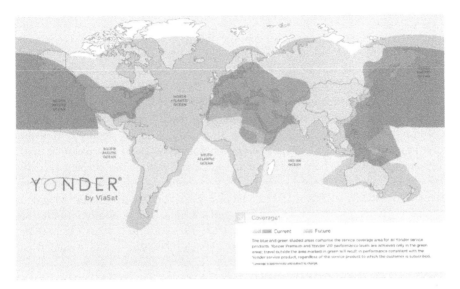

Figure 4.13 Yonder service coverage map. Courtesy of ViaSat.

TABLE 4.8 Yonder Service Packs

	Yonder	Yonder Premium	Yonder VIP
Suitable aircraft	Large cabin business	Large cabin business	Transport category
Supported applications	Web, VPN, Email, VoIP	Add streaming media	Add video, conference calls for more users
Maximum speed	1.5 Mbps	2 Mbps	4 Mbps
Data threshold	5 GB/month	15 GB/month	Unlimited

coverage map). Yonder and Yonder Premium use the VMT-1500 system, a very light, small-footprint Ku-band terminal. The VMT-1524 is designed for narrow and wide-body aircraft to support higher data rates and a larger number of users on Yonder VIP. In 2010, ViaSat announced a partnership agreement with JetBlue to deploy the first Ka-band commercial aviation broadband network on its more than 170 aircraft using ViaSat-1 in North America. In 2011, ViaSat announced that the YahSat subsidiary, Star Satellite Communications, will use YahSat 1B Ka-band capacity for the Yonder service. Also in 2011, ViaSat announced that it had entered into an agreement with Asia Broadcast Satellite to use Ka-band capacity in the Middle East using the ABS-7 satellite with 600 MHz of Ka-band capacity for fixed and MSSs as a part of ViaSat's expansion of its Yonder mobile satellite network [CAS201201] (see Table 4.8 for illustrative service plans).

Returning to the HTS-based IFC service, ViaSat has also developed a Ka-band terminal for aircraft that is designed for the high-capacity ViaSat-1 network. Just as ViaSat-1 and the associated Ka-band ground system technology fundamentally altered the economics of consumer-based satellite broadband services, the Ka terminal technology is expected to fundamentally alter the economics of aeronautical broadband by enabling service over ViaSat-1 at a lower cost per bit than previously possible. Because it leverages ViaSat's existing Ka-band network, the mobile application of this technology can be deployed immediately [VIA201201]. The antenna is a mechanically steered waveguide horn array antenna. As has been discussed in the earlier part of the chapter, recently ViaSat made an application to the US FCC for blanket authority to operate up to 4,000 technically identical transmit/receive earth stations to provide service in the United States using the 28.35–29.1 and 29.5–30.0 GHz portions of the Ka band for uplink communications and the 18.3–19.3 and 19.7–20.2 GHz portions for downlink communications. The terminals are targeted to be mounted on commercial and private aircraft and will be used to provide two-way, in-flight broadband communications, including Internet access, for passengers and flight crew. The terminals will communicate with Ka-band satellites ViaSat-1 at 115.1° W.L., WildBlue-1 at 111.1° W.L., and ANIK-F2 at 111.1° W.L. The terminals will communicate with (i) ViaSat-1 in the 18.3–19.3, 19.7–20.2, 28.35–29.1, and 29.5–30.0 GHz portions of the Ka-band; (ii) WildBlue-1 in the 19.7–20.2 and 29.5–30.0 GHz portions of the Ka-band; and (iii) ANIK-F2 in the 19.7–20.2 and 29.5–30.0 GHz portions of the Ka-band. Each of these satellites is already being used to provide service in the United States in these frequency bands.

4.3.2.3 Live TV LiveTV, LLC is a wholly-owned subsidiary of JetBlue Airways Corporation that provides IFE, voice communication, and connectivity services for commercial and general aviation aircraft. LiveTV was expected to be the first service provider actually using HTS Ka-band capacity (from ViaSat over the United States and from Eutelsat over Europe) for in-flight connectivity services. The first commercial flight broadcasting LiveTV took place in 2000; hence, the company has years of experience. LiveTV provides both IFE products, such as seatback solutions, wireless IFE, and global and regional live television solutions, in addition to connectivity services in L-, Ku-, and Ka-band. The new Ka-band service will have JetBlue and United as first customers in the United States and Aer Lingus as the first announced airline in Europe. The service will be operated throughout the ViaSat Exede network over North America and Eutelsat KA-Sat in Europe [RUI201301].

4.3.2.4 OnAir OnAir is a Switzerland-based company (in partnership between Airbus and SITA). OnAir provides IFC services in L-band through Inmarsat's Swift-Broadband services discussed earlier. The provider has reportedly signed agreements with Inmarsat to provide in-flight connectivity services to commercial and business aviation markets using GX. OnAir first launched GSM services and later on Internet access services. OnAir serves commercial airlines, the VIP/business jet market, and the maritime market. It partnered with technology specialists such as Thales and has roughly 350 agreements with mobile network operators in 80 countries in addition to agreements with regulatory bodies [RUI201301].

4.3.2.5 *Row 44*   Row 44 Inc. is a US-based company that provides connectivity services over Ku-band satellite; it also provides live television and video on demand. In early 2013, Row 44, which provides in-flight broadband services for Southwest Airlines and several international carriers, announced that it had installed its service on its 400th aircraft [FCC201301]. Satellite capacity is supported by Hughes' network. The company expects to have global coverage by 2015.

4.3.2.6 *AeroMobile*   AeroMobile is a UK-based company that entered the market in 2007 specializing in in-flight GSM communications and voice and data services on mobile phone and tablets. It operates over Inmarsat's L-band services and also in Ku-band (where more emphasis is expected in the future). Panasonic acquired a majority stake in 2012 in AeroMobile.

4.3.2.7 *Panasonic*   Panasonic Avionics, a key IFE manufacturer, also provides connectivity solutions. Panasonic makes use of Intelsat's Ku-band HTS; it signed a long-term agreement with Intelsat for over 1 Gbps of capacity on IS-29e and IS-33e (at this time the firm is not pursuing Ka-band solutions). Panasonic Avionics provides in-flight broadband services to United Airlines and several international airlines. In 2013, United Airlines announced it had outfitted a Boeing 747 with Panasonic Ku-band satellite technology, and that it planned to install satellite-based Wi-Fi on numerous aircraft models in its fleet [FCC201301]. The Panasonic in-flight Internet system makes use of a phased array antenna.

4.3.2.8 *AT&T*   AT&T (in addition to the DirecTV acquisition plan) is also planning to launch IFC services in combination with Honeywell by 2015. AT&T reportedly "sees a significant market opportunity" for IFC services [HOL201401], [ATT201401]. Some observers believe that AT&T's service could impact the satellite opportunity for a longer term, depending on how AT&T deploys its network, for example, how much spectrum it will deploy. For the US continental market, Gogo is currently spectrum-constrained with just 4 MHz of spectrum (only 3 MHz in use); hence, Gogo is in the process of deploying satellite-based connectivity to supplement and/or replace the ATG service over time. However, a wireless carrier could create a more competitive service if it is willing to allocate larger spectrum bands to the ATG service and make the investments necessary to support the business, including network operations, billing services, and customer care. This is reportedly AT&T's plan. The company aims to launch a nationwide ATG service in late 2015 using a number of alliances with the industry, including with Honeywell to provide hardware and service capabilities for a high-speed 4G LTE-based IFC. AT&T operates one of the major 4G LTE networks in the United States, and the company believes it

Figure 4.14   ViaSat aeronautical terminal. Courtesy of ViaSat.

has the capability "to transform airborne connectivity" by leveraging its expertise, spectrum, and financial strength.

4.3.2.9 *Technology Providers* Platforms: Platform vendors that supply the intrinsic on-the-ground technology include but are not limited to ViaSat, HNS, Gilat, NewTec, and Idirect.

End-systems: A number of vendors supply aeronautical terminals. Notable in that list is ViaSat Inc. The terminal is an integrated modem–antenna–radome set and is able to maintain a continuous link to aircraft in flight by performing seamless switching from beam-to-beam across spot beams from multiple Ka-band satellites. The in-flight Internet system is certified by the FAA on A320 and Boeing 737 aircraft and is licensed for mobile operation by the FCC (Figure 4.14).

REFERENCES

[ATT201401] CNBC online magazine http://www.cnbc.com/id/101621395, April 28, 2014. Article stated that "AT&T announced plans to launch a high-speed 4G LTE-based, in-flight connectivity service stating that "we are building on AT&T's significant strengths to develop in-flight connectivity technology unlike any other that exists today, based on 4G LTE standards."

[AZZ201001] T. Azzarelli, L-band Spectrum Process, ESA Presentation, 27 May 2010.

[BAL201401] E. Ballard, R. Wall, Wall Street Journal, June 5, 2014.

[BOW201301] S. Bowles, "Movie ticket prices jump to all-time high", USA TODAY, July 23, 2013.

[CAS201201] High Throughput Satellite Networks, White Paper, June 2012. CASBAA Executive Office 802 Wilson House 19–27 Wyndham Street Central, Hong Kong.

[COM201401] Comsat Laboratories, a division of ViaSat, Informational Material, 20511 Seneca Meadows Parkway, Suite 200, Germantown, MD 20876.

[FCC201301] Application of AC BidCo, LLC, Gogo Inc., and LiveTV, LLC, WT Docket No. 12–155, Federal Communications Commission, For Consent To Assign Commercial Aviation Air-Ground Radiotelephone (800 MHz band) License, Call Sign WQFX729, Memorandum Opinion And Order; Adopted March 29, 2013, Released March 29, 2013.

[HOL201401] M. Holmes, AT&T Exec Talks to Via Satellite About In-Flight Connectivity Plans, Via Satellite/Satellitetoday.com, May 23, 2014.

[INM201401] Inmarsat marketing materials on Global Xpress. http://www.inmarsat.com.

[INT201401] Intelsat marketing materials on EPIC. http://www.intelsat.com

[ITU201101] ITU-R, Report ITU-R S.2223: Technical and operational requirements for GSO FSS earth stations on mobile platforms in bands from 17.3 to 30.0 GHz (Question ITU-R 70-1/4) (2011)

[NSR201201] Mobile Satellite Services, 8th Edition, NSR, 2012.

[RUI201301] N. De Ruiter, B. Prokosh, *High Throughput Satellites: The Quest For Market Fit -- A vertical market analysis of major drivers, strategic issues and demand take up*, An Euroconsult Executive Report. November 2013. 86 Blvd. Sebastopol, 75003 Paris, France. www.euroconsult-ec.com.

[USP201001] US Patent US 7787819 B2, "Ground-based beamforming for satellite communications systems", Publication number US7787819 B2, Publication date Aug 31, 2010. John L. Walker, Rolando Menendez, Douglas Burr, Gilles Dubellay inventors.

[VIA201201] ViaSat, Inc. EXHIBIT A to FCC Filing for Ka band Aeronautical Antenna, Public Interest Statement and Waiver Requests. 2012, U.S. Federal Communications Commission.

[WLI201301] W. Li, R. Roithner, C. Fargier, Prospects for In-Flight Entertainment And Connectivity – Sector dynamics, analysis and forecasts addressing the IFEC market for commercial airlines and business aviation. An Euroconsult Executive Report. April 2013. 86 Blvd. Sebastopol, 75003 Paris, France. www.euroconsult-ec.com.

5

MARITIME AND OTHER MOBILITY SERVICES

This chapter[1] assesses emerging maritime and terrestrial mobility applications and services that can be provided with satellite-based systems.

5.1   APPROACHES TO MARITIME COMMUNICATION

There is a growing need for vessels to be connected to the institution's intranet, the Internet, or industry extranets in order to improve operational productivity through the transfer of real-time information, as well as regulatory communications requirements, and also support crew connectivity to family and friends. For example, ship personnel typically need to download the latest chart updates, engine monitoring and weather-routing applications, all from some centralized source, and distribute this information in real time to all the vessels in the fleet simultaneously. According to industry estimates there were approximately 112,500 ships in at the beginning of the decade with satellite services (68,000 merchant ships, 23,000 fishing vessels, 6,500 passenger ships, 6,000 large yachts, 9,000 government vessels), and also there were 8,500 oil and gas platforms.

---

[1] Services, service options, and service providers change over time (new ones are added and existing ones may drop out as time goes by); as such, any service, service option, or service provider referred to in this chapter is mentioned strictly as illustrative examples of possible examples of emerging technologies, trends, or approaches. As such, the mention is intended to provide pedagogical value. No recommendations are implicitly or explicitly implied by the mention of any vendor or any product (or lack thereof).

Innovations in Satellite Communications and Satellite Technology: The Industry Implications of DVB-S2X, High Throughput Satellites, Ultra HD, M2M, and IP, First Edition. Daniel Minoli.
© 2015 John Wiley & Sons, Inc. Published 2015 by John Wiley & Sons, Inc.

   Maritime broadband satellite services have seen growth over the past 10 years with the market being served by lower bandwidth L-band solutions (Inmarsat, Iridium) and regular C-and Ku-band Very Small Aperture Terminal (VSAT) services. While the coverage of these services is now practically global, these systems have limitations in terms of available bandwidth per vessel, particularly in high-traffic shipping routes in the Northern Atlantic as well as in the Pacific Ocean. As connectivity requirements in the maritime industry are increasing and as vessels are implementing a larger number of higher bandwidth applications such as video conferencing, real-time monitoring, route planning, cloud-computing applications, and crew connectivity, satellite operators have responded by including high-traffic maritime routes in the coverage areas of their upcoming High Throughput Satellite (HTS) systems. A growing inventory of HTS capacity along the most highly travelled maritime routes from operators such as Inmarsat (GX), Telenor (Thor 7), Intelsat (EPIC) and O3b was becoming available at press time In particular, operators that already provide regular capacity to maritime markets (Intelsat, Telesat, Telenor) are planning to overlay new HTS beams to enhance capacity within their "regular" satellite footprints [RUI201301].

   Table 5.1 depicts the most active maritime routes and Table 5.2 identifies some key operators and services; notice that some of these services were in the process of being rolled-out at press time as several HTS systems with oceans coverage were being launched (Inmarsat GX, O3b, Intelsat EPIC, among others). The expectation is that the lower cost of HTS capacity in conjunction with more extensive geographic coverage as new HTS spacecraft is launched and will accommodate the increasing demand for (affordable) service.

   Traditionally the maritime industry has made significant use of machine-to-machine (M2M) services on MSS bands, as described in Chapter 6. M2M services operate at lower speeds and support basic functions such as geo-location information, machine surveillance, surveillance of various conditions (e.g., temperature and

TABLE 5.1   Routes with Major Concentration of Shipping Traffic

Market ID	From	To (and Vice Versa)	Maritime Traffic
1	North America	North America	H
2	North America	South America	L
3	Pacific Ocean entry/exit to Panama	Atlantic Ocean entry/exit to Panama	H
4	Caribbean	Caribbean	H
	North America	Europe	VH
5	South America	Europe	M
6	Europe	Africa	L
7	Europe/Mediterranean	Indian Ocean (via Suez)	H
8	East Asia	Far East Asia	M
9	Far East Asia (north–south)	Far East Asia (north–south)	H
11	Far East Asia	Australia & Islands	M
12	Far East Asia	North America	H

Low = low; M = medium; H = high; VH = very high

TABLE 5.2   Key Maritime Satellite Services Operators (as of press time)

Provider	HTS Infrastructure	Band	Geography Focus	Application Focus
Inmarsat	Global Xpress	Ka	Worldwide	Merchant ships, cruises, offshore, oil/gas platforms/exploration
Intelsat	Epic	Ku, Ka	Northern hemisphere	Merchant ships, cruises, offshore, oil/gas platforms/exploration
O3b	8 spacecraft constellation	Ka	Worldwide (but within 45 L north and 45 L south)	Cruise ships
Telenor	Thor 7	Ka	Europe area	Merchant ships, cruises, offshore, oil/gas platforms/exploration
Telesat	Vantage	Ku	Caribbean and Europe area	Merchant ships, cruises, offshore, oil/gas platforms/exploration

speed), Supervisory Control and Data Acquisition (SCADA), and ship identification services used for Automatic Identification System (AIS) or Long-Range Identification and tracking (LRIT). LRIT was established as an international system in May 2006 by the International Maritime Organization (IMO). AIS is a collision-avoidance navigational aid system also mandated by the IMO that originally operated in the VHF radio band (see Chapter 1); while it was originally designed for short-range operation, it is now possible to receive such data by satellite in many (but not all) parts of the world.

Around 375,000 satellite communication terminals were in operation in the maritime market in 2014. The market is comprised of narrowband (less than 128 kbps) terminals, M2M terminals, and broadband terminals. At press time, about 71% of the maritime terminals were used for M2M and for safety communications (e.g., Inmarsat-C), 13% were narrowband terminals (e.g., Inmarsat-B, FleetPhone, Thuraya Seagull), and 16% were broadband terminals (e.g., FleetBroadband, Iridium OpenPort, Iridium Pilot).

With a projected (combined) CAGR of around 7% over the next few years, the number of maritime terminals is estimated to reach 650,000 by 2022-MSS terminals are still expected to account for the large majority (95%) of terminal deployment over the next few years, and the population should grow from around 375,000 terminals in 2014 to around 619,000 terminals by 2022 (with VSATs growing from 12,500 to 31,000). Over the same period the CAGR of broadband VSAT terminals is expected to be 13%, for M2M terminals 8%, and for narrowband terminals −7%. The VSAT deployment (which is a subset of the broadband pool) will grow at a more significant rate, especially with the deployment of Ka HTS: the VSAT terminals should grow from around 9,000 terminals in 2011 to around 30,100 terminals in 2022, with a CAGR over the 10-year period of 12%. Figure 5.1 shows terminal and revenue projections synthetized from industry sources, including [WLI201201]; Table 5.3 also depicts other forecast data. It should be noted that most of the deep sea-ships have multiple communication solutions for different shipboard functions. We saw in earlier chapters that Ka terminal are less expensive (as an initial first cost) compared with traditional VSAT equipment, typically costing between $30,000 and $70,000 per vessel. Traditionally service providers have had to purchase the terminal equipment and then lease it to end-users because of the high cost. The reduced Ka costs, due to the smaller size of the equipment, may fuel more wide-scale uptake of connectivity to vessels of all sizes in the next 5 years, to the end of the decade. According to market research forecasts for Euroconsult, the number of Ka HTS maritime VSAT terminals is expected to grow from a nascent market today, to about 13,000 terminals by 2022 (up from about 1,000 in 2015). Net additions of HTS terminals are expected to ramp up in 2016, averaging nearly 1,750 units/year until 2022, this being about 2.5 times the net VSAT terminal additions for regular C-band and Ku-band combined. Total maritime service revenues are projected to be around $1.7 billion in 2022.

The Atlantic Ocean represents about 50% of the total maritime market (in terms of terminals and traffic), owning to the concentration of shipping routes in the North Atlantic, the cruise ship business in the Caribbean, and Oil & Gas (O&G) installations in the Gulf of Mexico. The second largest region is the Pacific Ocean owning to the import business from China. Maritime VSAT service providers (including but not limited to MTN, KVH, Astrium Services/Airbus Group, and Harris Caprock)

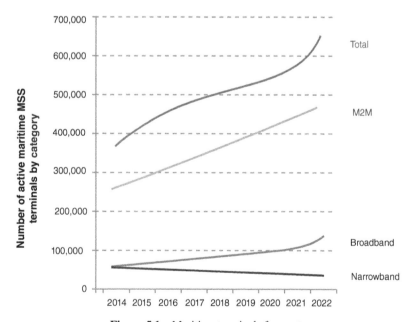

Figure 5.1   Maritime terminals forecast.

TABLE 5.3 Maritime Terminal and Revenue Forecasts

	2014	Details	2022	Details	Notes
MSS terminals Non-VSAT	362,500	See Note 1	619,000	See Note 2	*Note 1:* 260,000 M2M terminals 52,500 broadband 50,000 narrowband
					Note 2: 480,000 M2M terminals 130,000 broadband 40,000 narrowband
VSAT terminals	12,500		31,000		
Ka		500		13,000	
Ku		9,000		15,000	
C		3,000		3,000	
Total	375,000		650,000		
(a) VSAT revenue	$320M		$700M		
Ka		$ 10M		$100M	
Ku		$260M		$525M	
C		$ 50M		$ 75M	
(b) MSS Revenue	$430M		$500M		
(c) Wholesale bandwidth revenue	$450M	See Note 3	$525M	See Note 4	*Note 3:* $200M broadband $ 70M M2M $180M narrowband
					Note 4: $360M broadband $ 95M M2M $ 70M narrowband
Total (a) + (b)	$750M	See Note 5	$1,200M	See Note 6	*Note 5:* 65% merchant ships Each 7% for government, O&G, leisure boats, Passenger ships, fishing ships
					Note 6: 50% merchant ships Each 10% for government, O&G, leisure boats, passenger ships, fishing ships
Total (a) + (b) + (c)	$1,200M		$1,750M		

are expected to integrate HTS capacity into their current service plans, allowing them to offer higher throughput at current price levels or offer the current bandwidth packages at a reduced cost per bit in the future. Inmarsat, with an existing maritime business in L-band now reportedly plans to be offering worldwide coverage through a tessellation of high-power Ka-band spot beams, addressing global shipping companies with dispersed fleets and routes. Traditional providers may encourage their client bases to switch from legacy Ku-band satellites to their HTS Ka-band systems by offering appealing deals on the new services [RUI201301]. O3b also has a maritime non-GSO offering that focuses on the high-end of the market (large cruise ships, e.g., Royal Caribbean and off-shore O&G), which has large bandwidth requirements and an economic incentive to deploy the relatively large on-board equipment; the system architecture of the O3b constellation and ground terminal cost, however, makes it less suitable for smaller vessels.

5.2   KEY PLAYERS

The M2M focus of the maritime market is discussed in greater details in chapter 6. Here we provide a more general overview.

5.2.1   Inmarsat

Inmarsat offers high-availability VSAT services through a combination of both GX (Ka-band) and FleetBroadband. Inmarsat has served the market for over 30 years. Their network is global, thus facilitating the delivery of higher performance consistently – with a downlink up to 50 Mbps and an uplink to 5 Mbps. The all-IP network enables point-and-click capability with a consistent user experience worldwide: Inmarsat GX will enable seamless, real-time operations between vessels and land offices, in addition to enhanced communications and entertainment for the crew. Inmarsat GX is much simpler and quicker to install than existing Ku-band systems, minimizing downtime in the dock for vessels. There is the option to use a 60-cm terminal that does not require a crane for installation. GX terminals will be lower cost than those currently available for Ku-band, and services will be available for a fixed monthly fee.

5.2.2   ViaSat/KVH

ViaSat, in partnership with KVH Industries, provides reliable, high-speed Internet access to seagoing vessels over the vast majority of the busiest maritime corridors. Applications include cruise, commercial business, and offshore O&G exploration. Operating on the ViaSat Yonder® network, the mini-VSAT Broadband service offers cable modem-like speeds so customers can connect to a virtual private network, send and receive email with attachments, make VoIP phone calls, browse the Internet, and transfer files, just as they would from a land-based business office. The following KVH mini-VSAT Broadband service options are available:

- *TracPhone V7* – The first FCC-approved 24-inch VSAT antenna. With over 1000 V7 systems shipped in the last three years, it is the fastest growing maritime VSAT broadband system.

- *TracPhone V3* – The world's smallest and most affordable maritime VSAT System. The 14-inch antenna weighs just 25 pounds and is perfect for use on tuna towers.

- *TracPhone V11-IP*: Dual-mode C-/Ku-band antenna for global VSAT connectivity and a complete IT & C solution from a 1.1 meter (42.5-inch) antenna.

- *TracPhone V7-IP*: Enterprise-grade solution for broadband connectivity and onboard network management, from a 60 cm (24-inch) antenna with Ku-band service.

- *TracPhone V3-IP*: The world's smallest maritime VSAT, with fast data speeds and low airtime costs, from a 37 cm (14.5-inch) antenna with Ku-band service.

- *IP-MobileCast*: A fast, affordable content delivery service that utilizes multicasting technology to deliver entertainment and operations content to vessels at sea over the mini-VSAT Broadband network.

5.2.3 Intelsat

Intelsat offers the Global Mobility Network to provide broadband solutions for maintaining communications for vessels, crews and passengers across the commercial shipping, oil and gas, and cruise and yacht industries. Utilizing a hybrid satellite and fiber infrastructure, the firm is able to provide constant, high-speed IP access for converged voice, data, and Internet applications – all on one convenient global platform. Intelsat's support of a range of shipboard antenna sizes and modem solutions provides true flexibility in designing a mobility solution. Proven global C-band coverage, combined with global Ku-band mobility beams over the world's major shipping routes, ensures that maritime operators stay connected anywhere they are. Figure 5.2 depicts the global coverage of Intelsat's mobility services.

5.2.4 O3b

O3b, as a non-GSO operator, is targeting the maritime market (among other markets). O3b satellite beams will follow ship tracks on ship's normal route (the cruise operator provides O3b with normal ship course). The service will provide 350 Mbps Committed Information Rate (CIR) to each ship. The service entails beam tracking updates in real-time if the ship has to change course. O3b will maintain ship within beam center, but the ship needs to provide latitude/longitude updates on 2-h intervals via in-band or out-of-band channel (see Figure 5.3).

O3b states that the following advantages accrue from the use of their service [BUR201301]:

- Improve customer satisfaction;
- Attract new and more affluent passengers who need better connectivity;

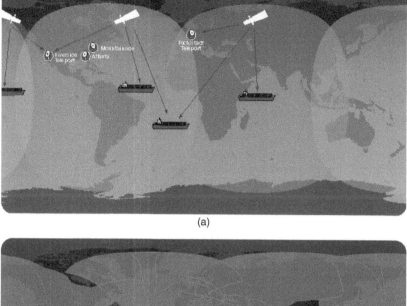

(a)

(b)

Figure 5.2   Global coverage of Intelsat's mobility services. (a) C-band. (b) Ku-band. Courtesy: Intelsat.

- Add new services to the passenger experience;
- Enable additional revenue streams from bandwidth services by selling daily and weekly Internet access packages to passengers;
- Better leverage social networking as an advertising channel;
- Attract and retain quality staff through improved crew welfare.

O3b also targets the scientific and military maritime segments. For example, O3b offers the ability to provision between 300–500 Mbps to a vessel or group of vessels

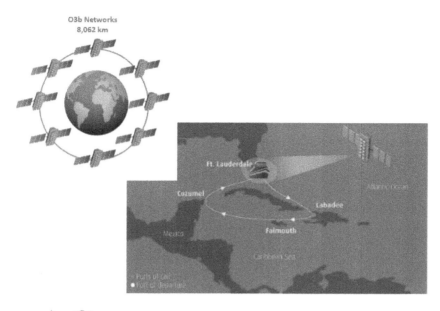

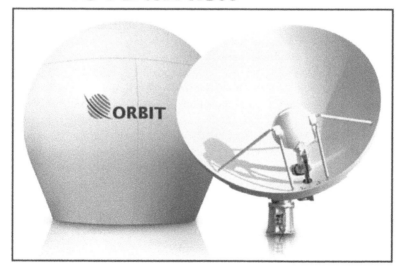

Figure 5.3 O3b maritime support. Courtesy: O3b and Orbit

as they maneuver. The options to either track vessels or to move beams based upon predetermined courses or route are available.

Key to the service is the terminal technology. O3b has partnered to offer a maritime terminal with the following features:

- OrBand™300 is the evolution of Orbit's OrBand portfolio into Ka-band;
- Compact maritime VSAT system that offers industry-standard RF performance equivalent to a 2.4 m/95″ dish in just a 2.7 m/106″ footprint;
- Takes up 40% less deck space than industry-standard 2.4 m/95″ and 3.8 m/150″ systems, and is more than 30% lighter than competitive solutions;
- Enables the most demanding maritime vessels and platforms to enjoy fiber-like broadband communication for high-speed internet services;
- Small enough to be shipped as a single, fully assembled and tested unit in a standard 20-foot container, the OrBand300 is designed for quick and simple single-day installation;
- OrBand300 can be installed while ships are on routine port calls, substantially driving down operational costs and eliminating the need for vessels to await dry dock.

5.3 COMMS-ON-THE-MOVE APPLICATIONS

Military and government missions typically require to be able to sustain communications while on the move, such communication is known as *Comms-On-The-Move*. These capabilities enable vehicles and other assets to achieve high-throughput, reliable, and secure communications. While somewhat-mundane communications can be achieved by the military (staff, vehicles) with 4G cellular services (where such services are available), even with VPN-based access to government networks, there are situations where satellite communication is required either because of a particular military-based application that entails satellite-oriented connectivity, or because 3G/4G cellular connectivity is not available (say in rural areas, battleground theaters, and so on). Applications also include nonterrestrial situations, for example, airborne missions for Intelligence, Surveillance, and Reconnaissance (ISR), Command & Control, and en-route connectivity for the delivery of data-rich content to warfighters in the theater and commanders around the world.

Advanced antenna design techniques have resulted in highly transportable antenna systems, including backpack terminals employed in DOD and military applications and highly mobile antennas used in *Comms-On-The-Move* applications. Phased array antennas have come into some use; for example, a new meta-materials flat panel Ka-band antenna has been tested and is nearing introduction into the marketplace by Kymeta, as mentioned in Chapter 3.

One illustrative example[2] of an approach to *Comms-On-The-Move* is ViaSat's systems. ViaSat's airborne satellite communications services offer worldwide access

---

[2]This material is based on company information.

and offer a range of service levels providing connectivity and performance options tailored to meet a variety of needs. Network performance is optimized for reliable, secure two-way mobile broadband, so that airborne operators have the capability to send full-motion video, make secure phone calls, conduct video conferences, access classified networks, and perform mission-critical communications while in flight. For airborne ISR missions, ViaSat private managed networks deliver critical intelligence such as High Definition (HD) full-motion video into theater, while military and political leaders stationed beyond the horizon can monitor a mission's progression throughout execution. ViaSat's private managed airborne networks are proven to support multiple simultaneous aircraft in-flight (termed simultaneous operations – SIMOPS). Dozens of compatible airborne satcom terminals, hubs and ground infrastructure – some located in DISA teleports – can access US Department of Defense and coalition data entry points around the world. Figure 5.4 depicts an example of the technology.

5.4 HTS/Ka-BAND TRANSPORTABLE SYSTEMS

Transportable systems have been available for years for military, news gathering, and sporting events. The "new angle" is to use an IP service over a Ka HTS satellite to stream content via an Internet-based VPN to the intranet of the content aggregator (e.g., a TV station). At Ka frequencies, antenna pointing to a specific supporting spacecraft can be somewhat difficult; hence, more sophisticated systems enable quick setup using built-in antenna pointing aid for rapid satellite acquisition.

According to some market research firms, the use of HTS Ka-band capacity for Satellite News Gathering (SNG) traffic represents a growth opportunity for operators; the assertion is that the multiplication of HTS systems in coming years will unlock growth opportunities in the SNG market, as HTS capacity becomes increasingly available, and coverage footprints are expanded globally. Usage of Ka-band will be driven by the cost of satellite distribution; such cost is expected to be up to 50% cheaper than Ku/C capacity. Ka-band is still new in the SNG market and offers opportunities to develop services: the share of HTS Ka-band in total SNG revenues is expected to grow from less than 0.5% in 2013 to a forecasted 18% in 2023. Avanti, Eutelsat, and ViaSat were the first operators to launch HTS Ka-band SNG services [BUC201401]. Advantages of Ka-band for SNG services include (i) the lower chance of adjacent satellite interference and (ii) the smaller risk of uplink mispointing to other spacecraft considering the plethora of existing Ku/C spacecraft now on station.

One illustrative example of Ka-band HTS-based transportable systems, especially for SNG applications is *Exede*® *Enterprise* service from ViaSat; the satellite-based service delivers live HD video from the field – no matter where the story happens. Exede helps news organizations get the story whether it is in a metropolitan area or in a natural disaster situation with no electricity, cell connection, or even vehicle access. With the "internet-on-the-go" organizations can get affordable HD video benefits by the following:

ViaSat Global Network & Services
* Worldwide Satcom for Mobile Broadband Applications
  * Global, Regional, and Private Networks
  * Assured Access with Secure Connectivity
* Bandwidth for the Most Demanding Applications
  * ISR with Real-Time HD Video Streaming
  * Command, Control, & Communications On-the-Move
  * Enroute Mission Planning
  * Office-in-the-Sky Voice, Video, & Internet Access
  * Data Rates from 512 kbps to 10 Mbps
* Service Plans Shaped to Meet Mission Requirements
  * Standard, Premium and Priority Access Levels
  * ViaSat Elite Service Plans for Government and
    Enterprise Customers
  * Optional ViaSat Yonder® Service Including
    Premium and VIP Plans
  * Annual Plans for Predictable Budget Planning
* Worldwide Technical Support
  * Tiered Plans with 24/7/365 Response
  * TS/SCI FSRs Worldwide
  * Ku- and Ka-band Network Overlays

Figure 5.4   Illustrative example of Comm-on-the-Move Hardware. Courtesy: ViaSat.

- Augmenting your existing Electronics News Gathering (ENG) and SNG vehicles.
- Replacing them with more affordable Exede SNG vans, trucks, and crew cars.
- Using our portable terminal – it fits in an airline-checkable suitcase.

The ViaSat SurfBeam 2 Pro Portable Terminal is designed for users requiring high throughput connectivity in a compact and portable package. Field reporters, remote medical and peace workers, and emergency responders benefit from high-speed Internet with the convenience of "near-instant" connectivity even in locations where no

Figure 5.5   Illustrative example of transportable (SNG) systems.   Courtesy: ViaSat.

other communications infrastructure is available. The form factor is ruggedized to support operation in harsh conditions and supports multiple configuration options to suit user needs. The complete system, including the satellite terminal and antenna, can be packed into a single case for transit and is lightweight enough to be carried by a single user. The unit enables fast web browsing and supports video streaming, file transfers, VPN connections, and bandwidth-intensive Internet applications. It is capable of delivering downstream rates up to 40 Mbps and upstream rates up to 20 Mbps, and provisioning tools enable the network operator to create different classes of service with configurable downstream and upstream rates. The modem has an embedded acceleration client that works with acceleration servers in the gateway to provide a faster, more responsive user experience. With four standard Ethernet connections, the terminal natively supports up to four IP devices, such as PCs, cameras, and routers as well as other user equipment. The antenna includes a collapsible satellite reflector and feeds, transmits, and receives electronics. The modem can be deployed attached to or located separately from the antenna. The compact SurfBeam 2 Pro Portable Terminal is designed for a quick and reliable user installation in less than 10 min by personnel with minimal satellite training (see Figure 5.5).

REFERENCES

[BUC201401] D. Buchs, M. Welinski, M. Bouzegaoui, Video Content Management And Distribution, Key Figures, Concepts and Trends, Euroconsult Report, March 2014, Euroconsult, 86 Blvd. Sebastopol, 75003 Paris, France. www.euroconsult-ec.com.

[BUR201301] D. Burr, MEO Satellite Applications to Support Mobility Colloquium on Satellite Services for Global Mobility, 14 October, 2013.

[RUI201301] N. De Ruiter, B. Prokosh, High Throughput Satellites: The Quest For Market
          Fit – A Vertical Market Analysis of Major Drivers, Strategic Issues and Demand Take
          Up, An Euroconsult Executive Report. November 2013. 86 Blvd. Sebastopol, 75003 Paris,
          France. www.euroconsult-ec.com.
[WLI201201] W. Li, Maritime Telecom Solutions by Satellite—Global Market Analysis &
          Forecasts. An Euroconsult Executive Report. March 2012. 86 Blvd. Sebastopol, 75003
          Paris, France. www.euroconsult-ec.com.

6

M2M DEVELOPMENTS AND SATELLITE APPLICATIONS

In this chapter, we first provide a generic overview of the emerging *Internet of Things* (IoT) and a contemporary perspective on machine-to-machine (M2M) communications, then we focus on satellite support of these important requirements for those environments where satellites can provide an optimal solution. According to industry watchers,[1] the IoT has the potential to sustain an added global economic value of $6 trillion annually by 2025; satellite's share is currently seen in being "a few billion dollars," however, the opportunity for a larger role for satellite-based services exists. With M2M technology that enables automated data exchanges between devices, organizations can track, monitor, and manage global assets; just-in-time inventories; enterprise fleets; energy grids; pipelines; mining operations; remote infrastructure; emergency operations; personnel deployments; and natural processes such as weather, ecology, farming, forestry, and water flows. The intrinsic satellite-based M2M value proposition is that "global businesses need access to global data." Satellite operators, for example, Iridium, make the point that global pole-to-pole connectivity services can extend M2M to the 90% of the earth not serviced by terrestrial networks.

---

[1]E.g, see reference [MCK201301].

Innovations in Satellite Communications and Satellite Technology: The Industry Implications of DVB-S2X, High Throughput Satellites, Ultra HD, M2M, and IP, First Edition. Daniel Minoli.
© 2015 John Wiley & Sons, Inc. Published 2015 by John Wiley & Sons, Inc.

6.1 A GENERAL OVERVIEW OF THE INTERNET OF THINGS AND M2M

The[2, 3] proliferation of an ever-growing set of devices that are able to be directly connected to a data network or to the Internet is leading to a new ubiquitous-computing paradigm: the opportunity now exists to connect all "things" and objects that have (or will soon have) embedded wireless (or wireline) connectivity to control systems that support data collection, data analysis, decision-making, and (remote) actuation [ASH200901]. In the IoT, commonly deployed devices and objects contain an embedded device or microprocessor that can be accessed by some communications mechanism, typically utilizing wireless links. "Things" include, but are not limited to, machinery, home appliances, vehicles, individual persons, pets, cattle, animals, habitats, habitat occupants, as well as enterprises. Interactions are achieved utilizing (i) a plethora of possibly different networks; (ii) computerized devices of various functions, form factors, sizes, and capabilities – such as iPads, smartphones, monitoring nodes, sensors, and tags; and (iii) a gamut of host application servers. This new paradigm seeks to enhance the traditional Internet (or intranet) into an ecosystem created around intelligent interconnections of diverse objects in the physical world. The IoT aims at closing the gap between objects in the material world, the "things," and their logical representation in information systems. The "things" are also variously known as "objects," "devices," "end nodes," "remotes," or "remote sensors," to list just a few commonly used terms.

The IoT generally utilizes low-cost information gathering and dissemination devices – such as sensors and tags – that facilitate fast-paced interactions in any place and at any time, among the objects themselves, as well as among objects and people; actuators are also part of the IoT. Hence, the IoT can be described as a new-generation information network that enables seamless and continuous M2M[4] and/or human-to-machine (H2M) communication. One of the initial goals of the IoT is to enable connectivity for the various "things," allowing some form of interrogation of the "thing's" status or some local variable; a next goal is to be able to have the "thing" provide back appropriate, application-specific telemetry, perhaps based on Artificial Intelligence (AI) principles; an intermediary next step is to provide a web-based interface to the "thing" (especially when human access is needed); the final step is to permit actuation by the "thing" (i.e., to cause a function or functions to take place either under external control or with intrinsic

---

[2]Some portion of this material are summarized from reference [MIN201301].
[3]Services, service options, and service providers change over time (new ones are added and existing ones may drop out as time goes by); as such, any service, service option, or service provider referred to in this chapter is mentioned strictly as illustrative examples of possible examples of emerging technologies, trends, or approaches. As such, the mention is intended to provide pedagogical value. No recommendations are implicitly or explicitly implied by the mention of any vendor or any product (or lack thereof).
[4]Some (e.g., 3GPP) also use the term Machine-Type Communications (MTC) to describe M2M systems.

AI mechanisms). Certain "things" are stationary, such as an appliance in a home; other "things" may be in motion, such as a car, a container, a carton (or even an item within the carton) in a supply chain environment (either end-to-end or while in an intermediary warehouse). Thus, the two fundamental components of the IoT are (i) intelligence (microprocessing) to be embedded into the various "things," and (ii) a network or a collection of networks (the first hop of which is almost invariably wireless) to connect the "things" to the rest of the world. In general, "things" have the following characteristics, among others [LEE201101]:

- Have the ability to sense and/or actuate (but not always)
- Are generally small (but not always)
- Have limited computing capabilities (but not always)
- Are energy/power limited (but not always)
- Are related to the physical world
- Are mobile (but not always)
- Sometimes have intermittent connectivity
- Managed by devices, not people (but not always)
- Of interest to people

For the purpose of this specific satellite-oriented discussion, objects ("things") tend to be cargo or an airplane/ship engine systems/subsystems; however, other "things" are not excluded.

As a point of observation, the Institute of Electrical and Electronics Engineers (IEEE) Computer Society stated that

> ... *The Internet of Things (IoT) promises to be the most disruptive technology since the advent of the World Wide Web. Projections indicate that up to 100 billion uniquely identifiable objects will be connected to the Internet by 2020, but human understanding of the underlying technologies has not kept pace. This creates a fundamental challenge to researchers, with enormous technical, socioeconomic, political, and even spiritual, consequences. IoT is just one of the most significant emerging trends in technology* ...
> [IEEE201301].

The European Commission recently made these observations, which we can employ in our discussion of the IoT [INF200801]:

> ... Considering the functionality and identity as central it is reasonable to define the IoT as "*Things having identities and virtual personalities operating in smart spaces using intelligent interfaces to connect and communicate within social, environmental, and user contexts*". A different definition, that puts the focus on the seamless integration, could be formulated as "Interconnected objects having an active role in what might be called the Future Internet". The semantic origin of the expression is composed by two

words and concepts: "Internet" and "Thing," where "Internet" can be defined as *"The world-wide network of interconnected computer networks, based on a standard communication protocol, the Internet suite (TCP/IP),"* while "Thing" is *"an object not precisely identifiable"* Therefore, semantically, "Internet of Things" means "a world-wide network of interconnected objects uniquely addressable, based on standard communication protocols" …

Figure 6.1 depicts the high-level logical partitioning of the interaction space; this figure illustrates human-to-human (H2H) communication, M2M communication, H2M communications, and machine in (or on) humans (MiH) communications [MiH devices may include chips embedded in humans, medical monitoring probes, and Global Positioning System (GPS) bracelets]. Recently, the IoT has been seen as an emerging "paradigm of building smart communities" through networking of various devices enabled by M2M technologies (but not excluding H2M), for which standards are now emerging. As a general concept, the IoT effectively eliminates time and space isolation between geographical space and virtual space, forming what proponents label as "smart geographical space" and creating new human-to-environment (and/or H2M) relationships. The latter implies that the IoT can advance the goal

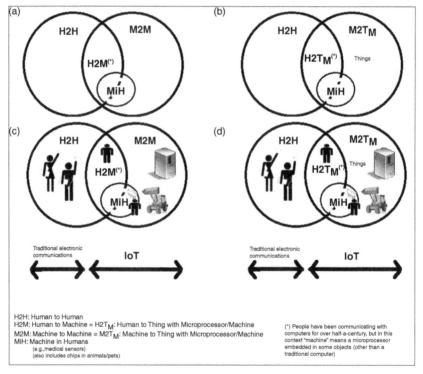

Figure 6.1 H2H, H2M, and M2M environment. (a) Interaction space partitioning showing humans and machines. (b) The target machine is shown explicitly to be embedded in the "thing." (c) Interaction space showing icons. (d) Embedded machine, icon view.

of integration of human beings with their surroundings. A smart environment can be defined as consisting of networks of federated sensors and actuators, and can be designed to encompass homes, offices, buildings, and civil infrastructure; from this granular foundation, large-scale end-to-end services supporting smart cities, smart transportation, and Smart Grids (SGs), among others, can be contemplated.

The IoT is a general, heterogeneous concept with somewhat open-ended definitions; M2M, on the other hand, is a tighter well-defined concept (a subset of the IoT), where standards have evolved and have been published by industry organizations such as European Telecommunications Standards Institute (ETSI). The focus of the IoT is on M2M, H2M, and MiH applications (also see Table 6.1). *M2M services* aim at automating decision and communication processes and support consistent, cost-effective interaction for ubiquitous applications (e.g., fleet management, smart metering, home automation, e-health). *M2M communications per se* is the communication between two or more entities that do not necessarily need direct human intervention: it is communication between remotely deployed devices with specific roles and requiring little or no human intervention. M2M communication modules are usually integrated directly into target devices, such as Automated Meter Readers (AMRs), vending machines, alarm systems, surveillance cameras, and automotive/aeronautical/maritime equipment, to list a few. These devices span an array of domains including (among others) industrial, trucking/transportation, financial, retail Point of Sales (POS), energy/utilities, smart appliances, and healthcare. The emerging standards allow both wireless and wired systems to communicate with other devices of similar capabilities; M2M devices, however, are typically connected to an application server via a mobile data communications networks.

At the "low end" of the spectrum, the thing's information is typically stored in a Radio Frequency Identification (RFID) electronic tag; and the information is uploaded by noncontact reading using an RFID reader. At the "mid range" of the spectrum, one finds devices with embedded intelligence (microprocessors) and embedded active wireless capabilities to perform a variety of data gathering and, possibly, control functions. Automatic on-body biomedical measurements, home appliance and power management, and industrial control are some examples of these applications. At the other end of the spectrum, more sophisticated sensors can also be employed in the IoT: some of these sensor approaches use distributed Wireless Sensor Networks (WSNs) systems that (i) can collect a wide variety of environmental data such as temperature, atmospheric and environmental chemical content, engine status, homeland security applications [e.g., geofencing, unmanned aerial vehicles (UAV)], or even low- or high-resolution ambient video images from geographically dispersed locations; (ii) can optionally preprocess some or all of the data; and (iii) can forward all this information to a centralized (or distributed/virtualized) site for advanced processing. These objects may span a city, region, or large distribution grid; or they may be mobile devices associated with airplane or ship engines or process control systems on oil well (on land or in the water.)

Eight specific vertical market segments recognized at press time included the following: utilities and SG; automotive and transportation; logistics; public safety; security and surveillance; retail and vending; healthcare; intelligent buildings and smart

TABLE 6.1 Taxonomy of "Things" in IoT

		H2M (Human to Machine)								
		Stationary access/connectivity			Local mobility access/connectivity			Full mobility access/connectivity		
Data integration point or person (DIPP) "thing"	H									
"Remote thing" also known as data end point (DEP)	M	Target device is stationary	Target device has local mobility	Target device has full mobility	Target device is stationary	Target device has local mobility	Target device has full mobility	Target device is stationary	Target device has local mobility	Target device has full mobility
Example		Access a home thermostat from an office PC	Access a monitor on a home-bound pet from an office PC	Access a GPS device on a teenager's car from an office PC	Access a home thermostat from a home, office, or hotspot wireless PC	Access a monitor on a home-bound pet from a home, office, or hotspot wireless PC	Access a GPS device on a teenager's car from a home, office, or hotspot wireless PC	Access a home thermostat from a smartphone	Access a monitor on a home-bound pet from a smartphone	Access a GPS device on a teenager's car from a smartphone

M2M (Machine to Machine)

DIPP "thing"	M1	Stationary access/connectivity			Local mobility access/connectivity			Full mobility access/connectivity		
Remote "thing"	M2	Target device is stationary	Target device has local mobility	Target device has full mobility	Target device is stationary	Target device has local mobility	Target device has full mobility	Target device is stationary	Target device has local mobility	Target device has full mobility
Example		Access a home electrical meter from an office/provider server	Access a monitor on a home-bound pet from an office/provider server	Access a GPS device on a person's car from an office/provider server Network Operation Center access to an airplane engine via satellite	Access a home electrical meter from a WLAN-based office/provider server	Access a monitor on a home-bound pet from a WLAN-based office/provider server	Access a GPS device on a person's car from a WLAN-based office/provider server	Access a home electrical meter from roaming-3G/4G based provider server	Access a monitor on a home-bound pet from a roaming-3G/4G based provider server	Access a GPS device on a person's car from a roaming-3G/4G based provider server Military in-theater access to an UAV from a vehicle via satellite

cities; and consumer electronics. More specifically, the following applications are of interest in the satellite context:

- Remote structures, such as water-deployed Oil & Gas (O&G) platforms, O&G pipelines, power transmission lines, sparse geofencing structures (e.g., sensors on national borders), and mining, forestry, and agricultural sensors. These structures include in-water oil wells throughout the world, land-based pipelines traversing long rural and sparsely inhabited regions, and dispersed power lines. Energy management functions fit under this category, as well as environmental protection activities, and homeland security endeavors. Often it is impossible to have cellular or other wireless coverage in many exurban or mountainous environments and satellite solutions are ideal.

- Land mobile platforms, including vehicles or structures at a fixed location or in motion. Vehicles can be automobiles, trucks, and military equipment. While cellular coverage may be available, vehicles traveling a long distance may find that a homogeneous solution that works end-to-end is preferable. Fleet management functions fit under this category.

- Maritime platforms, including vessels traveling on a body of water, especially in oceans. These vessels may include merchant ships, tankers, container ships, fishing vessels, cruise ships, ferries, pleasure yachts (>60 ft) and superyachts (>120 ft), Coast Guard ships, and buoys.

- Aeronautical platforms, including commercial and business airplanes, helicopters, UAVs, military planes (the military may have its own systems, but they also rely on commercially provided services).

- Civil infrastructure management (cities, traffic, resources) in Third World countries (and/or rural geographies) where modern connectivity systems may still be rudimentary.

Figures 6.2 and 6.3 (the last one from [PTC201401]) provide pictorial views of typical IoT applications; these figures only depict illustrative cases and are not exhaustive or normative. Note that personal communication devices (smartphones, pads, etc.) can be viewed as machines or just simply as end nodes; when personal communication devices are used for H2M devices where the human employs the smartphone to communicate with a machine (such as a thermostat or a home appliance), then we consider the personal communication devices part of the IoT (otherwise we do not). It has been estimated that in 2011 there were 7 billion people on earth and 60 billion machines worldwide. Market research firm *Frost & Sullivan* recently forecasted that total *cellular* M2M connections are expected to increase from around 24 million in 2010 to more than 75 million by 2017; worldwide, the expectation is that the number of M2M device connections will grow from around 60 million in 2010 to over 2 billion in 2020 [DUK201101]. Satellite-based M2M systems are currently in the range of 1–2 million by the number of deployed units. Other market research puts the worldwide M2M revenues at over $38 billion in 2012 [PEE201201]. Yet other market research companies project 15 billion connected

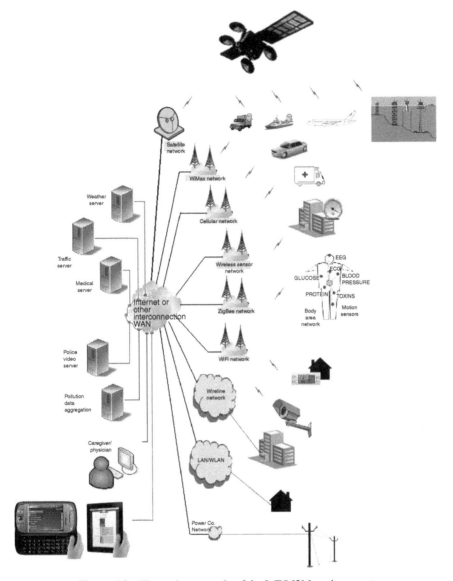

Figure 6.2   Illustrative example of the IoT/M2M environment.

devices moving 35 trillion gigabytes of data at a cost of $3 trillion annually by 2015 [KRE201201]. Other market research sees the wireless M2M market as accounting for nearly $196 billion in annual revenue by the end of 2020, following a Compound Annual Growth Rate (CAGR) of 21% during the 6-year period between 2014 and 2020; these market estimates also see the installed base of M2M connections (wireless and wireline) as growing at a CAGR of 25% between 2014 and 2020,

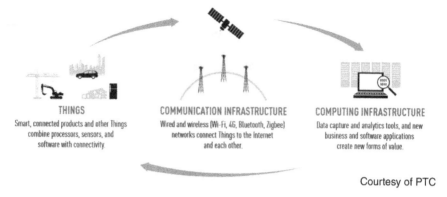

Courtesy of PTC

Figure 6.3   A simplified environment of the IoT.

eventually accounting for nearly 9 billion connections worldwide [SST201401]. These market data point to major development and deployment of the IoT technology in the next few years.

It should be noted that the characteristics of IoT/M2M communication are often different from other types of networks or applications. For example, cellular mobile networks are designed for human communication and communication is connection centric; this kind of communication entails interactive communication between humans (voice, video), or data communication involving humans (web browsing, file downloads, etc.); it follows that cellular mobile networks are optimized for traffic characteristics of human-based communication and applications. On the other hand, in M2M environment the expectation is that there are many devices, there will be long idle intervals, transmission typically entails small messages, there may be relaxed delay requirements, the data are coded in defined codepoints or are raw measurements, and the device's energy efficiency is, generally, paramount. Table 6.2 depicts some key properties and requirements of M2M applications.

A number of wireless technologies to support M2M are available, which are as follows:

- Personal Area Networks (PANs): Zigbee®, Bluetooth®, especially Bluetooth Low Energy (BLE), Near Field Communications (NFCs), and proprietary systems; specifically, there is interest in Low-power Wireless Personal Area Networks (LoWPANs); some of these PANs are also classified as Low-Rate Wireless Personal Area Networks (LR-WPANs);
- Wireless Local Area Networks (WLANs): Wi-Fi® IEEE Standard 802.11 (including vendor-specific implementations for low power[5]);
- Metropolitan Area Networks (MANs): WiMAX;

---

[5] In recent years, several improvements have been made to the Wi-Fi LAN standard; some of these improvements (including IEEE Standard 802.11v) are aimed at reducing its power consumption. Wi-Fi is optimized for traditional Office Automation (OA) of large data transfer where high throughput is needed; it is not generally intended for coin cell operation.

TABLE 6.2 Properties and Requirements of M2M Applications

	Land Mobile	Aeronautical	Maritime	eHealth	Surveillance	Smart Meters
Mobility	Vehicular medium/long distance	High-speed long distance	Low-speed long distance	Pedestrian/vehicular	None	None
Message size	Small - Medium	Small - Medium	Small - Medium	Medium	Large	Small (few kB)
Traffic pattern	Periodic/Irregular	Periodic/Irregular	Periodic/Irregular	Periodic/Irregular	Periodic	Periodic
Device density	Medium	Medium	Medium	Medium	Low	Very high
Latency requirements	Medium (s)	Medium (s)	Medium (s)	Medium (s)	Medium (<200 ms)	Low (up to hours)
Power efficiency requirements	Medium	Medium	Medium	High (battery power devices)	Low	High (battery powered meters)
Reliability	High	High	High	High	Medium	High
Security requirements	High	Very high	Very high	Very high	Medium	High

- WSN: application-specific technology, in general;
- 3G/4G cellular: Universal Mobile Telecommunications System (UMTS), General Packet Radio Service (GPRS), Enhanced Data rates for GSM Evolution (EDGE), Code Division Multiple Access (CDMA-2000/EV-DO), Wideband CDMA (W-CDMA/HSPA), and Long-Term Evolution (LTE);
- Satellite (GEO, MEO, LEO) networks for remote, widespread, hostile, or global environments; and
- Hybrid (e.g., satellite backhaul of cellular systems) for developing environments

While IoT/M2M connectivity might be achieved by wireline means, for example, PLC (Power Line Communication)-based grid management, some operators have used wireless technology for meter reading. Furthermore, although energy suppliers routinely utilize Supervisory Control and Data Acquisition (SCADA)-based systems to enable remote telemetry functions in the power grid, and although, traditionally, SCADA systems have used wireline networks to link remote power grid elements with a central operations center, at this time an increasing number of utilities are turning to public cellular networks to support these functions. Due to its global reach and the ability to support mobility in all geographic environments (including Antarctica), satellite communications can play a critical role in many broadly distributed M2M applications. While some background on other wireless technologies is provided at the end of the chapter, the core of this discussion is on satellite communication. In the satellite arena, M2M has generally been supported by Mobile Satellite Service (MSS) approaches. As we have seen in previous chapters, MSS deals with the delivery of services (data streams) to portable and mobile terminals with a variety of (small) antennas. These terminals are typically found on O&G, land mobile, maritime, and aeronautical platforms. Satellite M2M players include Iridium, Inmarsat, Orbcomm, and Globalstar. M2M is considered to be a value-added service and equipment market, rather than a simple capacity-only business, which some satellite operators seem to thrive on.

As noted, various technologies have indeed emerged in the past two decades that can be utilized for implementations including PANs, such as IEEE 802.15.4; WLANs; WSNs; 3G/4G cellular networks; satellite networks; metro-Ethernet networks; MultiProtocol Label Switching (MPLS); and Virtual Private Network (VPN) systems. Wireless access and/or wireless ad hoc mesh systems reduce the "last mile" cost of IoT applications for distributed monitoring and control applications. Satellite communication also can play a significant role for the five classes of application listed earlier (remote structures, land mobile, maritime, aeronautical, civil infrastructure in the Third World), especially for unconnected/undeveloped/underdeveloped environments. On a different vein, we believe that the fundamental technical advancement that will foster the deployment of the IoT is Internet Protocol version 6 (IPv6). *In fact, IoT may well become the "killer-app" for IPv6.* IoT is deployable using IP version 4 (IPv4) as has been the case in the recent past, but only IPv6 provides the proper scalability and functionality to make it economical, ubiquitous, and pervasive. There are many advantages in using IP for IoT, but we have to ascertain that the infrastructure and the supporting technology scale to meet the challenges. This is why there is broad agreement that IPv6 is critical for the deployment of the IoT.

The reader should realize, however, that despite significant technological advances in many subtending disciplines, difficulties associated with the evaluation of IoT solutions under realistic conditions in real-world experimental deployments still hamper their maturation and significant rollout. Obviously, with limited standardization, there are capability mismatches between different devices; also there are mismatches between communication and processing bandwidth. Although IoT systems can utilize existing Internet protocols, in a number of cases the power-, processing-, and capabilities-constrained IoT environments can benefit from additional protocols that help optimize the communications and lower the computational requirements. The M2M environment has been a fragmented space, but recent standardization efforts are beginning to show results. Standards covering many of the underlying technologies are critical because proprietary solutions fragment the industry. Standards are particularly important when there is a requirement to physically or logically connect entities across an interface. Device, network, and application standards can enable global solutions for seamless operations at reduced costs. Specifically proponents make the case that IPv6 is the fundamental optimal network communication technology to deploy IoT in a robust, commercial manner rather than just a preliminary desktop "science experiment" in some academic researcher's lab (layer 2 wireless technologies are also critical to IoT's end-to-end connectivity).

IoT standardization spans several domains including physical interfaces, access connectivity (e.g., low-power IEEE802.15.4-based wireless standards such as IEC62591, 6LoWPAN, and ZigBee Smart Energy (SE) 2.0, DASH7, ETSI M2M), networking (such as IPv6), and applications. Some studies have shown that for the home two wireless physical layer communications technologies that best meet the overall performance and cost requirements are Wi-Fi (802.11/n) and ZigBee (802.15.4) [DRA201001]. Examples of standardization efforts targeted for these environments include the initiatives known as "Constrained RESTful Environments (CoRE)," "IPv6 over Low power WPAN (6LoWPAN)," and "Routing Over Low power and Lossy (ROLL) networks," which have (and are being) studied by appropriate Working Groups of the IETF [IAB201101].

6.2   M2M FRAMEWORKS

This section elaborates on the concept, definition, and a usable framework of the IoT. We noted above that the IoT is an evolving type of Internet *application* that endeavors to make a thing's information (whatever that may be) securely available on a global scale if/when such information is needed by an aggregation point or points. M2M systems have enjoyed some (useful) standardization in recent years. Some applicable observations related to M2M include the following:

> The M2M ... *term is used to refer to machine-to-machine communication, i.e., automated data exchange between machines. ("Machine" may also refer to virtual machines such as software applications.) Viewed from the perspective of its functions and potential*

uses, M2M is causing an entire "Internet of Things," or internet of intelligent objects, to emerge … On closer inspection, however, M2M has merely become a new buzzword for demanding applications involving telemetry (automatic remote transmission of any measured data) and SCADA (Supervisory, Control and Data Acquisition). In contrast to telemetry and SCADA-based projects, the majority of M2M applications are broadly based on established standards, particularly where communication protocols and transmission methods currently in use are concerned. Telemetry applications involve completely proprietary solutions that, in some cases, have even been developed with a specific customer or application in mind. M2M concepts, meanwhile, use open protocols such as TCP/IP [Transport Control Protocol/Internet Protocol], which are also found on Internet and local company networks. The data formats in each case are similar in appearance …

[WAL200701].

… Order(s) of magnitude bigger than the Internet, no computers or humans at end-point, inherently mobile, disconnected, unattended … IoT is going to be an advanced network including normal physical objects together with computers and other advanced electronic appliances. Instead of forming ad hoc network, normal objects will be a part of the whole network so that they can collaborate, understand real time environmental data and react accordingly in need … The basic idea is that IoT will connect objects around us (electronic, electrical, non electrical) to provide seamless communication and contextual services provided by them. Development of RFID (radio-frequency identification) tags, sensors, actuators, mobile phones make it possible to materialize IoT which interact and co-operate each other to make the service better and accessible anytime, from anywhere … The "Internet of Things (IoT)" refers to the networked interconnection of everyday objects. An "IoT" means "a world-wide network of interconnected objects uniquely addressable, based on standard communication protocols" … In the IoT, "things" are very various such as computers, sensors, people, actuators, refrigerators, TVs, vehicles, mobile phones, clothes, food, medicines, books, etc. These things are classified as three scopes: people, machine (for example, sensor, actuator, etc) and information (for example clothes, food, medicine, books and so on). These "things" should be identified at least by one unique way of identification for the capability of addressing and communicating with each other and verifying their identities … if the "thing" is identified, we call it the "object" …

[BOT200901], [LEE201101].

… M2M describes devices that are connected to the Internet, using a variety of fixed and wireless networks and communicate with each other and the wider world. They are active communication devices. The term embedded wireless has been coined, for a variety of applications where wireless cellular communication is used to connect any device that is not a phone. This term is widely used by the GSM Association (GSMA) …

[OEC201201].

The International Telecommunications Union, Telecommunications (ITU-T) is in the process of identifying a common way to define/describe the IoT. So far, the ITU-T has not found "*a good definition to cover all aspects of IoT as the IoT has quite big scope not only the technological viewpoints but also other views … We recognized whatever we define, everyone cannot be happy*" [ITU201101]. One can view the

Internet as an *infrastructure* providing a number of technological capabilities or as a *concept* providing an array of data exchange and linkage services. The infrastructure perspective describes the Internet as a global system of interconnected computer networks (of many conceivable technologies) that use the TCP/IP suite to communicate; the networks comprise millions of private, public, business, academic, and governmental servers, computers, and nodes. One can view the Internet as a *concept.* The concept perspective sees the Internet as a worldwide logical interconnection of computers and networks that support the exchange of information among users, including but not limited to interlinked hypertext documents of the World Wide Web (WWW).

While the IoT can, in principle, be seen as a more encompassing concept than what is captured under the ETSI M2M standards and definitions, nonetheless, the M2M definitions can serve the purpose of adding some structure to the discussion of the general IoT environment. From Figure 6.1 it has been noted that a high-level logical partitioning of the entity-to-entity interaction space included H2H communication, M2M communication, H2M communications, and machine in (or on) humans (MiH) communications (MiH devices may include medical monitoring probes, and GPS

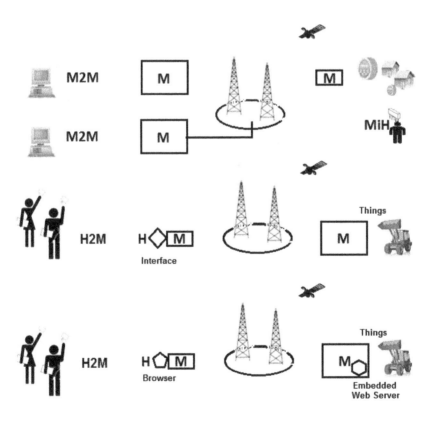

Figure 6.4 Classes of generic IoT arrangements.

bracelets). Figure 6.4 illustrates classes of generic IoT arrangements that are included in our discussion.

Intuitively, an M2M/H2M environment comprises three basic elements: (i) the data integration point (DIP); (ii) the communication network; and (iii) the data end point (DEP) (again, a Machine M) (see Figure 6.5, where the process (X) and application (Y) form the actual functional end points). Typically, a DEP refers to a microcomputer system, one end of which is connected to a process or to a higher level subsystem via special interfaces; the other end is connected to a communication network. However, the DEP can also be a Machine M in a Human H, as is the case in the MiH environment. Many applications have a large base of dispersed DEPs [WAL200701]. A DIP can be an Internet server, a software application running on a firm-resident the host, or an application implemented as a cloud service. Basic applications include, but are not limited to, smart meters, e-health, track-and-trace, monitoring, transaction, control, home automation, city automation, connected consumers, and automotive.

As noted earlier, at a macro level, an IoT comprises a remote set of sensing assets (Sensing Domain, also known as M2M Domain in a M2M environment), a Network Domain, and an Applications Domain. Figure 6.6 provides illustrative pictorial view of the domains.

A High-Level M2M System Architecture (HLSA) (Figure 6.7) is defined in the ETSI TS 102 690 V1.1.1 (2011-10) specification that is useful to the present discussion. We describe the HSLA next, summarized from [ETS201101]. The HSLA comprises the Device and Gateway Domain, the Network Domain, and the Applications Domain.

The *Device and Gateway Domain* is composed of the following elements:

1. *M2M Device.* A device that runs M2M Application(s) using M2M Service Capabilities. M2M Devices connect to Network Domain in the following processes:
   – *Case 1 "Direct Connectivity"*. M2M Devices connect to the Network Domain via the Access network. The M2M Device performs the procedures

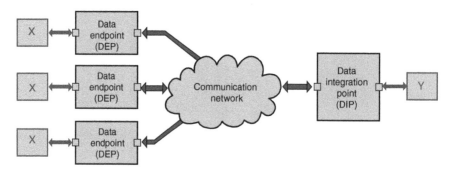

Figure 6.5   Basic elements of an M2M application.

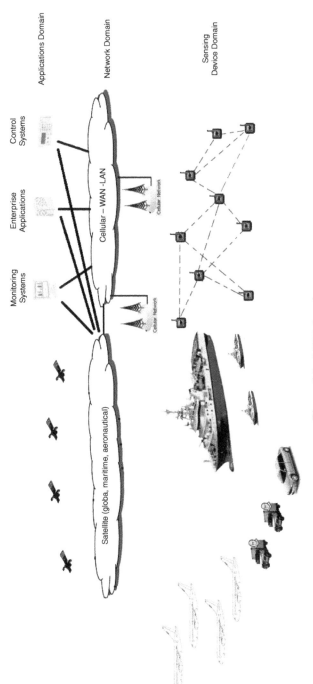

Figure 6.6 M2M domains.

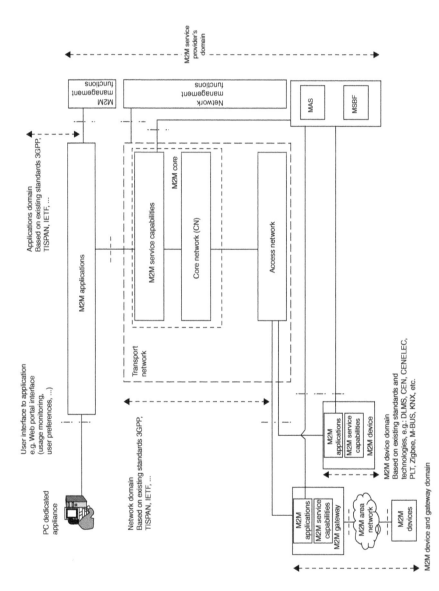

Figure 6.7 M2M high-level system architecture.

such as registration, authentication, authorization, management, and provisioning with the Network Domain. The M2M Device may provide service to other devices (e.g., legacy devices) connected to it those are hidden from the Network Domain.

- *Case 2 "Gateway as a Network Proxy"*. The M2M Device connects to the Network Domain via an M2M Gateway. M2M Devices connect to the M2M Gateway using the M2M Area Network. The M2M Gateway acts as a proxy for the Network Domain toward the M2M Devices that are connected to it. Examples of procedures that are proxied include authentication, authorization, management, and provisioning.

(M2M Devices may be connected to the Networks Domain via multiple M2M Gateways.)

2. *M2M Area Network*. Provides connectivity between M2M Devices and M2M Gateways. Examples of M2M Area Networks include PAN technologies such as IEEE 802.15.1, ZigBee, Bluetooth, IETF ROLL, ISA100.11a, among others, or local networks such as PLC, M-BUS, Wireless M-BUS, and KNX.[6]

3. *M2M Gateway*: A gateway that runs M2M Application(s) using M2M Service Capabilities. The Gateway acts as a proxy between M2M Devices and the Network Domain. The M2M Gateway may provide service to other devices (e.g., legacy devices) connected to it those are hidden from the Network Domain. As an example, an M2M Gateway may run an application that collects and treats various information (e.g., from sensors and contextual parameters).

The *Network Domain* is composed of the following elements:

1. *Access Network*. A network that allows the M2M Device and Gateway Domain to communicate with the Core Network (CoN). Access Networks include (but are not limited to) xDSL (Digital Subscriber Line), HFC (Hybrid Fiber Coax), satellite, GERAN (GSM/EDGE Radio Access Network), UTRAN (UMTS Terrestrial Radio Access Network), eUTRAN (evolved UMTS Terrestrial

---

[6]KNX (administered by the KNX Association) is an OSI-based network communications protocol for intelligent buildings defined in standards CEN EN 50090 and ISO/IEC 14543) (KNX is the follow-on standard built on the European Home Systems Protocol (EHS), BatiBUS, and the European Installation Bus (EIB or Instabus). Effectively, KNX uses the communication stack of EIB but augmented with the physical layers and configuration modes BatiBUS and EHS; thus, KNX includes the following PHYs:

- Twisted pair wiring (inherited from the BatiBUS and EIB Instabus standards). This approach uses differential signaling with a signaling speed of 9.6 kbps. Media access control is controlled with the CSMA/CA method;
- Powerline networking (inherited from EIB and EHS);
- Radio (KNX-RF);
- Infrared; and
- Ethernet (also known as EIBnet/IP or KNXnet/IP).

Radio Access Network), WLAN, and WiMAX (Worldwide Interoperability for Microwave Access).

2. *Core Network.* A network that provides the following capabilities (different CoNs offer different feature sets):
   - IP connectivity at a minimum, and possibly other connectivity means;
   - Service and network control functions;
   - Interconnection (with other networks);
   - Roaming;
   - CoNs include (but are not limited to) 3GPP CoNs, ETSI TISPAN CoN, and 3GPP2 CoN.

3. *M2M Service Capabilities.*
   - Provide M2M functions that are to be shared by different applications;
   - Expose functions through a set of open interfaces;
   - Use CoN functionalities;
   - Simplify and optimize application development and deployment through hiding of network specificities.

The "M2M Service Capabilities" along with the "CoN" is known collectively as the "M2M Core."

The *Applications Domain* is composed of the following elements:

1. *M2M applications.* Applications that run the service logic and use M2M Service Capabilities accessible via an open interface.

There are also management functions within an overall M2M Service Provider Domain, as follows:

1. *Network Management Functions.* Consists of all the functions required to manage the Access and CoNs; these functions include provisioning, supervision, and fault management.

2. *M2M Management Functions.* Consists of all the functions required to manage M2M Service Capabilities in the Network Domain. The management of the M2M Devices and Gateways uses a specific M2M Service Capability.
   - The set of M2M Management Functions include a function for M2M Service Bootstrap. This function is called M2M Service Bootstrap Function (MSBF) and is realized within an appropriate server. The role of MSBF is to facilitate the bootstrapping of permanent M2M service layer security credentials in the M2M Device (or M2M Gateway) and the M2M Service Capabilities in the Network Domain.
   - Permanent security credentials that are bootstrapped using MSBF are stored in a safe location, which is called M2M Authentication Server (MAS). Such a server can be an AAA server. MSBF can be included within MAS, or may communicate the bootstrapped security credentials to MAS, through an

appropriate interface (e.g., the DIAMETER protocol defined in IETF RFC 3588) for the case where MAS is an AAA server).

The H2M portion of the IoT could theoretically make use of these same mechanisms and capabilities, but the information flow would likely need to be front-ended by an access layer (which can also be seen as an application in the sense described above) that allows the human user to interact with the machine using an intuitive interface. One such mechanism can be an HTML/HTTP-based browser that interacts with a suitable software peer in the machine (naturally this requires some higher level capabilities to be supported by the DEP/machine in order to be able to run an embedded web server software module.) (When used in embedded devices or applications, web servers must assume that they are secondary to the essential functions the device or application must perform; as such, the web server must minimize its resource demands and should be deterministic in the load it places on a system.[7])

6.3 M2M APPLICATIONS EXAMPLES AND SATELLITE SUPPORT

This section looks at some IoT/M2M applications, in general, and then focuses on satellite-based support of the space. Related to IoT applications, proponents make the observation that [LEE201101]:

… there are so many applications that are possible because of IoT. For individual users, IoT brings useful applications like home automation, security, automated devices monitoring, and management of daily tasks. For professionals, automated applications provide useful contextual information all the time to help on their works and decision making. Industries, with sensors and actuators operations can be rapid, efficient and more economic. Managers who need to keep eye on many things can automate tasks connection digital and physical objects together. Every sectors energy, computing, management, security, transportation are going to be benefitted with this new paradigm. Development of several technologies made it possible to achieve the vision of Internet of things. Identification technology such as RFID allows each object to represent uniquely by having unique identifier. Identity reader can read any time the object allows real time identification and tracking. Wireless sensor technology allows objects to provide real time environmental condition and context. Smart technologies allow objects to become more intelligent which can think and communicate. Nanotechnologies are helping to reduce the size of the chip incorporating more processing power and communication capabilities in a very small chip.

---

[7] As an illustrative example of an embedded web server, Oracle's GoAhead WebServer is a simple, portable, and compact web server for embedded devices and applications; it runs on dozens of operating environments and can be easily ported and adapted. The GoAhead WebServer is a simple, compact web server that has been widely ported to many embedded operating systems. Appweb is faster and more powerful – but requires more memory. If a device requires a simple, low-end web server and has little memory available, the GoAhead WebServer is ideal; if the device needs higher performance and extended security, then Appweb is the right choice. As one of the most widely deployed embedded web servers, Appweb is being used in networking equipment, telephony, mobile devices, consumer and office equipment as well as hosting for enterprise web applications and frameworks. It is embedded in hundreds of millions of devices. The server runs equally well stand-alone or in a web farm behind a reverse proxy such as Apache [EMB201101].

6.3.1   Examples of General Applications

This section provides a sample of applications than can be provided with/by IoT, although any such survey is invariably incomplete and is limited in the temporal domain (with new applications being added on an ongoing basis). We look at applications that are already emerging and/or have a lot of current industry interest. Typically, in a general context, IoT applications range widely from energy efficiency to logistics and from appliance control to "smart" electric grids. Indeed, there is an increasing interest in connecting and controlling, in real time, all sorts of devices for personal healthcare (patient monitoring fitness monitoring), building automation [also known as building automation and control (BA&C) – for example, security devices/cameras; Heating, Ventilation, and Air-Conditioning (HVAC); AMRs]; residential/commercial control (e.g., security HVAC, lighting control, access control, lawn, and garden irrigation); consumer electronics (e.g., TV, DVRs); PC and peripherals (e.g., mouse, keyboard, joystick, wearable computers); industrial control (e.g., asset management, process control, environmental, energy management); and supermarket/supply chain management (this being just a partial list.)

Vertical industries in arenas such as automotive and fleet management, telehealth (also called telecare by some) and Mobile Health (m-Health – when mobile communications are used), energy and utilities, public infrastructure, telecommunications, security and defense, consumer telematics, Automated Teller Machines (ATMs)/kiosk/POS, and digital signage are in the process deploying IoT services and capabilities. Proponents make the claim that IoT will usher in a wide range of smart applications and services to cope with many of the challenges individuals and organizations face in their everyday lives. For example, remote healthcare monitoring systems could aid in managing costs and alleviating the shortage of healthcare personnel; intelligent transportation systems (ITSs) could aid in reducing traffic congestion and the issues caused by congestion such as air pollution; smart distribution systems from utility grids to supply chains could aid in improving the quality and reducing the cost of their respective goods and services; and tagged objects could result in more systematic recycling and effective waste disposal [GLU201101]. These applications may change the way societies function and, thus, have a major impact on many aspects of people's lives in the years to come. Many of today's home entertainment and monitoring systems often offer a web interface to the end-user; the IoT aims at greatly extending those capabilities to many other devices and many other applications.

A list of (early) applications includes the following (also see Table 6.3):

- Things on the move
  - Retail;
  - Logistics;
  - Pharmaceutical;
  - Food;

TABLE 6.3   The Scope of IoT

Service Sector	Application Group	Location (partial list)	Devices ("Things") of Interest (partial list)
Real estate (Industrial)	Commercial/ institutional	Office complex, school, retail space, hospitality space, hospital, medical site, airport, stadium	UPS, generator, HVAC, fire and safety (EHS), lighting, security monitoring, security control/access
	Industrial	Factory, processing site, inventory room, clean room, campus	
Energy	Supply providers/ consumers	Power generation, power transmission, power distribution, energy management, AMI (Advanced Metering Infrastructure)	Turbine, windmills, UPS, batteries, generators, fuel cells
	Alternative energy systems	Solar systems, wind system, cogeneration systems	
	Oil/gas operations	Rigs, well heads, pumps, pipelines, refineries	
Consumer and home	Infrastructure	Home wiring/routers, home network access, home energy management	Power systems, HVAC/thermostats, sprinklers, MID, dishwashers, refrigerators, ovens, eReaders, washer/dryers, computers, digital videocameras, meters, lights, computers, game consoles, TVs, PDRs
	Safety	Home fire safety system, home environmental safety system (e.g., CO_2), home security/intrusion detection system, home power protection system, remote telemetry/video into home, oversight of home children, oversight of home-based babysitters, oversight of home-bound elderly	

(*continued*)

TABLE 6.3　*(Continued)*

Service Sector	Application Group	Location (partial list)	Devices ("Things") of Interest (partial list)
	Environmentals	Home HVAC, home lighting, home sprinklers, home appliance control, home pools and jacuzzis	
	Entertainment	TVs, PDRs	
Healthcare	Care	Hospitals, ERs, mobile POC, clinic, labs, doctor's office	MRIs, PDAs, implants, surgical equipment, BAN devices, power systems
	In vivo/home	Implants, home monitoring systems, body area networks (BANs)	
	Research	Diagnostic lab, pharmaceutical research site	
Industrial	Resource automation	Mining sites, irrigation sites, agricultural sites, monitored environments (wetlands, woodlands, etc.)	Pumps, valves, vets, conveyors, pipelines, tanks, motors, drives, converters, packaging systems, power systems
	Fluids management	Petrochemical sites, chemical sites, food preparation site, bottling sites, wineries, breweries	
	Converting operations	Metal processing sites, paper processing sites, rubber/plastic processing sites, metalworking site, electronics assembly site	
	Distribution	Pipelines, conveyor belts	
Transportation	Nonvehicular	Airplanes, trains, buses, ships/boats, ferries	Vehicles, ships, planes, traffic lights, dynamic signage, toll gates, tags
	Vehicles	Consumer and commercial vehicle (car, motorcycle, etc.), construction vehicle (e.g., crane)	
	Transportation subsystems	Toll booths, traffic lights and traffic management, navigation signs, bridge/tunnel status sensors	

TABLE 6.3 *(Continued)*

Retail	Stores	Supermarkets, shopping centers, small stores, distribution centers	POS terminals, cash registers, vending machines, ATMs, parking meters
	Hospitality	Hotel, restaurants, café', banquet halls, shopping malls	
	Specialty	Banks, gas stations, bowling, movie theaters	
Public safety and security	Surveillance	Radars, military security, speed monitoring systems, security monitoring systems	Vehicles, ferries, subway trains, helicopters, airplanes, video cameras, ambulances, police cars, fire trucks, chemical/radiological monitors, triangulation systems, UAVs
	Equipment	Vehicles, ferries, subway trains, helicopters, airplanes	
	Tracking	Commercial trucks, postal trucks, ambulances, police cars	
	Public infrastructure	Water treatment sites, sewer systems, bridges, tunnels	
	Emergency services	First responders	
IT systems and networks	Public networks	Network facilities, central offices, data centers, submarine cable, cable TV headends, telco hotels, cellular towers, poles, teleports, ISP centers, lights-off sites, NOCs	Network elements, switches, core routers, antenna towers, poles, servers, power systems, backup generators
	Enterprise networks	Data centers, network equipment (e.g., routers)	

- Ubiquitous intelligent devices
  - Smart appliances;
  - Efficient appliances via the use of eco-aware/ambient-aware things;
  - Interaction of physical and virtual worlds; executable tags, intelligent tags, autonomous tags, collaborative tags;

- Intelligent devices cooperation;
- Ubiquitous readers;
- Ambient and assisted living
  - Health;
  - Intelligent home;
  - Smart living;
  - Security-based living;
  - Transportation;
- Education and Information
- Environmental aspects/resource efficiency
  - Energy and resource conservation;
  - Advanced Metering Infrastructure (AMI);
  - Energy harvesting (biology, chemistry, induction);
  - Power generation in hash environments;
  - Energy recycling;
  - Pollution and disaster avoidance.

A longer, but far from complete, list of applications includes the following:

- *Public Services and Smart Cities*:
  - Telemetry: for example, smart metering, parking metering, and vending machines
  - ITSs and traffic management
  - Connecting consumer and citizens to public infrastructure (such as public transportation)
  - In-building automation, municipal and regional Infrastructure
  - Metropolitan operations (traffic, automatic tolls, fire, and so on)
  - Electrical grid management at a global level; SGs
  - Electrical Demand Response (DR) at a global level
- *Automotive, Fleet Management, Asset Tracking*:
  - e-Vehicle: for example, navigation, road safety, and traffic control
  - Driver safety and Emergency Services
  - Fleet management systems: hired-car monitoring, goods vehicle management
  - Backseat Infotainment device integration
  - Next-gen GPS services
  - Tracking such as asset tracking, cargo tracking, and order tracking
- *Commercial Markets*:
  - Industrial monitoring and control; for example, industrial machines and elevator monitoring

- Commercial building and control
- Process control
- Maintenance automation
- Home automation
- Wireless AMR/Load Management (LM)
- Homeland security applications: chemical, biological, radiological, and nuclear wireless sensors
- Military sensors
- Environmental (land, air, sea) and agricultural wireless sensors
- Finance: POS terminals, ticketing
- Security: Public surveillance, personal security
- *Embedded Networking Systems in the Smart Home and Smart Office*:
  - Smart appliances: for example, AC power control, lighting control, heating control, and low-power management
  - Automated home: remote media control
  - Smart meters and energy efficiency: efficiencies obtained by exploiting the potential of the SG
  - Telehealth (e-health): Assisted living and in-home m-health services (including remote monitoring, remote diagnostic)
  - Security and emergency services: integrated remote services

Figure 6.8, partially inspired by [SCA201201], depicts a grouping of applications, particularly in the M2M context. As should be clear by now, some of the possible short-term applications include the following: building automation and remote control (facilitating efficient commercial spaces); SE (supporting office building/home energy management); healthcare (providing health and fitness monitoring); home automation (giving rise to smart homes); and retail services (enabling smart shopping).

Table 6.4 provides some examples by category as defined in 3GPP Machine-Type Communications (MTCs) documentation [3GP201001] (MTC is the term used in 3GPP to describe M2M systems.)

In recent years, ETSI has published a number of use cases for IoT (specifically for M2M) applications in the following documents:

- ETSI TR 102 691: *"Machine-to-Machine communications (M2M); Smart Metering Use Cases."*
- ETSI TR 102 732: *"Machine to Machine Communications (M2M); Use cases of M2M applications for eHealth."*
- ETSI TR 102 897: *"Machine to Machine Communications (M2M); Use cases of M2M applications for City Automation."*
- ETSI TR 102 875: *"Access, Terminals, Transmission and Multiplexing (ATTM); Study of European requirements for Virtual Noise for ADSL2, ADSL2plus and VDSL2."*

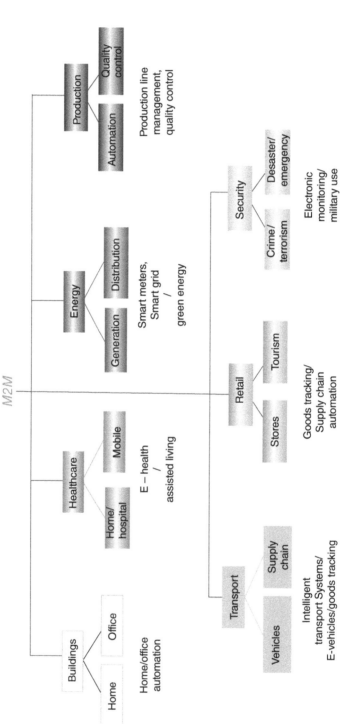

Figure 6.8 Grouping of applications in the M2M context.

TABLE 6.4 Examples of MTC Applications as Defined in 3GPP TS 22.368 Release 10

Category	Specific Example
Consumer devices	Digital camera Digital photo frame eBook
Health monitoring vital signs	Remote diagnostics Supporting the aged or handicapped Web access telemedicine points
Metering	Gas Grid control Heating Industrial metering Power Water
Payment	Gaming machines Point of sales Vending machines
Remote maintenance/control sensors	Elevator control Lighting Pumps Valves Vehicle diagnostics Vending machine control
Service Area MTC applications	Backup for landline Car/driver security Control of physical access (e.g., to buildings) Security surveillance systems
Tracking and tracing fleet management	Asset tracking Navigation Order Management Pay as you drive Road tolling Road traffic optimization/steering Traffic information

- ETSI TR 102 898: *"Machine to Machine Communications (M2M); Use cases of Automotive Applications in M2M capable networks."*
- ETSI TS 102 412: *"Smart Cards; Smart Card Platform Requirements Stage 1 (Release 8)."*

Some of these (ETSI-covered) applications are discussed in the sections that follow.

6.3.1.1   City Automation   Some applications in this domain include but are not limited to the following:

- Traffic Flow Management System in combination with Dynamic Traffic Light Control
- Street Light Control
- Passenger Information System for Public Transportation
- Passive surveillance

Generic city sensors include environmental sensors and activity sensors. Environmental sensors include devices capable of thermal, hygrometric, anemometric, sound, gas, particles light or other EM spectrum, and seismic measurements. Activity sensors include devices capable of pavement/roadway pressure, vehicle and pedestrian detection, and parking space occupancy measurements.

ETSI TR 102 897: "*Machine to Machine Communications (M2M); Use cases of M2M applications for City Automation*" provides the following description of these applications [ETS201002]:

- *Use Case 1: Traffic Flow Management System in Combination with Dynamic Traffic Light Control.* The flow of road traffic within cities depends on a number of factors such as the number of vehicles on the road, the time and the day, the current or expected weather, current traffic issues and accidents, as well as road construction work. Traffic flow sensors provide key traffic flow information to a central traffic flow management system; the traffic flow management system can develop a real-time traffic optimization strategy, and, thus, endeavor to control the traffic flow. The traffic control can be achieved by dynamic information displays informing the driver about traffic jams and congested roads; traffic signs can direct the traffic to utilize less used roads. The Traffic Flow Management system can also interact with controllable traffic lights to extend or to reduce the green light period to increase the vehicle throughput on heavily-used roads; dynamically changeable traffic signs can lead to an environment where the vehicular traffic is managed more efficiently, thus enabling cities to reduce fuel consumption, air pollution, congestions, and the time spent on the road.

- *Use Case 2: Street Light Control.* Street lights are not required to shine at the same intensity to accomplish the intended safety goal. The intensity may depend on conditions such as moonlight or weather. Adjusting the intensity helps to reduce the energy consumption and the expenditures incurred by a municipality. The street light controller of each street light segment is connected (often wirelessly) with the central street light managing and control system. Based on local information measured by local sensors, the control system can dim the corresponding street lights of a segment remotely or is able to switch street lights on and off.

- *Use Case 3: Passenger Information System for Public Transportation.* Public transportation vehicles, such as buses, subways, and commuter trains operate on a schedule that may be impacted by external variables and, thus, have a degree of variability compared with a baseline formal schedule. Passengers need to know when their next connection is available; this information also allows passengers to select alternative connections in the case of longer delays. In this application, the current locations of the various public transport vehicles are provided to the central system that is able to match the current location with the forecasted location at each time or at specific check points. Based on the time difference, the system is able to calculate the current delay and the expected arrival time at the upcoming stops. The vehicle location can be captured via check points on the regular track or via GPS/GPRS tracking devices that provide the position information in regular intervals. Two approaches are possible:

- With a Check Point based approach, the line number (of the bus or the street car) is captured at each station where the vehicle stops regularly, or at defined check point in between. Because of the fact that the sensor at a specific station is able to provide the data to the central system, the expected delay can be calculated by comparing the information of the scheduled arrival time and the actual arrival time. This change can be added to the arrival time displayed at each following station. Each vehicle must be equipped with a transponder (variously based on infrared, RFID, short-range communication, or optical recognition). In addition, each station has to be equipped with one or more check point systems that are able to readout or to receive the line number information of the vehicle. In case of larger stations with several platforms, multiple systems are needed.

- With a GPS/GPRS-based approach, each vehicle has to be equipped with a GPS/GPRS tracking device that provides beside the current position also the information that can be directly or indirectly matched to the serviced line number. Based on the "regular" position/time pattern, the system is able to calculate the actual time difference and provide the expected time on the passenger display.

A combination of check point based and GPS/GPRS-based solution can be used to integrate railed vehicles (such as subways and street cars) and road vehicles (such as buses).

6.3.1.2 *Automotive Applications*

IoT/M2M automotive and transportation applications focus on safety, security, connected navigation, and other vehicle services such as, but not limited to, insurance or road pricing, emergency assistance, fleet management, electric car charging management, and traffic optimization. These applications typically entail IoT/M2M communication modules that are embedded into the car or the transportation equipment. Some of the technical challenges relate to mobility management and environmental hardware considerations. A brief description of applications follows from reference [ETS201003] (on which the next few paragraphs are based).

- *bCall (Breakdown Call).* A bCall sends the current vehicle position to a roadside assistance organization and initiates a voice call. The bCall trigger is usually a switch that is manually pushed by the user in order to activate the service. An "enhanced" bCall service allows current vehicle diagnostic information to be transmitted in addition to the vehicle position.

- *Stolen Vehicle Tracking.* A basic application for automotive M2M communications is tracking of mobile assets – either for purposes of managing a fleet of vehicles or to determine the location of stolen property. The goal of a Stolen Vehicle Tracking (SVT) system is to facilitate the recovery of a vehicle in case of theft. The SVT service provider periodically requests location data from the Telematics Control Unit (TCU) in the vehicle and interacts with the police. The TCU may also be capable of sending out automatic theft alerts based on vehicle intrusion or illegal movement. The TCU may also be linked to the Engine Management System (EMS) to enable immobilization or speed degradation by remote command. Vehicles contain embedded M2M devices that can interface with location determination technology and can communicate via a mobile cellular network to an entity (server) in the M2M core. The M2M devices will communicate directly with the telecommunications network; the M2M devices will interface with location-determination technology such as stand-alone GPS, or network-based mechanisms such as assisted GPS, Cell ID, and so on. For theft-tracking applications, the M2M device is typically embedded in an inaccessible or inconspicuous place so that it may not be easily disabled by a thief. The Tracking Server is an entity located in the M2M core and owned or operated by the asset owner or service provider to receive, process, and render location and velocity information provided by the deployed assets. The Tracking Server may trigger a particular M2M device to provide a location/velocity update, or the M2M devices may be configured to autonomously provide updates on a schedule or upon an event-based trigger.
- *Remote Diagnostics.* Remote diagnostic services can broadly be grouped into the following categories:
  - Maintenance minder – when the vehicle reaches a certain mileage (e.g., 90% of the manufacturer's recommended service interval since the previous service), the TCU sends a message to the owner or the owner's named dealership, advising the owner (or the dealership) that the vehicle is due for service.
  - Health check – either on a periodic basis or triggered by a request from the owner, the TCU compiles the vehicle's general status using inbuilt diagnostic reporting functions, and transmits a diagnostic report to the owner, the owner's preferred dealership, or to the vehicle manufacturer.
  - Fault triggered – when a fault [a Diagnostic Trouble Code (DTC)] is detected with one of the vehicle systems, this triggers the TCU to send the DTC code and any related information to the owner's preferred dealer, or to the vehicle manufacturer.
  - Enhanced bCall – when a manual breakdown call is initiated by the owner, the TCU sends both position data and DTC status information to the roadside assistance service or to the vehicle manufacturer.
- *Fleet Management.* The fleet owner wishes to track the vehicles – that is, to know, over time, the location and velocity of each vehicle – in order to plan and

optimize business operations. A fleet management application assumes that a fleet of vehicles has been deployed with M2M devices installed that are able to

- Interface with sensors on the vehicle that measure velocity;
- Interface with devices that can detect position;
- Establish a link with a mobile telecommunication network using appropriate network access credentials, such as a USIM (Universal Subscriber Identity Module).

A server in the fleet owner's employ receives, aggregates, and processes the tracking data from the fleet and provides this information to the fleet owner. Devices could be configured to autonomously establish communication with the Server via a cellular network either at regular intervals, at prescheduled times, or based on some event such as crossing a geographic threshold. Alternatively, the M2M devices could be commanded by the M2M Server to report their location/velocity data (see Figure 6.9 for an illustrative example).

- *Vehicle-to-Infrastructure Communications.* A European Intelligent Transport Systems Directive[8] seeks the implementation of eSafety applications in

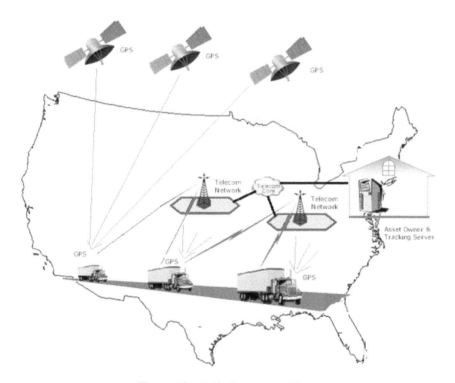

Figure 6.9   Vehicular asset tracking.

[8]A directive is a legislative act of the European Union requiring member states to achieve a stated result but without mandating the means of achieving that result.

vehicles. Some vehicle manufacturers have begun to deploy vehicle-to-vehicle communication, for example, in the context of Wireless Access in Vehicular Environments (WAVEs). On the other hand, vehicle to roadside applications are less well developed; in this case, vehicles have embedded M2M devices that can interface with location-determination technology and can communicate via a mobile telecommunications network to an entity (server). This application assumes that vehicles have been deployed with M2M devices installed that are able to:

- Interface with sensors on the vehicle that measure velocity, external impacts
- Interface with devices that can detect position
- Establish a link with a mobile telecommunication network using appropriate network access credentials, such as a USIM
- Upload or download traffic and safety information to a Traffic Information Server

Devices could be configured to establish communication with the Server via the cellular network based on some event triggered by a vehicle sensor such as external impact, motor failure, and so on. For example, the traffic information server pushes roadside or emergency information out to vehicles based on location (cell location or actual location). Or, vehicle information is pushed to the traffic information server based on external sensor information, internal sensor information, or subscription basis.

- *Insurance Services.* Pay-As-You-Drive (PAYD) schemes offer insurers the opportunity to reduce costs based on actual risk, and provide more competitive products to the end-user based on getting feedback from the vehicle as to when, where, how, or how far the vehicle is being driven (or a combination of these factors).

6.3.2   Satellite Roles, Context, and Applications

As noted, the M2M satellite applications to date have gravitated to *MSS* services and to the L-band [some also call these satellite applications as LDR (Low Data Rate) applications]. The satellite industry has referred to many of the MSS-based data communication applications as being M2M applications; while we *can* accept that description, we tend to look at M2M applications as being more "futuristic" in the sense of being more related to IoT-inspired applications, such as those described earlier. We realize that the revenue stream for the satellite operators is now more in the categories we first introduced in Section 6.1 rather than those in Section 6.3.1, but as time goes by these other applications may come into play.

Satellite service providers perceive M2M communications as an approach to the global demand for uninterrupted and seamless data connectivity across a mixture of urban, suburban, exurban, rural, and oceanic environments: satellite-based M2M can facilitate the delivery of relatively small quantities of information to and from anywhere in the world (consistent with the requirements of Table 6.2). Applications include civil government, environmental monitoring and climate analysis, Police and

Coast Guard, off-shore oil drilling, and mining. Many other examples of IoT applications can be cited and many more will evolve in the future. For example, SCADA applications are now also being extended to be supported over satellite links. Some propose satellite-based M2M as a way to achieve "Global SCADA."

Some M2M services offered are only one-way communication; others services support two-way communications; clearly two-way communication is a lot more flexible allowing data collection and monitoring of customer assets. Companies that utilize remote locations, such as oil and gas wells, mining sites, water management systems, environmental/weather sites, pollution detection systems, early warning systems, and remote security will benefit from cost-effective, two-way M2M services for asset monitoring solutions.

Observers are noticing an increased demand for satellite services from several companies associated with finance, energy, and maritime industries. Although at press time satellite-based services were only a small share of the M2M market, which is largely dominated by cellular systems (around 2% in terms of volume and 6% of revenue in 2011), M2M is a growing segment for the satellite industry: forecasts say the global satellite M2M market will reach $3 billion by 2016. The region with the highest rate of progress will be the Asia-Pacific with developments in countries such as China, Indonesia, Vietnam, and India [IDA201201]. Proponents make the case that "M2M Market represents an interesting and potentially huge revenue stream for the satellite industry with opportunities in many markets, particularly vertical ones" [STA201101]; perhaps the opportunity is "significant" rather than "huge." Some industry observers see growth opportunities in fixed asset tracking while others see it in mobile asset tracking. Iridium and Globalstar reportedly expect growth in the mobile M2M market, whereas SkyWave and Inmarsat reportedly believe that their future growth would come mainly from fixed asset monitoring [FAR201301].

Figure 6.10 (partially based on reference [FAR201201]) shows how the solutions from different providers are positioned relative to one another in terms of their optimal amount of data needed to be transferred each month. Globalstar and Orbcomm are the most direct competitors at the low end of the market. Iridium has focused on satellite solutions designed to carry 10 KB plus of data each month; the recent introduction of ISatDataPro by Inmarsat now provides an alternative at those usage rates to services by Iridium.

6.3.3 Antennas for Satellite M2M Applications

M2M Antenna types include the following [STA201401]:

- *Embedded Antennas.* Most M2M products employ multiband internal antennas (planar inverted F) that are galvanically connected to the radiating ground plane of the product. These antenna elements are tailored to match the radiating properties of the product itself, so the product as a whole radiates efficiently over the frequency bands specified. Each product has unique electromagnetic properties owing to size, shape, and interference.

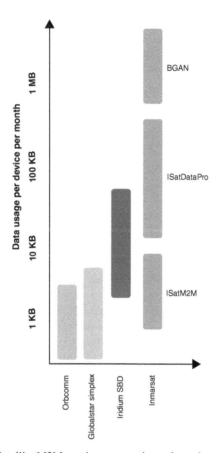

Figure 6.10   Satellite M2M services supporting volume-based requirements.

- *Stubby Antennas.* These antennas are external to the product but need to be carefully tuned to the electromagnetic requirements of the product in order to achieve optimal performance over the frequency bands specified.
- *(Mini) Very Small Aperture Terminal (VSAT) Antennas.* These are tracking antennas used for Ku-band and Ka-band that for M2M-related applications (especially in the broadband range) are typically 1 m or smaller.

Figures 6.11 and 6.12 depict some typical antennas.

6.3.4   M2M Market Opportunities for Satellite Operators

Earlier we noted that the market can be classified into the following arenas: remote structures such as O&G platforms; land mobile platform include vehicles; maritime platforms; aeronautical platform; and civil infrastructure management. We provide some assessment of these segments below.

Figure 6.11 Stubby or other M2M antennas for GNSS (GPS, GLONASS, COMPASS, Galileo); Iridium, Inmarsat, and Thuraya satellites; and terrestrial M2M, MSS, and 4G LTE applications. Courtesy: MAXTENA

6.3.4.1 Maritime In spite of the expected growth of the VSAT systems in the Ku- and Ka-band (as discussed in Chapter 3 and 4), the MSS operators as an aggregate still account for over 95% of the maritime satellite communication market (in terms of terminals). All the major MSS operators – Inmarsat, Iridium, Orbcomm, Globalstar, and Thuraya (Dubai – UAE) – serve the maritime market; however, only Inmarsat, Iridium, and Thuraya have a direct focus on the maritime sector. Inmarsat is the maritime/MSS market leader. At present, MSS services, and in particular the legacy Inmarsat services, are widely utilized across the maritime world for a plethora of mission-critical communications. As noticed in Chapter 5, at press time about 71% of the maritime terminals were M2M and safety communications (e.g., Inmarsat-C), 13% were narrowband terminals (e.g., Inmarsat-B, FleetPhone, Thuraya Seagull), and 16% were broadband terminals (e.g., FleetBroadband, Iridium OpenPort, Iridium Pilot). We also noted in Chapter 5 that merchant ships constitute 65% of the maritime M2M market: government, O&G, leisure boats, passenger ships, and fishing ships each comprises about 7% of the market.

FSS operators tend to focus on the provision of satellite capacity while MSS operators are typically involved in service delivery; in fact, MSS operators provide a menu of services, including medium data rate M2M services and voice/fax, safety-support communications, narrowband data (<128 kbps) services; but they also

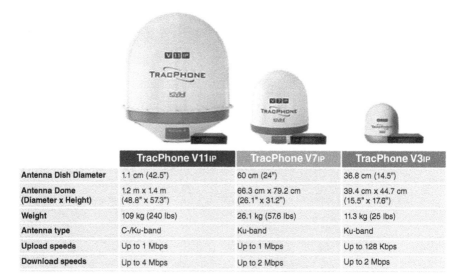

	TracPhone V11ᴵᴾ	TracPhone V7ᴵᴾ	TracPhone V3ᴵᴾ
Antenna Dish Diameter	1.1 cm (42.5")	60 cm (24")	36.8 cm (14.5")
Antenna Dome (Diameter x Height)	1.2 m x 1.4 m (48.8" x 57.3")	66.3 cm x 79.2 cm (26.1" x 31.2")	39.4 cm x 44.7 cm (15.5" x 17.6")
Weight	109 kg (240 lbs)	26.1 kg (57.6 lbs)	11.3 kg (25 lbs)
Antenna type	C-/Ku-band	Ku-band	Ku-band
Upload speeds	Up to 1 Mbps	Up to 1 Mbps	Up to 128 Kbps
Download speeds	Up to 4 Mbps	Up to 2 Mbps	Up to 2 Mbps

Figure 6.12   Mini-VSAT Broadband antennas. Courtesy: KVH

provide broadband services, such as Inmarsat FleetBroadband and Iridium Pilot. A number of MSS products are compliant with maritime safety communications requirements. The leading two MSS systems, Iridium and Inmarsat, provide global coverage or near-global coverage (global for Iridium, while the Inmarsat system does not cover the polar caps.)

M2M communication, in general, and asset-tracking solutions, in particular, are gaining popularity in the maritime sector as the subtending technologies become cheaper and easier to deploy and manage. In addition, an increasing number of value-added services are being offered through low-data-rate terminals, enabling vessel owners to increase safety, security, and operational efficiency. Typical applications include geolocation information, machine surveillance, surveillance of various conditions (e.g., temperature and speed), SCADA, and ship identification services used for Automatic Identification System (AIS) or Long-Range Identification and tracking (LRIT).

Container tracking could, in theory, be an MSS/M2M application; however, ship-board techniques using other (short range) wireless networks are generally needed in addition to one satellite-enabled gateway, since each container does not have direct line-of-sight to the satellite (see Figure 6.13).

While the majority of the maritime terminals as of press time were still LDR-type terminals used principally for safety communications and also used for other related basic M2M applications, broadband Internet access and streaming video have been the driver of sector growth in recent years; however, evolving IoT/M2M as described earlier and in [MIN201301] are not limited to low-speed-only applications.

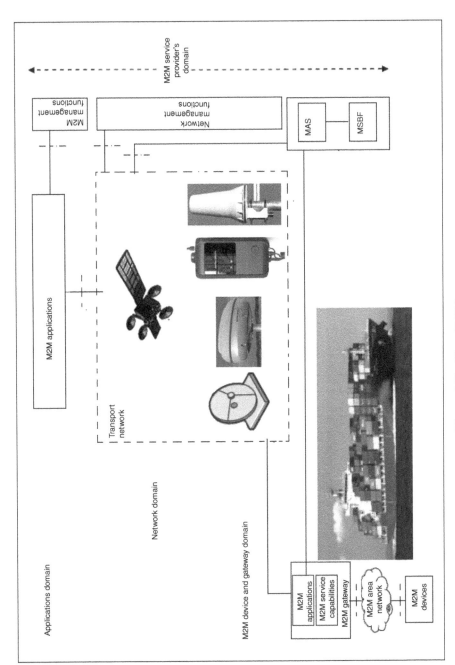

Figure 6.13 Use of M2M gateways.

In 2014, there were an estimated 52,000 broadband terminals, 50,000 narrowband terminals, 260,000 M2M terminals, and 12,500 VSAT terminals in use, as noted in Table 4.11. LDR terminals (used for regulation and security) represent the bulk of MSS terminals; these systems also support data loggers. While the deployment of MSS broadband services terminals enjoys a growth trajectory that results in an estimated 130,000 terminals in use by 2022, the number of terminals with legacy narrowband service, typically delivered over the Inmarsat-3 constellation, has been declining due to migration to MSS broadband and VSAT services. By 2022, it is estimated that there will be 130,000 broadband terminals, 40,000 narrowband terminals, 480,000 M2M terminals, and 31,000 VSAT terminals in use.

In terms of specific vertical application within the maritime industry cited above (merchant ships, government, O&G, leisure boats, passenger ships, and fishing ships), some points of observation follow [WLI201201].

- *Fishing vessels.* The need to control the catches of fishing vessels is driving a number of countries to require the mandatory adoption of electronic logbooks, which in turn requires maritime connectivity.
- *Passenger ships.* Passenger ships were equipped with an estimated 11,000 satellite terminals at press time. The growth rate for satcom terminals in passenger ships is expected to be relatively low with a CAGR of approximately 4% over the next few years, since the high-end market is already well-served. Although migrations to a new generation of services, specifically broadband Internet-access Ku/Ka VSAT-based services, still present an opportunity, these substitutions will have relatively limited impact on the growth in terminals counts; however, the revenue potential will grow.
- *O&G.* Offshore O&G activity encompasses exploration development, production, and decommissioning. Maritime satellite communications usage is particularly intensive in the exploration phase by both survey vessels, mobile rigs, and supporting vessels. For offshore operations away from nearby terrestrial connectivity, VSAT and MSS services continue to make up the predominant networks.
- *Government.* An increasing number of government users need to survey, track, and monitor their assets, including at sea; as a consequence, the increased use of low-data-rate M2M devices in a government maritime environment to track cargo, to increase safety and situational awareness, and to support logistics is seen.

6.3.4.2 Land Mobile Land-mobile low-data-rate services in the M2M space, including asset tracking and fleet management application, are expected to be utilized by about 2.5 million units by 2022 (from a base of about 1.5 million in 2014, with a CAGR of about 7%, under some high-growth scenario assumptions – but not unreasonable assumptions). It represents the largest market for mobility and has the largest number of deployed systems. It is comprised of commercial and military vehicles, and emergency response vehicles. Retail revenues were forecast to grow from about $2.2B to about $3.5B in 2022 (with a CAGR of 6%) and wholesale revenues to grow from about $09B to $1.5B [NSR201201]. Retail revenues include retail equipment revenues (8%) and retail services revenues (92%). Both narrowband and broadband (MSS-based) applications will continue to see deployment. These services support both the communications-on-the-move (COTM) and communications-on-the-pause (COPT), for example, when the land mobile devices become stationary (say overnight). Service capabilities include portable units that provide VoIP and video (for the broadband service options); other capabilities include personal locator beacons and location-based services.

This market is typically served by low earth orbit (LEO) operators such as Iridium (with the 9602 transceiver) and Globalstar (with the SPOT service for emergency calls from remote locations, as well as by Inmarsat ISAT M2M providers (e.g., Skywave). It is generally recognized that cellular 3G/LTE services offer competitive alternatives to the satellite market. In 2012, Inmarsat also launched the BGAN M2M data service for real-time applications (supporting 1–50 MB data transfers per month) for applications such as smart metering, SCADA, remote asset monitoring, meteorological and environmental data collection, telemetry for utilities, O&G platform operators, and government agencies.

For a press time example, the Iridium 9602 SBD transceiver (Figure 6.14), designed to be integrated into a wireless data application with other host system hardware and software, provides a complete solution for a specific application or vertical market. The Iridium 9602 is ideal for M2M solutions, including tracking of maritime vessels, equipment monitoring, and automatic vehicle location. Another typical press time example is the Quake Global's Quake Q4000 dual-mode satellite/GSM M2M unit (using the Iridium chipset) that allows asset tracking and remote two-way communication for land mobile, maritime, and aeronautical applications (Figure 6.15). The device supports satellite network (GPS, Iridium, Orbcomm) as well as terrestrial 3G cellular systems; the M2M modems meet the demanding

Courtesy: Iridium

Figure 6.14   Iridium 9602 transceiver.

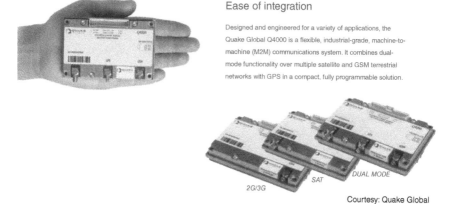

Ease of integration

Designed and engineered for a variety of applications, the Quake Global Q4000 is a flexible, industrial-grade, machine-to-machine (M2M) communications system. It combines dual-mode functionality over multiple satellite and GSM terrestrial networks with GPS in a compact, fully programmable solution.

DUAL MODE
SAT
2G/3G

Courtesy: Quake Global

Figure 6.15 Quake Global Q4000 modems.

quality requirements found in the heavy equipment and mining industries, among other industries.

6.3.4.3 Aeronautical A number of M2M applications exist in the aeronautical space, particularly from companies such as Inmarsat. Having covered aeronautical environments in Chapter 4, we only make note of the Automatic Dependent Surveillance-Broadcast (ADS-B) system here.

With ADS-B (and GPS), aircraft can transmit their location and altitude to other nearby aircraft and to air traffic control (ATC). On May 27, 2010, the US Federal Aviation Administration (FAA) issued a final rule mandating ADS-B equipage. Effective on January 1, 2020, any aircraft operating in airspace where a Mode C transponder is required today will also be required to carry an ADS-B OUT transmitter. ADS-B will redefine the paradigm of communications, navigation, and surveillance in Air Traffic Management today. Already proven in the field and internationally certified as a viable low-cost replacement for conventional radar, ADS-B allows pilots and air traffic controllers to "see" and control aircraft with more precision and over a far larger percentage of the earth's surface than has ever been possible before. Essentially, ADS-B is a one-for-one replacement for Radar, except that ADS-B is more accurate, reliable, and far more cost effective than Radar. ADS-B capable aircraft use an ordinary GNSS (GPS) receiver to derive their precise position from the GNSS constellation, and then combine that position with any number of aircraft variables, such as speed, heading, altitude, type of aircraft, and flight number. This information is then simultaneously broadcast line-of-sight to ATC ADS-B receivers on the ground and to other aircraft equipped to receive it. A space-based enhancement to ADS-B is an over-the-horizon air traffic management system capable of delivering an uncorrupted ADS-B digital message every second from virtually any ADS-B avionics system to ATC. In essence, this extends ADS-B into areas where a conventional line-of-sight connection between the aircraft and a terrestrial ADS-B ground station is either impossible or impractical to maintain. When an ADS-B equipped aircraft is over the horizon from an ADS-B ground station, an MSS constellation provides an alternate delivery path for the ADS-B signal. Features include [GRA201301]:

- *Automatic* – "Always ON" – No operator attention is required
- *Dependent* – Relies on very accurate GNSS position data
- *Surveillance* – Provides aircraft position, altitude, speed, heading, identification, and other data
- *Broadcast* – Requires no interrogation or triggering by other stations – data are broadcast to any aircraft or ground station equipped to receive the data link signal

One example is the utilization of the Globalstar Constellation; the Globalstar's system is called ADS-B Link Augmentation System (ALAS) by the firm.

6.3.4.4 Utility Market Although the near-term outlook for M2M satellite communications solutions is low in the utility market, a demand exists in both developed and developing markets. The developed markets in Europe and North America have seen some deployments of the services, but in the future Latin America and Asia hold potential. With high natural gas prices, abundant renewable energy sources such as hydroelectric power, and in the case of China, widespread air pollution, an ongoing emphasis on distributed generation sources and grid optimization efforts is almost certain. The market is still dominated by MSS forms factors that offer easy satellite and terrestrial integration into small footprints. From a sizing perspective the market for utility satellite communications services remains a market for in-service units, rather than service revenues for bandwidth demand (see Figure 6.16). Although new applications continue to emerge across the industry – from remote video monitoring to supporting remote workers in semi-manned substations or generating locations – overall bandwidth needs remain minimal [GRA201401]. As such, the introduction of GEO-HTS offerings will find their niche in supporting OA applications and M2M centric services taking a distant second place.

6.3.5 Key Satellite Industry Players and Approaches

This section provides some information on some of the major suppliers for illustrative purposes; this discussion is not intended to be exhaustive in any manner.

6.3.5.1 Inmarsat Inmarsat is the maritime/MSS market leader; it has a long history and experience in maritime communications. Inmarsat has been able to maintain its leading position among MSS operators in the maritime segment due to (i) limited competition for deep-sea communications in the 1990s, (ii) the company's globally established service provision and distribution network, and (iii) its continuous product innovations. Inmarsat's maritime products include Inmarsat-B, Inmarsat-C, Inmarsat-M, Inmarsat Fleet, and Inmarsat FleetBroadband. Related specifically to M2M, services include Inmarsat D+, IsatM2M, ISatData Pro (developed by Sky-Wave), and BGAN (Broadband Global Area Network) M2M. Two of the leading

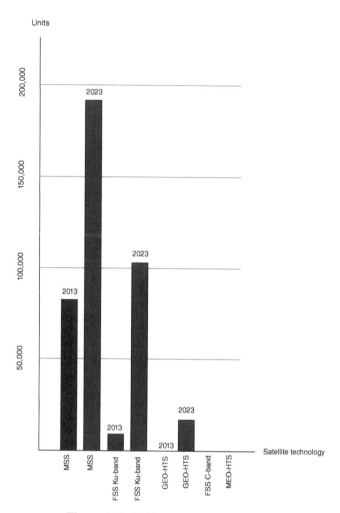

Figure 6.16   Utility market, M2M satcom.

Inmarsat system-based M2M service providers are SkyWave and EMS Global Tracking. Among all Inmarsat-related maritime businesses, it has been estimated that the operator accounted for more than 70% of MSS maritime terminals and more than 80% of MSS maritime wholesale revenue in 2011 [WLI201201]. Inmarsat is also pursuing asset monitoring applications, such as SG applications (on the other hand, Iridium, Globalstar, and Orbcomm reportedly see little demand for fixed asset monitoring and SCADA-type applications – their growth to be almost entirely driven by mobile assets, particularly in the transportation segment). In the US, utilities (which are one of the most obvious opportunities for fixed asset monitoring) are governed by rate-of-return legislation that (effectively) encourages investing in capital equipment, giving them an incentive to invest in unlicensed wireless networks, or even wired solutions, even if these are more expensive, rather than incurring ongoing costs for data transmission over cellular or satellite network. In contrast, Inmarsat's business is more global, and Inmarsat is focused on growth opportunities in BRICA (Brazil, Russia, India, China, Africa) countries, although few MSS low-data-rate solutions are used in at present, and thus it may take time to grow awareness [FAR201201].

Key applicable Inmarsat services include the following (according to the company), listed here for illustrative purposes:

- *BGAN M2M.* Two-way global IP data service for long-term monitoring and control of fixed assets in unmanned, remote locations.
- *IsatM2M.* A worldwide two-way, low-data-rate messaging service for tracking, monitoring, and control of remote assets.
- *IsatData Pro.* Two-way, short-burst data exchange for tracking and monitoring fixed or mobile assets globally.

BGAN M2M is a reliable, global, two-way IP data service designed for long-term M2M management of fixed assets. It connects monitoring and control applications in remote, unmanned locations, giving the user full visibility and management of the dispersed assets across an entire operational area. BGAN M2M is targeted to customers with data volume requirements ranging from megabytes to gigabytes, such as real-time surveillance or high-volume metering and telemetry. It is a 3G satellite network service and provides full IP data connectivity supported by remote terminal management, debugging, and configuration options. Using robust and lightweight hardware, BGAN M2M enables a wide range of M2M applications. Some features are as follows:

- *Global Coverage.* BGAN M2M is accessible across the globe except in the extreme polar regions.
- *Performance and Latency.* Send data using BGAN Standard IP at a rate of up to 448 kbps with a low latency from 800 ms, assuring real-time visibility of critical data.
- *Easily Integrated.* Simple for field teams to set up and integrate with specific applications, and to maintain without technical expertise or training.
- *Robust Terminals.* BGAN M2M is accessed through a robust terminal that can be housed in a sturdy, weatherproof unit for long-term unmanned deployment.
- *Reliable Network.* Operates over our global satellite and ground network, with 99.9% availability and an operational lifespan expected well into the 2020s.
- Affordable low hardware costs with subscription-based price plans, no minimum connection fee and minimum billing increments.
- Specific applications
  - IP SCADA for data backhaul
  - Asset tracking
  - Fixed monitoring – remote surveillance

- Fixed monitoring – remote telemetry
- Fixed monitoring – remote tracking
- Friendly force tracking
- Mobile monitoring – remote surveillance
- Mobile monitoring – remote telemetry
- Mobile monitoring – remote tracking
- Oil well head telemetry and monitoring
- Railway track and crossings
- Remote control of assets
- Remote personnel tracking
- Road signs
- Secure and encrypted ATM/PoS solution
- Smart grid
- Smart metering
- Telemetry – SCADA
- Weather, environmental monitoring
- General applications
  - ECDIS
  - GPS location data look-up-and-send
  - IP SCADA
  - SCADA
  - Surveillance

Inmarsat only had a few thousand BGAN terminals deployed at press time, but the expectation was for greater deployment in the future.

As an illustrative example, the Hughes 9502 IP satellite terminal provides reliable connectivity over the Inmarsat BGAN for IP SCADA and M2M applications. The Hughes terminal delivers affordable, global, end-to-end IP data connectivity enabling applications in industry sectors such as environmental monitoring, smart grid, pipeline monitoring, compressor monitoring, well site automation, video surveillance, and out-of-band management to primary site communications. The low power consumption (<1 W idle) of the Hughes 9502 makes it possible to provide end-to-end IP connectivity to sites that are off the grid; this equipment provides end-to-end IP connectivity to power-challenged locations that rely upon solar battery arrays involving sensitive power budgets. Some features are as follows:

- Satellite transmit frequency: 1626.5–1675 MHz
- Satellite receive frequency: 1518–1559 MHz
- GPS frequency: 1574.42–1576.42 MHz
- IDU weight: < 1.5 kg
- IDU dimensions: 15 cm × 20 cm × 45 cm

IsatM2M is a global, store-and-forward low-data-rate messaging to and from remote assets for tracking, monitoring, and control operations. The service supports critical applications such as transport vehicle security, industrial equipment monitoring, and marine tracking, giving companies visibility and control of fixed or mobile assets. IsatM2M is a two-way burst messaging service that enables a wide range of M2M applications for tracking and monitoring remote fixed or mobile assets on a global basis – whether on land, at sea, or in the air. This next-generation satellite telematics service is based on Inmarsat D+ and offers fast data forwarding rates, quicker responses to polling requests, and shorter time to first transmission. Some features are as follows:

- *Global Coverage.* IsatM2M is available across the globe, apart from the extreme polar regions.
- *Performance and Latency.* Speeds of 10.5 or 25.5 bytes in the send direction and 100 bytes in the receive direction, with latency typically between 30 and 60 s.
- Specific applications
  - Asset tracking
  - Fixed monitoring – remote surveillance
  - Fixed monitoring – remote telemetry
  - Fixed monitoring – remote tracking
  - Friendly force tracking
  - Mobile monitoring – remote surveillance
  - Mobile monitoring – remote telemetry
  - Mobile monitoring – remote tracking
  - Remote personnel tracking
- General applications
  - GPS location data look-up-and-send
  - SCADA
  - Short message email

IsatData Pro is a global two-way short message service for M2M communication. It enables companies to track and monitor their fixed or mobile assets, giving them increased visibility of business operations, enhanced efficiency, and greater safety and security for their assets, cargo, and drivers – while lowering operational costs. It is a low-data-rate service ideal for remote management of fixed assets including tracking and telemetry, IsatData Pro operates in near real-time anywhere in the world. With burst-mode communication and a gateway for store-and-forward messaging, IsatData Pro also offers a convenient web-based portal for adjusting settings. Applications can be run on the terminal to reduce data sent over the air. Suitable for mission-critical applications, it offers a wide range of protocols for data collection. One can send 6,400 bytes and receive 10,000 bytes, with a latency of 15–60 s

Figure 6.17 Iridium global constellation of 66 cross-linked low earth orbit (LEO) satellites.

depending on message size. The service operates over the I–4 global satellite network, with an availability of 99.9% and expected lifespan into the 2020s.

6.3.5.2 Iridium Communications Inc. Iridium is major player in the M2M/MSS market, and the company has significant revenue stream from these services. The firm supports the handheld telephony and maritime markets, in addition to many other services and markets. The Iridium constellation consists of 66 active polar-orbit satellites and six additional in-orbit spares that provide global voice and data services to satellite phones and to/from integrated transceivers (the company also has nine ground spares for additional backup support). Iridium owns and operates the constellation and resells equipment and access to its MSS services. The satellites are in LEO, operating at a height of about 500 miles above the earth (Figure 6.17). The satellites are "cross-linked" and communicate with neighboring satellites via Ka-band intersatellite links (each satellite can have up to four intersatellite links: two to neighbors fore and aft in the same orbital plane, and two to satellites in neighboring planes to either side). With Iridium NEXT, expected to begin launching in 2015, Iridium is on schedule to fully replace its current satellite constellation over time and introduce new enhancements and innovations through higher data speeds and increased bandwidth. Iridium M2M will also offer greater transmitting power, able to penetrate buildings and other structures, smaller antennas and transceivers, and lower latency. It will offer backward capability with existing Iridium-connected applications.

A recent listing of the services they offer includes the following (Figure 6.18):

- Satellite phones
- Iridium Wi-Fi solutions
- Iridium Pilot®
- Modems and modules
- Broadband services
- Voice services
- Data services
- LRIT (Long-range identification and tracking)
- Tracking and monitoring
- Personal tracking and monitoring

Phones Iridium satellite phones provide rugged handsets connected to both fixed and mobile voice terminals, supporting global communications for marine, land, and air applications over Iridium's network.

Iridium Wi-Fi Solutions Iridium has developed a suite of Wi-Fi-enabling accessory and software solutions, giving users the ability to use smartphones, tablets, and laptops beyond the reach of existing terrestrial networks by having satellite links behind the locally deployed access points. Combined with your Iridium satellite phone, Iridium AxcessPoint creates a Wi-Fi hotspot, wherever the user is in the world. Usage charges are billed per minute once the Internet connection is established, until the user disconnect your Iridium AxcessPoint.

Iridium Pilot® Iridium provides an integrated antenna, modem, and receiving station to handle broadband data and high-quality voice in one low-cost platform. Options include the following:

- Iridium OpenPort® (being phased out after 2016, *and is been replaced by its second generation maritime solution, Iridium Pilot™. Product support will be available until 2016*).
- Iridium Pilot
- Iridium Pilot Land Station

Iridium Pilot is positioned by the vendor as a next-generation global maritime communications. The outdoor antenna is engineered to perform in blazing sun, frigid cold, or high wind conditions. The ship-borne equipment supported by the Iridium OpenPort broadband is packet switched service that offers highly reliable broadband voice and data communications priced on a number of different pricing options. OpenPort broadband offers the largest and only truly global commercial communication network, providing pole-to-pole coverage for ships at sea, it works everywhere on

Figure 6.18 Iridium equipment and services (press time). Courtesy: Iridium.

Voice Services	Data Services	LRIT	Tracking and Monitoring	Personal Tracking and Monitoring	Voice and Data Terminals
Iridium GoChat®	Iridium Burst	Iridium LRIT	TRIG	TRIG	
Iridium Prepaid	Iridium SBD	BlueTraker LRIT	TracerTrak	NALSHOUT	
Iridium Postpaid	Iridium SMS	THORIUM LRIT	GSatTrack	Spider S3	
Netted Iridium	Iridium RUDICS	WatchDog 750 LRIT	GeoProWeb	MOB Guardian	
	Netted Iridium		Two10degrees GAP	GeoPro	
	SkyTrac ISAT-200R		SATTRANS Navigator	EPSILON Tracking	
	WebSentinel		FMS MLT-400i	NALSHOUTnano	
			Datalink Tracker i50	Yellowbrick 3	
			ECT Data Trans Sys	InReach	
			M3i Buoy	Briartek Cerberus	
			BlueTraker VMS	HawkEye PT	
			Latitude SkyNode S100	Track24 Whisper	
			Latitude SkyNode S200	inReach SE	
			Latitude GeoNode	T200G Remote Monitor	
			Satcourier	Spidertracks	
			ZUNIBAL V77	Field Tracker 2000	
			GeoPro	Field Tracker 2100	
			Mobiltex RMU3		
			Connectport X5		
			Yellowbrick 3		
			WebSentinel		
			InReach		
			SkyBitz GTP1000		
			SkyBitz GTP1100		
			Guardian 5		
			AFIRS228B		
			Track24 C4i Platform		
			HawkEye 5200		
			SkyRouter		
			iSLDMB		
			GIT X-Track™		
			GTTS-3000		
			IridiumLink		
			v2track V5		
			GIT GAT100L		
			SkyTrac ISAT-200A		
			SkyTrac Cockpit Interface		
			Cartoweb		
			Text Anywhere		
			SkyWeb		
			Blue Traker® SSAS		
			Proximity T°		
			GTTS-2000B Tracking		
			Blue Sky D2000		
			Blue Sky D2000A		
			Blue Sky D1000C		
			Blue Sky D10000A		
			NAL 9522B SatTracker		
			iTrac VMS		
			iSVP Drifter		
			iSPHERE Tracking		
			GMN XTracker		
			FLYHT afirs UpTime		
			waySmart 820 RTS		

Iridium LRIT

WatchDog 750 LRIT System

BlueTraker LRIT

THORIUM LRIT Terminal CLS-TST 100

Figure 6.18 *(Continued)*

the planet (for ships traveling in the A4 region, Iridium OpenPort broadband service is the only broadband option available). The system and service can also work as a backup to a larger VSAT service that the ship may have. The linked LEO satellites provide inherent advantages over GEO satellite constellations, supporting low latency links; furthermore, because the Iridium OpenPort broadband service operates in the L-band spectrum, it is mostly impervious to inclement weather. Many customers with mission-critical communications requirements rely on the Iridium OpenPort Service to back-up to VSAT, or other communications on their vessel. The Iridium OpenPort broadband service can substantially reduce the satellite communications costs compared to mini-M, Fleet, or even FleetBroadband, while providing better reliability and throughput. The Iridium OpenPort broadband service offers a wide range of customizable voice and data plans, allowing the users to choose what best suits their needs and their budget.

Some features of the equipment package include the following:

Antenna
- Height: 9.06 in. (230 mm)
- Diameter: 22.44 in. (570 mm)
- Weight: 24.25 lb (11 kg)

Below-Decks Unit
- Height: 7.78 in. (200 mm)
- Width: 9.84 in. (250 mm)
- Diameter: 2.17 in. (55 mm)
- Weight: 2.98 lb (1.35 kg)

Data
- Up to 134 kbps bidirectional with Iridium OpenPort broadband service

Telephony
- Three independent phone lines

Coverage
- Global, pole-to-pole

Iridium Pilot Land Station is targeted to the land mobile market; also it can be used in fixed environment in remote locations. The equipment is engineered for durability in harsh environments and delivers reliable communications performance and value in a variety of fixed and mobile applications – everywhere on the planet; features include the following:

- Supports data applications such as email, web browsing, or social media
- Powers up to three simultaneous voice calls, even during data transmission
- Simply mounts on buildings for quick deployment in remote locations

Modems and Modules   Iridium has developed an array of transceivers, modems, and receivers that can be integrated by other partners into products they may be developing. For example, available to registered Iridium partners, the Iridium 9522A Satellite Transceiver sends and receives data from equipment anywhere on the planet. It functionally supports all of Iridium's voice and data services and easily integrates into a wide variety of applications through the RS232 serial interface and AT command set. Some of the products include the Quake Q4000i, the Quake Q9612 Modem, and the SkyBitz GLS9602 Modem.

Data services include the following:

- Iridium Burst
- Iridium SBD
- Iridium SMS
- Iridium RUDICS
- Netted Iridium®
- SkyTrac ISAT-200R
- Latitude WebSentinel

A brief description of some M2M-relevant services follows.

- Iridium Burst supports a one-to-many global data broadcast service, outbound to remote devices. Delivers data to an unlimited number of enabled devices within a targeted geographic region at a fraction of the cost of comparable services. Iridium Burst service receivers are small, light, and meet many environmental standards.
- Low-cost Short Burst Data (SBD) is an example of an M2M service from Iridium. This service enables user devices to send periodic time-based messages, GPS coordinates, and external sensor data automatically; the use sets up trigger thresholds for the sensors that initiate reports, send alerts, or data transmissions. The service is geared to low-bandwidth applications including telnet sessions. Applications include environmental remote monitoring systems; large animal/herd tracking systems; remote equipment tracking; networks of remote stations for monitoring waste water systems, power network systems, oil delivery systems; and agriculture analysis and monitor system. These systems also allow computer-to-satellite-to-Internet communication, voice communication (e.g., Voice Iridium Privacy Handset), analog voice communication, chat SMS, GPS data acquisition, mail weather blog, and Grab & Go, and sensors connectivity.
- Router-based Unrestricted Digital Internetworking Connectivity Solution (RUDICS) is an example of an M2M service from Iridium. It provides an enhanced gateway termination and origination capability for circuit switched data calls across the Iridium satellite network. When a user places a call to the RUDICS Server located at the Iridium Gateway, the RUDICS Server connects

the call to a predefined IP address allowing an end-to-end IP connection between the Host Application and the user.

- The Netted^{SM} Iridium concept combines voice communication and positioning services with less than two-second push-to-talk capability – all in a portable, handheld device. It is currently available only through the Department of Defense's Iridium gateway, but we believe it has benefits to offer to a wide range of organizations.

LRIT (Long-Range Identification and Tracking)   Iridium's low-latency LRIT solutions provide complete coverage for all seas, including polar sea areas. It allows naval operators to improve maritime safety with tracking, monitoring control, and surveillance of ships, and gather the information needed to maintain optimal operations. Service elements include the following:

- Iridium LRIT
- BlueTraker LRIT
- THORIUM LRIT Terminal CLS-TST 100
- WatchDog 750 LRIT System

Tracking and Monitoring   Iridium offers an array of tracking and monitoring services and hardware for both "machines" and "people." From shipping to delivery, these devices are designed to provide organizations with end-to-end tracking. By knowing where items are when they are on the move, asset utilization and productivity are improved and theft/recovery costs are reduced. Figure 6.18 listed some of the equipment used for these services, and the interested reader should consult the provider's web site.

These capabilities fit well under the M2M rubric, which is the topic of this chapter.

Iridium M2M   "Iridium M2M" is an integrated capability offered by the firm. Iridium M2M leverages the power of the Iridium's network to extend the high value of intelligent M2M data flows beyond the barriers of terrestrial networks. The company states that the Iridium satellite network offers truly global M2M service with the lowest latency in the industry – extending the value of intelligent data far beyond the 10% of the earth serviced by terrestrial networks. With an ecosystem of over 240 partners leveraging Iridium's core technology to innovate and advance M2M data exchanges, Iridium M2M is the catalyst for cutting-edge solutions across a wide range of markets. Iridium M2M provides the following:

- 100% truly global coverage
- Low latency; near real-time data delivery
- Higher transmission speeds than terrestrial networks
- One-price global coverage
- Low-cost next-generation ultralight transceivers

As noted earlier, Iridium Burst^{SM} provides cost-effective one-to-many M2M communication. Iridium Burst is the world's first one-to-many global data broadcast service. Available as an on-demand or dedicated service, Iridium Burst offers secure, simple, cost-effective, and low-latency one-to-many transmission of data to an unlimited number of enabled devices within a targeted geographic region – at a fraction of the cost of comparable services. From command and control of troops and deployed assets to software, firmware and electronic billboard updates, to warnings about tsunamis, fires, earthquakes, tornados, and hurricanes, Iridium Burst transmits highly actionable, customer-specific critical information anywhere on the planet.

6.3.5.3 Orbcomm Orbcomm offers M2M products and services, including satellite and cellular connectivity (in conjunction with global network partners), asset tracking and monitoring devices, satellite modems, RFID tags, and web reporting applications. Orbcomm makes use of a cadre of Value Added Resellers (VARs) to place their M2M products into the market, but they are in the process of rationalizing that model through the acquisition of a number of VARs (approximately half-a-dozen in recent years). Orbcomm generate a significant portion of their revenue (approximately 50%) from sales of wholesale satellite capacity on its own network.

The company launched six next-generation OG2 satellites from Cape Canaveral Air Force Base in Florida aboard a dedicated SpaceX Falcon 9 rocket in June 2014. Eventually, 17 second-generation satellites will be launched to enable the company to deliver adequate service quality. The OG2 satellites are more advanced than its current OG1 satellites and provide existing customers with enhancements, such as faster message delivery, larger message sizes, and better coverage at higher latitudes, while drastically increasing network capacity. In addition, the OG2 satellites are equipped with an AIS payload to receive and report transmissions from AIS-equipped vessels for ship tracking and other maritime navigational and safety efforts, increasing asset visibility and the probability of detection for Orbcomm's AIS customers.

As an illustrative example, around press time Orbcomm announced that its partner, Savi Technology (Savi), has been awarded a 5-year US Department of Defense contract as the sole provider of the RFID-IV program. The RFID-IV program offers a range of technology solutions, including newly upgraded RFID, satellite-based Enhanced In-Transit Visibility (EITV), and Advanced Intrusion Tracking Detection (ATID) products and services to a wide range of government customers for global asset planning and tracking of personnel, equipment, and sustainment cargo worldwide. In support of the RFID-IV program, Orbcomm and Savi will offer satellite and RFID tags as well as other sensor technologies using Orbcomm's global communications networks, which will enhance the visibility and security of government cargo in transit. Orbcomm's tracking and monitoring solutions will provide government customers with the current location of their cargo and send real-time alerts for security breaches and other anomalies such as cargo entering or exiting a geozone outside of predetermined parameters for immediate resolution. US government customers, including the Army, Navy, Air Force, USTRANSCOM, Special Forces, and Defense Logistics Agency, can now seamlessly track and secure their cargo anywhere in the world and access near-real-time operational data through existing government EITV systems or secure, dedicated web portals to meet specific mission needs.

The firm offers a range of devices to track and monitor assets anywhere in the world with complete global coverage. Orbcomm's dual-mode devices utilize both satellite and cellular communications to manage remote and mobile assets for optimal connectivity, cost, and global coverage (see Table 6.5 for illustrative examples of M2M devices).

6.3.5.4 Globalstar Globalstar is a provider of mobile satellite voice and data services. Globalstar offers these services to commercial and recreational users in more than 120 countries around the world. The firm's products include mobile and fixed satellite telephones, simplex and duplex satellite data modems, and flexible service packages. Many land-based and maritime industries benefit from

TABLE 6.5   Illustrative Examples of M2M Devices (Orbcomm)

Device	Features
GT 500	The ruggedized GT 500 is an environmentally hardened asset tracking solution for the transportation, oil and gas, and other industrial markets. This one-way satellite-based device is fully waterproof and completely self-contained. The GT 500 provides accurate position information and motion detection capabilities for assets worldwide.
GT 600	The GT 600 is a self-contained, reliable intrusion detection and tracking solution for the transportation, oil and gas, and other industrial markets. Utilizing one-way satellite communications, the GT 600 supports a security cable that can be used to wrap around container locking bars or other latching mechanisms for maximum cargo security.
GT 1000	The GT 1000 is a cellular-enabled electronic bolt seal that sets the standard for enhanced cargo security and visibility. Codeveloped with the world's largest seal company, Tyden Brooks, the GT 1000 is targeted for theft prevention and recovery as well as virtual warehouse and in-transit visibility applications for intermodal containers in transit around the world.
GT 1100	The award-winning GT 1100 is a self-powered, solar recharging GPS trailer tracking and monitoring solution with low power consumption, long service life in the field and efficient messaging via cellular communications. Its small form factor and flexible design make it easy to install on trailers or in the groove of intermodal containers to enable greater asset visibility and utilization.
GT 2300	The GT 2300 is a container tracking device with cellular communications and GPS, providing visibility into the status, movement, loading, and unloading transactions of dry intermodal container operations. By leveraging powerful telematics capabilities integrated with intermodal information systems, domestic container operators can optimize movement and accelerate utilization.

(continued)

TABLE 6.5 *(Continued)*

Device	Features
RT 6000+	RT 6000+ is a two-way reefer monitoring and control device that provides comprehensive temperature, fuel management, maintenance, and logistical applications services for temperature-controlled cargo. The solution provides visibility, control, and decision rules to transportation companies worldwide, providing maximum compliance, efficiency, and return on investment.
GT 2000	The GT 2000 is an economical dual-mode cargo security and monitoring device that enables location tracking and intrusion detection of cargo containers and other assets. The device's low-profile enclosure facilitates mounting beneath container or trailer locking bars. Security cables can also be threaded through the holes in the metal plate and around the locking bars for added security.
GT 3000	The GT 3000 is a dual-mode cargo tracking, security, and situational monitoring device that facilitates near-real-time asset management throughout the supply chain. The device's highly reliable sensor suite detects intrusion and tampering for maximum container security and visibility.
GT 3100	The GT 3100 is a fuel sensor and volume monitoring solution that provides a significant deterrent against fuel theft and pilferage. By providing valuable operations data and reporting from fuel sensors mounted in the tanker, the device enables continuous and reliable monitoring of fuel levels to maximize the security and accountability of fuel transport even in the most remote and hostile areas of the world.
HE 4000	The HE 4000 is a ruggedized, dual-mode tracking, and monitoring solution that can used for a broad range of heavy equipment types and sizes in the construction, rail, and utility industries. The device provides accurate and timely status and position information along with key operational metrics, so OEMs, dealers, and end-users can proactively manage their fleet anywhere in the world.

Globalstar services available in remote areas beyond cellular and landline service (Figure 6.19). Global customer segments include oil and gas, government, mining, forestry, commercial fishing, utilities, military, transportation, heavy construction, emergency preparedness, and business continuity as well as individual recreational users. Globalstar data solutions are ideal for various asset and personal tracking, data monitoring, and SCADA applications.

Globalstar successfully launched its final batch of second-generation satellites in early 2013. Once placed in service, Globalstar will be positioned as the first MSSs provider to complete the deployment of a second-generation constellation of LEO satellites. Combined with the company's affordable and award-winning suite of consumer retail SPOT products (e.g., see Figure 6.20), Globalstar will be uniquely positioned to offer the world's most extensive lineup of highly reliable and lowest priced MSSs to the broadest range of customers around the globe.

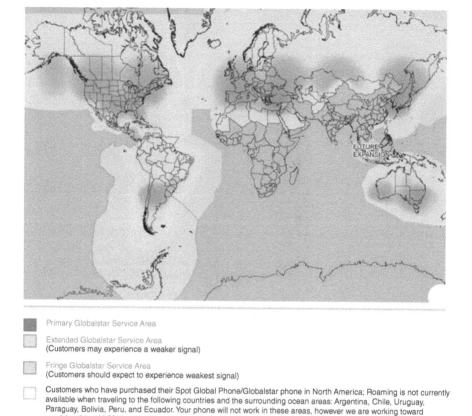

■ Primary Globalstar Service Area

▦ Extended Globalstar Service Area
(Customers may experience a weaker signal)

▦ Fringe Globalstar Service Area
(Customers should expect to experience weakest signal)

☐ Customers who have purchased their Spot Global Phone/Globalstar phone in North America; Roaming is not currently available when traveling to the following countries and the surrounding ocean areas: Argentina, Chile, Uruguay, Paraguay, Bolivia, Peru, and Ecuador. Your phone will not work in these areas, however we are working toward resolving this mid-2014.

☐ Coverage in Venezuela, Colombia, French Guiana, Guyana, Suriname, Trinidad and Tobago is currently experiencing technical issues impacting connectivity and service levels. Issues are identified, and we are working to resolve them. Full restoration of service expected 2Q 2014.

Coverage may vary. Map denotes coverage for satellite two-way voice and duplex data only.

Central America, Nigeria and Singapore coming to service date are subject to change.

Figure 6.19   Globalstar coverage. Courtesy: Globalstar.

6.3.5.5 SkyWave In contrast to Iridium and Globalstar, SkyWave reportedly sees considerable opportunity for ISatDataPro in SCADA applications, and now has a large share of its subscribers using dual-mode satellite-cellular devices. In particular, the SCADA segment has a rather higher ARPU (and is less price sensitive) than Sky-Wave's existing ISatM2M subscriber base, especially those using SkyWave solutions as an antitheft device in cars, so even if equipment volumes are slightly lower, Sky-Wave anticipates substantial revenue growth based on its close relationship with key systems integrators. Given that a key differentiator between existing BGAN M2M and ISatDataPro terminals, even in the fixed asset monitoring market, is that ISatDat-aPro operates with a much smaller omnidirectional antenna, if omni-BGAN can come close to ISatDataPro on equipment pricing, while maintaining current per Mbyte price levels ($5–$20 per Mbyte, compared to several hundred dollars per Mbyte for

Courtesy: Globalstar

Figure 6.20   SPOT examples.

ISatDataPro), then omni-BGAN may end up being in a strong position to compete for SkyWave customers [FAR201201].

6.3.5.6 SkyBitz SkyBitz provides global remote asset management solutions, providing real-time information on the location and status of assets. More than 700 enterprises rely on SkyBitz technology to achieve total asset visibility, improved security, lower operating and capital expenses, and enhanced customer service. SkyBitz delivers its solution via SkyBitz Insight, a secure web-based application that is fully customizable and requires no software downloads. SkyBitz specializes in real-time decision-making tools for companies with unpowered assets such as tractor-trailers, intermodal containers, rail cars, power generators, heavy equipment, and other assets. SkyBitz's asset tracking solution is delivered to commercial, transportation, military, and public safety customers, including sensitive shipment haulers of Arms, Ammunition and Explosives (AA&E) cargos.

As illustrative examples, the SkyBitz Galaxy series GTP1000 and GTP1100 are state-of-the-art, two-way global tracking and monitoring solutions built on the SkyBitz GLS9602 modem platform. The SkyBitz GLS9602 modem combines SkyBitz patented Global Locating System (GLS) technology with the Iridium® 9602 SBD transceiver and can be integrated into a wide variety of asset tracking solutions. These can then provide connectivity to even the most remote locations on earth via the global satellite communications network operated by Iridium Communications Inc. The new SkyBitz Galaxy series delivers operations manager's total asset and cargo visibility everywhere in the world – even if those assets are unpowered. The solution delivers the visibility required to optimize operations within transportation, oil and gas, chemical, intermodal, government, and other markets that rely on heavy equipment or remote assets.

Founded in 1992 and commercially operated since 2002, SkyBitz has gained valuable operational experience with over 800 large public, private, and public safety customers and hundreds of thousands of assets tracked on a daily basis. SkyBitz leveraged funding from DARPA (Defense Advanced Research Projects Agency) to develop its GLS system and an efficient satellite communications protocol. This protocol provides a more secure platform than traditional GPS and satellite communications solutions. The result is a product that excels in power efficiency, accuracy, and reliability. To save battery power and manage resources more effectively, the SkyBitz GLS-based communicator provides the position of an asset through the computing power of the SkyBitz Data Center, rather than in the device.

6.3.5.7 ARINC ARINC[9] (now a division of Rockwell) is a provider of aviation connectivity services including flight planning and high-speed data. Their services are captured under the GLOBALink banner of voice and data communications. ARINC developed the Aircraft Communications Addressing and Reporting System (ACARS)™ designed to transmit maintenance data back to the ground. Aeronautical communication typically spans the following elements, among others: Ku/Ka satellite connectivity, Iridium-based L-band solutions, ACARS®, and expanded capabilities of VDLM2.

While the services suite encompasses voice and general data transmission, we cover these ARINC services under an M2M rubric because many of the telemetry functions are intrinsically M2M functions, and will likely evolve to M2M-oriented solutions in the future.

The ARINC GLOBALink networks include the VHF ACARS, the VHF Data Link (VDL) system, the High Frequency Data Link (HFDL) system, the Aeronautical Satellite Communications (SATCOM) network, the Iridium Satellite, and GateFusion, an ARINC implementation of the GateLink system characterized in the AEEC standards. VHF data link extends throughout all of North and Central America as well as most of Europe and Asia and is 10 times faster than standard ACARS. Inmarsat satellite service expands VHF capabilities to include real-time reporting of flight and weather information. HF covers remote polar regions, while Iridium is used to fill in intermittent oceanic and polar coverage. Each of these communication subnetworks includes a series of RF channels, multiple ground stations, and a central processor as shared resources for all airline data link avionics.

ACARS is a trusted, proven, reliable communications system used throughout the aviation world. Over 300 airlines and 15,000 aircraft around the world rely on ACARS and the GLOBALink[SM] infrastructure for these critical communications. The technology has been in use for over 30 years. The system delivers more than 7.5 billion messages a year worldwide; it provides global coverage including North America, Central America, most of Europe and Asia, oceanic, polar, and remote regions

VDLM2 expands on ACARS capabilities to offer even greater bandwidth and a more extensive suite of flight information, operational control, and ATC applications than ever before. With a capacity more than 10 times that of ACARS, pilots can use the weather graphics, electronic charts, and aircraft monitoring programs now readily

---

[9]This information is based on vendor's materials.

available to boost flight efficiency and safety. ARINC Direct data link communications capabilities include the following:

- Support for all avionics types
- Uplink flight plans and weather to the FMS
- Uplink flight plans from ARINC and third-party providers
- Uplink Coded Departure Routes (CDRs) as needed
- Send and receive emails, faxes, and text (SMS) messages
- Message via Type B (IATA) and ICAO (AFTN) networks
- Concierge and message delivery services
- Text and graphical weather
- METARs/TAFs/PIREPs/SIGMETs/NOTAMs, etc.
- Uplink winds and temperature aloft
- Aircraft-to-aircraft messaging
- Email to aircraft

ARINC Direct SATCOM communications allow operators of private aircraft continuous coverage anywhere. It also provides real-time information on departures, destinations, movement times, engine parameters, delays, positioning, maintenance, and winds aloft. ARINC's satellite communications solutions include the following:

- SATCOM voice and data services
- Inmarsat solutions including support for Aero-H/H+/I, Mini-M, Swift64
- Solutions for legacy networks such as Aero-C
- Iridium data link solutions
- Satellite connectivity to FSS and oceanic ATC facilities
- Satellite graphical weather support
- Single-number fleet dialing through ARINC Direct Dial™
- Automatic detection of incoming faxes
- Online interface to view real-time activity
- Seamless L-band and Ku-band switching (vice versa)
- Managed service with no minimum contracts
- Online interface to view real-time activity
- Simplified pricing and billing

As part of a total communications solution, ARINC Direct delivers its customers global data link services using Iridium Data Link Solutions, thus ensuring the timely, accurate delivery of aircraft operational control (AOC) messages. Two-way text messaging, flight movement data, text and graphical weather, Notice to Airmen (NOTAM) alerts, and in-flight route planning are just a few of the applications made possible by Iridium satellite services around the world. Iridium also provides redundancy for satellite services while requiring minimal equipage or upgrade costs, creating a cost effective and vital communications service for your aircraft.

Pilots need to be able to access critical operational and safety information as quickly and efficiently as possible. ARINC offers ATS Data Link Services that makes this possible. ARINC's ATS solutions minimize cockpit workload while providing accurate real-time information. Data Link ATS services include the following:

- Predeparture Clearances (PDC and DCL)
- Oceanic Clearance Delivery (OCD)
- FANS, ADS, and CPDLC capability (approved avionics)
- D-ATIS, TWIP, flow reports
- North Atlantic Track (NAT) messages

ARINC Direct also supports flight deck weather graphics for a variety of avionics types, including next-generation flight decks with Honeywell Primus Epic® or Rockwell Collins Pro Line 21® avionics suites.

6.4 COMPETITIVE WIRELESS TECHNOLOGIES

This section surveys basic wireless technologies to support IoT/M2M applications, as it appears that many such implementations will entail wireless connectivity (rather than wireline connectivity) at the PHY/MAC layer.

Available wireless networks[10] that can be utilized for IoT/M2M applications include the ones listed in Table 6.6. Many of these technologies are PAN technologies and are not directly applicable to a satellite-focused discussion; however, because 3G/4G cellular (UMTS, GPRS, EDGE, and LTE) can be an alternative to satellite in land mobile environments, a brief description of some of these technologies follows.

Developers of IoT/M2M applications that are geographically dispersed over a city, region, or nation may find cellular networks to be the practical connectivity technology in lieu of the development of specialized new wireless networks. In the near future, M2M applications are expected to become important sources of traffic (and revenues) for cellular data networks. For example, energy suppliers routinely utilize SCADA-based systems to enable remote telemetry functions in the power grid. Traditionally, SCADA systems have used wireline networks to link remote power grid elements with a central operations center; however, at this time and increasing number of utilities are turning to public cellular networks to support these functions. Naturally, reliability and security are key considerations; endpoints typically will support VPN built on IPSec mechanisms in addition to other embedded firewall capabilities.

In starting the discussion about terrestrial mobile networks, one should keep in mind that, as implied in Table 6.2, IoT/M2M traffic has specific characteristics that relate to the priority of the data being communicated, the size of the data,

---

[10]Some refer to the entire "wireless networks" field as Wireless Information and Communication Technology (WICT).

the real-time streaming needs on one end of the requirements spectrum to the extremely high delay tolerance of the data on the other end of the requirements spectrum, and varying degrees of mobility; cellular/mobile networks are characterized by varying capacity, bandwidth, link conditions, and link utilization, and overall network load, that affect their ability to reliably transfer such M2M data [IEEE201201]. These details have to be reconciled in order to be able to

TABLE 6.6   Key Wireless Technology and Concepts Supporting IoT/M2M Applications

Technology/Concept	Description
3GPP (3rd generation partnership project)	3GPP unites six telecommunications standards bodies, known as "Organizational Partners" and provides their members with a stable environment to produce the highly Reports and Specifications that define 3GPP technologies. These technologies are constantly evolving through – what have become known as – Generations of commercial cellular/mobile systems. 3GPP was originally the standards partnership evolving Global System for Mobile Communication (GSM^{TM}) systems toward the 3rd Generation. However, since the completion of the first Long Term Evolution (LTE) and the Evolved Packet Core specifications, 3GPP has become the focal point for mobile systems beyond 3G. From 3GPP Release 10 onwards – 3GPP is compliant with the latest ITU-R requirements for Internation Mobile Telecommunications (IMT)-Advanced "Systems beyond 3G." The standard now allows for operation at speeds up to 100 Mbps for high mobility and 1 Gbps for low mobility communication. The original scope of 3GPP was to produce Technical Specifications and Technical Reports for a 3G Mobile System based on evolved GSM core networks and the radio access technologies that they support (i.e., Universal Terrestrial Radio Access (UTRA) both Frequency Division Duplex (FDD) and Time Division Duplex (TDD) modes). The scope was subsequently amended to include the maintenance and development of the GSM^{TM} Technical Specifications and Technical Reports including evolved radio access technologies (e.g., General Packet Radio Service (GPRS) and Enhanced Data rates for GSM Evolution (EDGE)) [3GP201201]. The term "3GPP specification" covers all GSM (including GPRS and EDGE), Wideband Code Division Multiple Access (W-CDMA), and LTE (including LTE-Advanced) specifications. The following terms are also used to describe networks using the 3G specifications: UMTS Terrestrial Radio Access Network (UTRAN), Universal Mobile Telecommunications System (UMTS) (in Europe), and FOMA (in Japan).

(continued)

TABLE 6.6   *(Continued)*

Technology/Concept	Description
3GPP2 (third-generation partnership project 2)	3GPP2 is a collaborative third-generation (3G) telecommunications specifications-setting project comprising North American and Asian interests developing global specifications for ANSI/TIA/EIA-41 Cellular Radiotelecommunication Intersystem Operations network evolution to 3G and global specifications for the radio transmission technologies (RTTs) supported by ANSI/TIA/EIA-41. 3GPP2 was born out of the International Telecommunication Union's (ITU) "IMT-2000" initiative, covering high speed, broadband, and Internet Protocol (IP)-based mobile systems featuring network-to-network interconnection, feature/service transparency, global roaming, and seamless services independent of location [3G2201201].
ANT/ANT+	ANT™ is a low-power proprietary wireless technology introduced in 2004 by the sensor company Dynastream. The system operates in the 2.4 GHz band. ANT devices can operate for years on a coin cell. ANT's goal is to allow sports and fitness sensors to communicate with a display unit. ANT+™ extends the ANT protocol and makes the devices interoperable in a managed network. ANT+ recently introduced a new certification process as a prerequisite for using ANT+ branding [SMI201201].
Bluetooth	Bluetooth is a Personal Area Network (PAN) technology based on IEEE 802.15.1. It is a specification for short-range wireless connectivity for portable personal devices initially developed by Ericsson. The Bluetooth Special Interest Group (SIG) made their specifications publicly available in the late 1990s at which time the IEEE 802.15 Group took the Bluetooth work and developed a vendor-independent standard. The sublayers of IEEE 802.15: (i) RF layer; (ii) Baseband layer; (iii) the Link manager; and (iv) the Logical Link Control and Adaptation Protocol (L2CAP). Bluetooth has evolved through four versions; all versions of the Bluetooth standards maintain downward compatibility. Bluetooth Low Energy (BLE) is a subset to Bluetooth v4.0 with an entirely new protocol stack for rapid buildup of simple links. BLE is an alternative to the "power management" features that were introduced in Bluetooth v1.0–v3.0 as part of the standard Bluetooth protocols. (Bluetooth is a trademark of the Bluetooth Alliance, a commercial organization that certifies the interoperability of specific devices designed to the respective IEEE standard.)
DASH7	A long-range low-power wireless networking technology, with the following features:

TABLE 6.6   *(Continued)*

Technology/Concept	Description
	• Range: dynamically adjustable from 10 m to 10 km; • Power: <1 mW power draw; • Data Rate: dynamically adjustable from 28 to 200 kbps; • Frequency: 433.92 MHz (available worldwide); • Signal Propagation: penetrates walls, concrete, water; • Real-time locating precision: within 4 m; • Latency: configurable, but worst case is less than 2 s; • P2P messaging; • IPv6 support; • Security: 128-bit AES, public key; • Standard: ISO/IEC 18000-7; advanced by the DASH7 Alliance.
EDGE (enhanced data rates for global evolution)	An enhancement of the GSM™ radio access technology to provide faster bit rates for data applications, both circuit switched and packet switched. As an enhancement of the existing GSM physical layer, EDGE is realized via modifications of the existing Layer 1 specifications rather than by separate, stand-alone specifications. Other than providing improved data rates, EDGE is transparent to the service offering at the upper layers, but is an enabler for High-Speed Circuit-Switched Data (HSCSD) and Enhanced GPRS (EGPRS). By way of illustration, the General Packet Radio Service (GPRS) can offer a data rate of 115 kbps, whereas EDGE can increase this to 384 kbps. This is comparable with the rate for early implementations of W-CDMA, leading some parties to consider EDGE as a 3G technology rather than 2G (a capability of 384 kbps allows EDGE systems to meet the ITU's IMT-2000 requirements). EDGE is generally viewed as a bridge between the two generations: a sort of 2.5G [ETS201201].
General packet radio service (GPRS)	Packet-switched functionality for GSM, which is essentially circuit switched. GPRS is the essential enabler for always-on data connection for applications such as web browsing and Push-to-Talk over Cellular. GPRS was introduced into the GSM specifications in Release 97, and usability was further approved in Releases 98 and 99. It offers faster data rates than plain GSM by aggregating several GSM time slots into a single bearer, potentially up to eight, giving a theoretical data rate of 171 kbps. Most operators do not offer such high rates, because obviously if a slot is being used for a GPRS bearer, it is not available for other traffic. Also, not all mobiles are able to aggregate all

TABLE 6.6   *(Continued)*

Technology/Concept	Description
	combinations of slots. The "GPRS Class Number" indicates the maximum speed capability of a terminal, which might be typically 14 kbps in the uplink direction and 40 kbps in the downlink, comparable with the rates offered by current wireline dial-up modems. Mobile terminals are further classified according to whether or not they can handle simultaneous GSM and GPRS connections: class A = both simultaneously, class B = GPRS connection interrupted during a GSM call, automatically resumed at end of call, class C = manual GSM / GPRS mode switching. Further data rate increases have been achieved with the introduction of EDGE [ETS201201].
GSM EDGE radio access network (GERAN)	GSM EDGE Radio Access Network (GERAN) is a radio access network architecture based on GSM/EDGE radio access technologies. GERAN is the term given to the second-generation digital cellular GSM radio access technology, including its evolutions in the form of EDGE and, for most purposes, GPRS. The GERAN is harmonized with the UTRAN through a common connectivity to the UMTS core network, making it possible to build a combined network for GSM/GPRS and UMTS. GERAN is also the name of the 3GPP™ Technical Specification Group responsible for its development. Technical Specifications that together comprise a 3GPP system and a GSM/EDGE radio access network are listed in 3GPP TS 41.101.
IEEE 802.15.4	IEEE Standard for Local and metropolitan area networks--Part 15.4: Low-Rate Wireless Personal Area Networks (LR-WPANs). IEEE 802.15.4-conformant devices support a wide range of industrial and commercial applications. The amended MAC sublayer facilitates industrial applications such as Process Control and Factory Automation in addition to the MAC behaviors that support the Chinese Wireless Personal Area Network (CWPAN) standard.
IEEE 802.15.4j (TG4j) medical body area networks	The purpose of Task Group 4j (TG4j) is to create an amendment to 802.15.4, which defines a physical layer for IEEE 802.15.4 in the 2360–2400 MHz band and complies with Federal Communications Commission (FCC) MBAN rules. The amendment may also define modifications to the MAC needed to support this new physical layer. This amendment allows 802.15.4- and MAC-defined changes to be used in the MBAN band [KRA201101].

TABLE 6.6 *(Continued)*

Technology/Concept	Description
Infrared data association (IrDA®)	IrDA is a Special Interest Group (SIG) consisting of about 40 members at press time. The SIG is pursuing a 1 Gbps connectivity link; however, this link only operates over a distance of less than 10 cm. One of the challenges with infrared (IR) signaling is its requirement for line-of-sight requirement. In addition, IrDA is also not very power efficient (power per bit) when compared with radio technologies.
ISA100.11a	ISA SP100 standard for wireless industrial networks developed by the International Society of Automation (ISA) to address all aspects of wireless technologies in a plant. The ISA100 Committee addresses wireless manufacturing and control systems in the areas of the (i) environment in which the wireless technology is deployed; (ii) technology and life cycle for wireless equipment and systems; and (iii) application of Wireless technology. The wireless environment includes the definition of wireless, radio frequencies (starting point), vibration, temperature, humidity, EMC, interoperability, coexistence with existing systems, and physical equipment location. ISA100.11a Working Group Charter addresses [ISA201201]: • Low energy consumption devices, with the ability to scale to address large installations; • Wireless infrastructure, interfaces to legacy infrastructure and applications, security, and network management requirements in a functionally scalable manner; • Robustness in the presence of interference found in harsh industrial environments and with legacy systems; • Coexistence with other wireless devices anticipated in the industrial work space; • Interoperability of ISA100 devices.
Long-term evolution (LTE)	LTE is the 3GPP initiative to evolve the Universal Mobile Telecommunications System (UMTS) technology toward a fourth generation (4G). LTE can be viewed as an architecture framework and a set of ancillary mechanisms that aims at providing seamless IP connectivity between User Equipment (UE) and the packet (IPv4, IPv6) data network without any disruption to the end-users' applications during mobility. In contrast to the circuit-switched model of previous generation cellular systems, LTE has been designed to support *only* packet-switched services.

(continued)

TABLE 6.6 *(Continued)*

Technology/Concept	Description
Near-field communication (NFC)	A group of standards for devices such as PDAs, smartphones, and tablets that support the establishment of wireless communication when such devices are in immediate proximity of a few inches. These standards encompass communications protocols and data exchange formats; they are based on existing radio-frequency identification (RFID) standards including ISO/IEC 14443 and FeliCa (a contactless RFID smart card system developed by Sony, for example, utilized in electronic money cards in use in Japan). NFC standards include ISO/IEC 18092 as well as other standards defined by the NFC Forum. NFC standards allow two-way communication between endpoints (earlier generation systems were one-way systems only. Unpowered NFC-based tags can also be read by NFC devices; therefore, this technology can substitute for earlier one-way systems. Applications of NFC include contactless transactions.
NIKE+	Nike+® is a proprietary wireless technology developed by Nike and Apple to allow users to monitor their activity levels while exercising. Its power consumption is relatively high, returning only 40 days of battery life from a coin cell. It is a proprietary radio that only works between Nike and Apple devices. Nike+ devices are shipped as a single unit: processor, radio, and sensor [SMI201201].
Radio frequency for consumer electronics (RF4CE)	RF4CE is based on ZigBee and was standardized in 2009 by four consumer electronics companies: Sony, Philips, Panasonic, and Samsung. Two silicon vendors support RF4CE: Texas Instruments and Freescale Semiconductor, Inc. RF4CE's intended use is as a device remote control system, for example, for television set-top boxes. The intention is that it overcomes the common problems associated with infrared: interoperability, line-of-sight, and limited enhanced features [SMI201201].
Satellite systems	As discussed in this textbook (and others [MIN200901]), and specifically in this chapter, satellite communication plays a key role in commercial, TV/media, government, and military communications because of its intrinsic multicast/broadcast capabilities, mobility aspects, global reach, reliability, and ability to quickly support connectivity in open-space and/or hostile environments.
UMTS terrestrial radio access network (UTRAN)	A collective term for the NodeBs (base stations) and Radio Network Controllers (RNC) that comprise the UMTS radio access network. NodeB is the equivalent to the base transceiver station (BTS) concept used in GSM. The UTRAN allows connectivity between the UE (user equipment) and the core network.

TABLE 6.6 *(Continued)*

Technology/Concept	Description
Universal mobile telecommunications system (UMTS)	UMTS is a third-generation (3G) mobile cellular technology for networks supporting voice and data (IP) based on the GSM standard developed by the 3GPP (3rd-Generation Partnership Project).
Wi-Fi	Wireless Local Area networks (LANs) based on the IEEE 802.11 family of standards, including 802.11a, 802.11b, 802.11g, and 802.11n [MIN200201]. (Wi-Fi is a trademark of the Wi-Fi Alliance, a commercial organization that certifies the interoperability of specific devices designed to the respective IEEE standard.)
WiMAX	WiMAX is defined as Worldwide Interoperability for Microwave Access by the WiMAX Forum, formed in June 2001 to promote conformance and interoperability of the IEEE 802.16 standard. The WiMAX Forum describes WiMAX as "a standards-based technology enabling the delivery of last mile wireless broadband access as an alternative to cable and DSL."
Wireless M-BUS	The Wireless M-BUS standard (EN 13757-4:2005) specifies communications between water, gas, heat, and electric meters and is becoming widely accepted in Europe for smart metering or Advanced Metering Infrastructure (AMI) applications. Wireless M-BUS is targeted to operate in the 868 MHz band (from 868 to 870 MHz); this band enjoys good trade-offs between RF range and antenna size. Typically, chip manufacturers, for example, Texas Instruments, have both single chip (SoC) and two chip solutions for Wireless M-BUS.
Wireless sensor network (WSN)	A sensor network is an infrastructure comprising sensing (measuring), computing, and communication elements that gives the administrator the ability to instrument, observe, and react to events and phenomena in a specified environment. Typically, the connectivity is by wireless means, hence the term WSN (see reference [MIN200701] for an extensive treatment of this topic).
WirelessHART (aka IEC 62591)	WirelessHART is a wireless sensor networking technology based on the Highway Addressable Remote Transducer Protocol (HART). In 2010, WirelessHart was approved by the International Electrotechnical Commission (IEC) as IEC 62591 as a wireless international standard. IEC 62591 entails operation in the 2.4 GHz ISM band using IEEE 802.15.4 standard radios and makes use of a time-synchronized, self-organizing, and self-healing mesh architecture. WirelessHART/IEC 62591 was defined for the requirements of process field device networks. It is a global IEC-approved standard that specifies an interoperable self-organizing mesh technology in which field devices form wireless networks that dynamically mitigate

(continued)

TABLE 6.6 *(Continued)*

Technology/Concept	Description
	obstacles in the process environment. This architecture creates a cost-effective automation alternative that does not require wiring and other supporting infrastructure [IEC201101].
ZigBee RF4CE specification	The specialty-use driven ZigBee RF4CE was designed for simple, two-way device-to-device control applications that do not require the full-featured mesh networking capabilities offered by ZigBee 2007. ZigBee RF4CE offers lower memory size requirements, thereby enabling lower cost implementations. The simple device-to-device topology provides easy development and testing, resulting in faster time to market. ZigBee RF4CE provides a multivendor interoperable solution for consumer electronics featuring a simple, robust, and low-cost communication network for two-way wireless connectivity. Through the ZigBee Certified program, the Alliance independently tests platforms implementing this specification, and has a list of ZigBee Compliant Platforms offering support for ZigBee RF4CE [ZIG201201].
ZigBee specification	The core ZigBee specification defines ZigBee's smart, cost-effective, and energy-efficient mesh network based on IEEE 802.15.4. It is a self-configuring, self-healing system of redundant, low-cost, very low-power nodes that enable ZigBee's unique flexibility, mobility, and ease of use. ZigBee is available as two feature sets, ZigBee PRO and ZigBee. Both feature sets define how the ZigBee mesh networks operate. ZigBee PRO, the most widely used specification, is optimized for low power consumption and to support large networks with thousands of devices [ZIG201201]. (ZigBee is a trademark of the ZigBee Alliance, a commercial organization that certifies the interoperability of specific devices designed to the respective IEEE standard).
Z-Wave	Z-Wave is a wireless ecosystem that aims at supporting connectivity of home electronics and the user via remote control. It uses low-power radio waves that easily travel through walls, floors, and cabinets. Z-Wave control can be added to almost any electronic device in the home, even devices that one would not ordinarily think of as "intelligent," such as appliances, window shades, thermostats, smoke alarms, security sensors, and home lighting. Z-Wave operates around 900 MHz (the band used by some cordless telephones but avoids interference with Wi-Fi devices). Z-Wave was developed by Zen-Sys, a Danish startup around 2005; the company was later acquired by Sigma Designs. The Z-Wave Alliance was established in 2005; it comprises about 200 industry leaders dedicated to the development and extension of Z-Wave as the key enabling technology for "smart" home and business applications.

cost-effectively utilize cellular technologies for a broad set of applications (while some applications may be less sensitive to cost consideration, many more applications will indeed require optimized connectivity cost metrics). Initial 3GPP efforts have focused on the ability to differentiate MTC-type devices, allowing operators to selectively handle such devices in congestion/overload situations. Specifically, low priority indicator has been added to the relevant UE (User Equipment)-to-network procedures; with this, overload and congestion control are done on both CoN and radio access network based on this indicator [RAO201201].

There are different opinions as to which cellular technologies are practical and/or ideal for M2M. Some proponents claim that many developers are concentrating on 4G products. However, the cost of 4G modules is two times more expensive than 3G modules, and three times more expensive than 2G modules; hence, some proponents only recommend a 4G device if it is going to be deployed in an urban setting, and the cost of connectivity was unimportant. Others argue that if a service provider or organization wanted to deploy an inexpensive system with a short lifespan of 1 or 2 years, they could go with 2G; but if a service provider or organization wanted to build a device to have longevity of around 10 years, then they should consider using 3G [PRI201201].

6.4.1   Universal Mobile Telecommunications System (UMTS)

UMTS is a third-generation (3G) mobile cellular technology for networks supporting voice and data (IP) based on the GSM standard developed by the 3GPP (3rd Generation Partnership Project). UMTS is a component of the ITU IMT-2000 standard set and is functionally comparable with the CDMA2000 standard set for networks based on the competing cdmaOne technology. UMTS can carry many traffic types from real-time Circuit Switched to IP-based Packet Switched.

Universal Terrestrial Radio Access Network (UTRAN) is a collective term for the NodeBs (base stations) and Radio Network Controllers (RNCs) that comprise the UMTS radio access network. NodeB is equivalent to the base transceiver station (BTS) concept used in GSM. The UTRAN allows connectivity between the UE and the CoN. As seen in Figure 6.21, UTRAN contains the base stations, which are called NodeBs, and the RNC; the RNC provides control functionalities for one or more NodeBs.

6.4.2   Long-Term Evolution (LTE)

LTE is the 3GPP initiative to evolve the UMTS technology toward a fourth generation (4G). LTE can be viewed as an architecture framework and a set of ancillary mechanisms that aims at providing seamless IP connectivity between UE and the packet (IPv4, IPv6) data network without any disruption to the end-users' applications during mobility. In contrast to the circuit-switched model of previous generation cellular systems, LTE has been designed to support *only* packet-switched services.

System Architecture Evolution (SAE) is the corresponding evolution of the GPRS/3G packet CoN evolution. LTE/SAE standards are defined in 3GPP Rel. 8

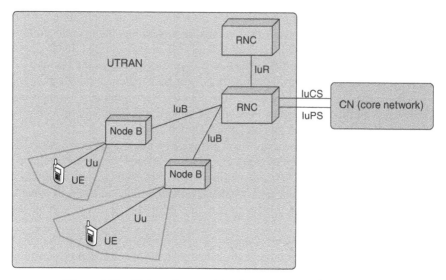

Figure 6.21   Universal terrestrial radio access network (UTRAN).

specifications. Colloquially, the term LTE is typically used to represent both LTE and SAE.

The key element provided by LTE/SAE is the EPS (Evolved Packet System), that is, together LTE and SAE comprise the EPS. EPS provides the user with IP connectivity to a packet data network for accessing the Internet, as well as for supporting services such as streaming video. Figure 6.22 shows the overall network architecture, including the network elements and the standardized interfaces. The EPS consists of the following:

- New air interface E-UTRAN (Evolved UTRAN) and
- The Evolved Packet Core (EPC) network.

Hence, while the term "LTE" encompasses the evolution of the UMTS *radio access* through the E-UTRAN, it is accompanied by an evolution of the *non-radio aspects* under the term SAE, which includes, as just noted, the EPC network.

In principle, LTE promises the following benefits:

- Simplified network architecture (Flat IP based);
- Efficient interworking;
- Robust Quality of Service (QoS) framework;
- Common evolution for multiple technologies;
- Real-time, interactive, low-latency true broadband;
- Multisession data;
- End-to-end enhanced QoS management (see below);

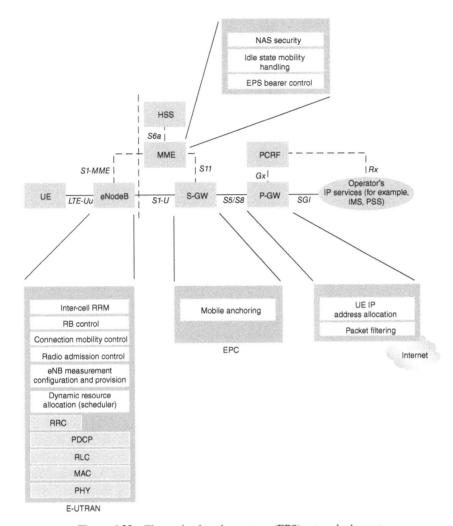

Figure 6.22   The evolved packet system (EPS) network elements.

- Policy control and management;
- High level of security.

The EPS uses the concept of *bearers* to route IP traffic from a gateway in the packet data network to the UE. A bearer is an IP packet flow with a defined QoS between the gateway and the UE. The E-UTRAN and EPC together set up and release bearers as required by applications. An EPS bearer is often associated with a QoS. Multiple bearers can be established for an end-user in order to provide different QoS streams or connectivity to different packet data networks or applications reachable via that network. For example, a user might be engaged in watching a video clip while at the same time performing web browsing or FTP download; a video bearer would provide the necessary QoS for the video stream, while a best-effort bearer would be suitable for the web browsing or file transfer session. This is achieved by means of several EPS network elements that have different roles.

REFERENCES

[3G2201201] Third Generation Partnership Project 2 Organization, http://www.3gpp2.org

[3GP201001] 3rd Generation Partnership Project, Technical Specification Group Services and System Aspects; Service Requirements for Machine Type Communications (MTC); Stage 1 (Release 10); Technical Specification 3GPP TS 22.368 V10.1.0 (2010-06).

[3GP201201] 3rd Generation Partnership Project (3GPP) Organization, www.3gpp.org

[ASH200901] K. Ashton, "That 'Internet of Things' Thing", RFID Journal, 22 July 2009.

[BOT200901] M. Botterman, "*Internet of Things: An Early Reality of the Future Internet*", Workshop Report, European Commission Information Society and Media, May 2009.

[DRA201001] J. Drake, D. Najewicz, W. Watts, "Energy Efficiency Comparisons of Wireless Communication Technology Options for Smart Grid Enabled Devices", White Paper, General Electric Company, GE Appliances & Lighting, December 9, 2010.

[DUK201101] R. Duke-Woolley, "Wireless Enterprise, Industry & Consumer Apps for the Automation Age", *M2M Zone Conference* at the International CTIA Wireless 2011 March 22-24, 2011, Orange County Convention Center, Orlando Florida.

[ETS201002] ETSI TR 102 897: "Machine to Machine Communications (M2M); Use Cases of M2M Applications for City Automation". (2010-01). ETSI, 650 Route des Lucioles F-06921 Sophia Antipolis Cedex – France.

[ETS201003] ETSI TR 102 898: "Machine to Machine Communications (M2M); Use cases of Automotive Applications in M2M Capable Networks". (2010-09). ETSI, 650 Route des Lucioles F-06921 Sophia Antipolis Cedex – France.

[ETS201101] Machine-to-Machine communications (M2M); Functional Architecture Technical Specification, ETSI TS 102 690 V1.1.1 (2011-10), ETSI, 650 Route des Lucioles F-06921 Sophia Antipolis Cedex – France.

[ETS201201] ETSI Documentation, ETSI, 650 Route des Lucioles F-06921 Sophia Antipolis Cedex – FRANCE.

[FAR201201] T. Farrar, "*MSS Industry Perspectives*", Telecom Media and Finance Associates, Inc. Market Report, November 30, 2012. TMF Associates, 3705 Haven Avenue, Suite 113 Menlo Park, CA 94025.

[FAR201301] T. Farrar, "*MSS Industry Perspectives*", Telecom Media and Finance Associates, Inc. Market Report, March 30, 2013. TMF Associates, 3705 Haven Avenue, Suite 113 Menlo Park, CA 94025.

[GLU201101] A. Gluhak, S. Krco, et al. "A Survey on Facilities for Experimental Internet of Things research", Communications Magazine, IEEE, November 2011, Volume: 49 Issue: 11, Page(s): 58–67.

[GRA201301] B. Grantham, Real Space-Based ADS-B, Press Release, June 11, 2013 Globalstar Inc., www.globarstar.com

[GRA201401] B. Grady, "NSR—The Potentials Of The Utility Market – Reports", 2014.

[IAB201101] Internet Architecture Board, Interconnecting Smart Objects with the Internet Workshop 2011, 25th March 2011, Prague.

[IDA201201] IDATE, *"The Satellite M2M Market 2012-2016"*, Report, IDATE Consulting & Research, April 23, 2012, London, UK.

[IEC201101] Emerson Process Management, *IEC 62591 WirelessHART, System Engineering Guide, Revision 2.3*, Emerson Process Management, 2011.

[IEEE201201] IEEE WoWMoM 2012 Panel, San Francisco, California, USA June 25–28, 2012.

[IEEE201301] IEEE Computer, "The Internet of Things: The Next Technological Revolution", Special Issue, February 2013.

[INF200801] *Internet of Things in 2020 – Roadmap For The Future,* INFSO D.4 Networked Enterprise & RFID, INFSO G.2 Micro & Nanosystems in co-operation with the Working Group RFID Of The ETP EPOSS. (European Commission – Information Society and Media). Version 1.1 – 27, May, 2008.

[ISA201201] ISA, 67 Alexander Drive, P.O. Box 12277, Research Triangle Park, NC 27709, info@isa.org

[ITU201101] International Telecommunications Union, Telecommunication Standardization Sector Study Period 2009-2012, "IoT-GSI – C 44 – E", August 2011.

[KRA201101] R. Krasinski, P. Nikolich, R. F. Heile, IEEE 802.15.4j Medical Body Area Networks Task Group PAR, IEEE P802.15 Working Group for Wireless Personal Area Networks (WPANs), January 18, 2011.

[KRE201201] K. Kreisher, "Intel: M2M Data Tsunami Begs for Analytics, Security", Online Magazine, October 8, 2012, http://www.telecomengine.com

[LEE201101] G. M. Lee, J. Park, N. Kong, N. Crespi, "The Internet of Things – Concept and Problem Statement," July 2011. Internet Research Task Force, July 11, 2011, draft-lee-iot-problem-statement-02.txt.

[MCK201301] McKinsey Global Institute, *"Disruptive Technologies: Advances that will Transform Life, Business, and the Global Economy"*, Report, May, 2013.

[MIN200201] D. Minoli, *Hotspot Networks: Wi-Fi for Public Access Locations*, McGraw-Hill, New York, NY, 2002.

[MIN200701] D. Minoli, *Wireless Sensor Networks* (co-authored with K. Sohraby and T. Znati), Wiley, Hoboken, NJ, 2007.

[MIN200901] D. Minoli, *Satellite Systems Engineering in an IPv6 Environment*, Francis and Taylor, Boca Raton, FL, 2009.

[MIN201301] D. Minoli, *Building the Internet of Things with IPv6 and MIPv6*, Wiley, New York, 2013.

[NSR201201] *"Mobile Satellite Services"*, Market Report, 8th Edition, NSR, 2012.

[OEC201201] OECD, "Machine-to-Machine Communications: Connecting Billions of Devices", OECD Digital Economy Papers, No. 192, 2012, OECD Publishing. doi: 10.1787/5k9gsh2gp043-en.

[PEE201201] S. Peerun, *"Machine to Machine (M2M) Revenues Will Reach $38.1bn in 2012"*, Visiongain Report, Visiongain, 230 City Road, London, EC1V 2QY, United Kingdom. 2012.

[PRI201201] B. Principi, "CTIA: Should M2M skip 3G and go right to 4G?", May 9, 2012, Online Article, http://www.telecomengine.com.

[PTC201401] The Internet of Things, PTC Whitepaper, 2014, PTC Inc. Paper J3220–IoT-eBook–EN–214. PTC Corporate Headquarters, 140 Kendrick Street, Needham, MA 02494.

[RAO201201] Y. S Rao, F. Pica, D. Krishnaswamy, "3GPP Enhancements for Machine Type Communications Overview", IEEE WoWMoM 2012 Panel, San Francisco, California, USA June 25–28, 2012.

[SCA201201] E. Scarrone, D. Boswarthick, "Overview of ETSI TC M2M Activities", March 2012, ETSI, 650 Route des Lucioles F-06921 Sophia Antipolis Cedex – France.

[SMI201201] P. Smith, "Comparing Low-Power Wireless Technologies", Tech Zone, Digikey Online Magazine, Digi-Key Corporation, 701 Brooks Avenue, South Thief River Falls, MN 56701 USA.

[STA201101] Staff, "Smart Cards, Mobile Telephony and M2M at the Heart of e-health Services", CARTES & IDentification Conference, Parc des Expositions Paris-Nord Villepinte, November, 2011.

[STA201401] Staff, "M2M Antenna Types", M2M Antennas, Inc., 8145 Ronson Road Suite A, San Diego, CA,US,92111, http://www.m2mantennas.com.

[SST201401] Signals and Systems Telecom, "The Wireless M2M & IoT Bible: 2014 - 2020 - Opportunities, Challenges, Strategies, Industry Verticals and Forecasts (Report)", May 2014. Reef Tower, Jumeirah Lake Towers, Sheikh Zayed Road, Dubai, UAE.

[WAL200701] K. -D. Walter, "Implementing M2M Applications via GPRS, EDGE and UMTS", Online Article, August 2007, http://m2m.com. M2M Alliance e.V., Aachen, Germany.

[WLI201201] W. Li, *Maritime Telecom Solutions by Satellite—Global Market Analysis & Forecasts*, March 2012, A Euroconsult Research Report. 86 Blvd. Sebastopol, 75003 Paris, France. www.euroconsult-ec.com.

[ZIG201201] ZigBee Alliance, http://www.zigbee.org/.

7

ULTRAHD VIDEO/TV AND SATELLITE IMPLICATIONS

This chapter assesses evolving Ultra High Definition video in general (known as UltraHD) and Ultra High Definition Television (UHDTV) services in particular; we refer to both as "UltraHD." UltraHD affords what has been called an immersive viewing experience. The chapter looks at the opportunities for satellite operators in providing distribution services, as well as challenges and issues. Some of the largest satellite operators get 75% of their revenues from video distribution, Direct to Home (DTH) in particular; hence, the topic can be important.

Satellite operators hope that the commercial service prospects for UltraHD are better than those that revolved around 3DTV. Three issues held back the 3DTV market, which are not applicable to, or impact, UltraHD. First, the distribution of 3DTV required only a relatively modest increment in additional bandwidth (around 30%), except for the high-end multiangle applications (which did not even take off the ground), thus putting an upper bound on the net bandwidth (transponder) growth requirements; on the contrary, UltraHD intrinsically requires significant bandwidth augmentation in the transport network (satellite-based or terrestrial) to be delivered, thus offering, in principle, more business potential for the operators. However, this could also act as a retardant to the deployment of the UltraHD technology because of the more onerous set of resources needed and (possibly) the intrinsic cost involved. Second, the 3DTV required not only a new screen but also active or passive glasses to be worn by all content watchers; the latter issue does not impact UltraHD, although the former does. Third, the production of 3DTV video material can be difficult, even when the stereoscopic cameras are available, especially for live programs such as

Innovations in Satellite Communications and Satellite Technology: The Industry Implications of DVB-S2X, High Throughput Satellites, Ultra HD, M2M, and IP, First Edition. Daniel Minoli.
© 2015 John Wiley & Sons, Inc. Published 2015 by John Wiley & Sons, Inc.

sporting events; by contrast, the production of UltraHD video material is technically simpler. On a passing note, we remain convinced that 3DTV-like services may have market potential either on autostereoscopic devices such as smartphones or game consoles, or with direct projection of the 3D content on glass lenses from miniature projectors built into the glasses (kind of a version of Google glasses), obviating the need for a far-end screen and eliminating the angle-of-view restrictions that are present in the far-screen approach to 3D (although glasses are needed in either approaches, the suggestion of simply-projecting-the-content from built-in miniature projectors onto the glasses' lenses is more elegant and can provide crisp video at all viewing angles as was possible with the simple View Master toy of old although that did not involve projecting/reflecting the images). Another approach would be to project the content directly into the eyes, again using miniature projectors built into the glasses, but now pointing towards the eyes themselves.

7.1 H.265 IN THE ULTRAHD CONTEXT

Emerging UltraHD provides video quality that is the equivalent of 8–16 HDTV screens (33 million pixels, for the 7680 × 4320 resolution option, compared to a maximum 2 million pixels for 1920 × 1080 resolution) for the current highest quality HDTV service, which obviously requires a lot more transmission/storage bandwidth, up to 16 × more. Figure 7.1 depicts the pixel density for various schemes, including UltraHD. UltraHD does not obligatorily require a new video coding scheme, but considering the data rates required, it is certainly advantageous to introduce a video compression scheme that provides the desired quality while at the same time being more efficient than existing schemes. The International Telecommunication Union Telecommunications/International Organization for Standardization (ITU-T/ISO) specification H.265 is one such scheme, which we describe next.

High Efficiency Video Coding (HEVC) is a newly developed standard (first stage approval was granted in January 2013) for video compression developed by the ITU-T and ISO; it is a joint publication of ISO/IEC and ITU-T, formally known as ISO/IEC 23008–2 (ISO-MPEG H Part 2) and ITU-T Recommendation H.265. HEVC has the same general structure as MPEG-2 and H.264/AVC, but it affords more efficient compression; namely, it can provide the same picture quality as the predecessor but with a lower data rate (and storage requirement), or it can provide better picture quality than its predecessor but at the same data rate (and storage requirement). There is an extensive published literature on this topic; we provide only a high-level summary in this discussion.

H.265 is seen as a successor to H.264/MPEG-4 AVC. The work was started around 2006 by some of the pertinent committees; the worked picked up steam in 2010, and a stable working draft was developed by 2012 (see Figure 7.2). Key developers of the standard include but are not limited to ATEME, Comcast, CableLabs, Motorola, Sony, Mitsubishi, Harmonic, NBC, Cisco, NEC, LG, Microsoft, and DirecTv.

HEVC is (yet another) hybrid codec, with its main structure being similar to conventional hybrid codecs, but offering enhanced flexibility. The design goals were to

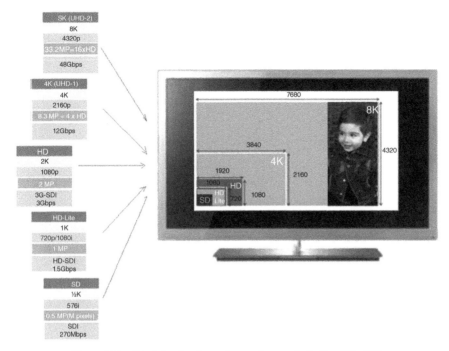

Figure 7.1 Pixel density for various schemes, including UltraHD.

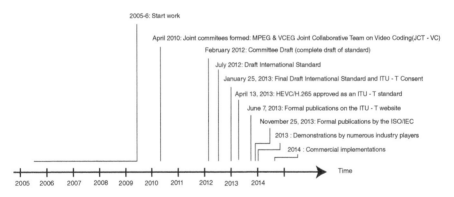

Figure 7.2 Development timetable for H.265.

achieve a compression gain of 50% over H264/AVC, sustaining a maximum 10-fold increase in complexity for the encoder and a maximum 2/3rd increase in complexity for the decoder. The operative feature is that video can be compressed into a smaller size or bit rate; actual savings ranging from 30% to 50% have been cited in the literature (up to $2\times$ better compression efficiency compared to the baseline H.264/AVC algorithm); however, it also has increased computational complexity compared to H.264, requiring more advanced chip sets on all equipment utilizing this technology.

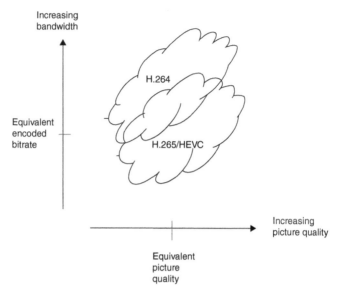

Figure 7.3 Graphical view of metrics related to H.264 and H.265.

Figure 7.3 depicts the value proposition for HEVC. Many demonstrations and simulations were developed in recent years, especially in 2013, and commercial-grade products are expected in the 2014–2015 timeframe, just in time for UltraHD applications. In fact, HEVC is designed to cover a broad range of applications for video content including, but not limited to, the following [BRO201301]:

- Broadcast (cable TV on optical networks/copper, satellite, terrestrial, etc.)
- Camcorders
- Content production and distribution
- Digital cinema
- Home cinema
- Internet streaming, download, and play
- Medical imaging
- Mobile streaming, broadcast, and communications
- Real-time conversational services (videoconferencing, videophone, telepresence, etc.)
- Remote video surveillance
- Storage media (optical disks, digital video tape recorder, etc.)
- Wireless display

Since HEVC provides superior video quality and up to twice the data compression as the previous standard (H.264/MPEG-4 AVC) [X26201401] [ANG201301], the implications for satellite operators are that users (e.g., the Dish Network, DirectTV) may, in principle, require less transponder bandwidth to support the necessary video quality. Or that, considering the ever-increasing bandwidth growth driven by evolving applications such as UltraHD, users will either increase the video quality and retain the same transponder bandwidth, or possibly need more bandwidth when UltraHD services are implemented.

Video frames have intrinsic natural structures and/or repetitive patterns. Hence a pixel's color can be predicted from the color of its neighbors within the same frame (intraframe, also known as "intra") or from recent frames (interframe, also known as "inter") – this is especially the case when motion is involved. The basic approach is to encode a block of pixels as a prediction mode plus a residual, or delta, from that prediction; the delta information is typically smaller (requiring fewer bits) than coding pixel values directly, thus achieving signal compression. Three patterns that are common in video frames are flat regions, smooth gradients, and straight edges. An advanced algorithm deals with these elements in an efficient manner.

Coded video content conforming to the HEVC recommendation uses a common syntax. To achieve a subset of the complete syntax, flags, parameters, and other syntax elements are included in the bitstream that signal the presence or absence of syntactic elements that occur later in the bitstream. The coded representation specified in the syntax is designed to enable a high compression capability for a desired image or video quality. The algorithm is typically not lossless, as the exact source sample values are typically not preserved through the encoding and decoding processes. A number of techniques may be used to achieve highly efficient compression; encoding algorithms (not specified *per se* in the standard) may select between inter and intra coding for block-shaped regions of each picture. Inter coding uses motion vectors for block-based inter prediction to exploit temporal statistical dependencies between different pictures. Intra coding uses various spatial prediction modes to exploit spatial statistical dependencies in the source signal for a single picture. Motion vectors and intra prediction modes may be specified for a variety of block sizes in the picture. The prediction residual may then be further compressed using a transform to remove spatial correlation inside the transform block before it is quantized, producing a possibly irreversible process that typically discards less important visual information while forming a close approximation to the source samples. Finally, the motion vectors or intra prediction modes may also be further compressed using a variety of prediction mechanisms, and, after prediction, are combined with the quantized transform coefficient information and encoded using arithmetic coding [BRO201301]. HEVC can support 8K UltraHD video, with a picture size up to 8192×4320 pixels. Figure 7.4 depicts a HEVC-based coder.

The algorithm provides a recursive quadtree structure for frame partitioning, larger block transforms, more efficient motion compensation and motion vector prediction, additional sample adaptive offset filtering, and an enhancement to Context-Adaptive Binary Arithmetic Coding (CABAC) called *Syntax-B*ased Context-Adaptive Binary Arithmetic *C*oder (SBAC) (CABAC is used in H.264/AVC [MAR200301]). Although improvements in data rate and/or picture quality are achieved, all these improvements significantly increase the encoding complexity and to some degree the decoding complexity; 1080p encodes are expected to be $5-10$ times more taxing, while 4K video

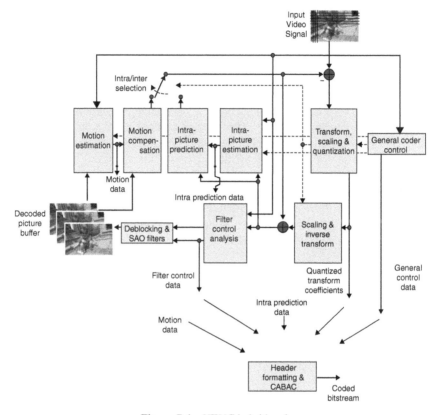

Figure 7.4   HEVC hybrid coder.

multiplies those demands by another 4 to 16 ×; it should be noted that the encoding processing can be parallelized, allowing a way to bring the computing time into an acceptable range (for both stored programming encoding and decoding, and for real-time encoding of live events) [SHA201301], [ANG201301]. Some of the key features of HEVC are [RIC201301]:

- More flexible partitioning
- Increased flexibility in prediction mode and transform block sizes
- More sophisticated interpolation and deblocking filters
- More refined prediction and signaling of modes and motion vectors
- Ability to support parallel processing

Figure 7.5 depicts the basic elements of the process in a satellite environment. As can be seen in the figure, the H.265 encoder starts the process by partitioning the picture on a frame into a group of units. This is followed by a step of predicting each unit using interpolation or intrapolation prediction and then subtracting the

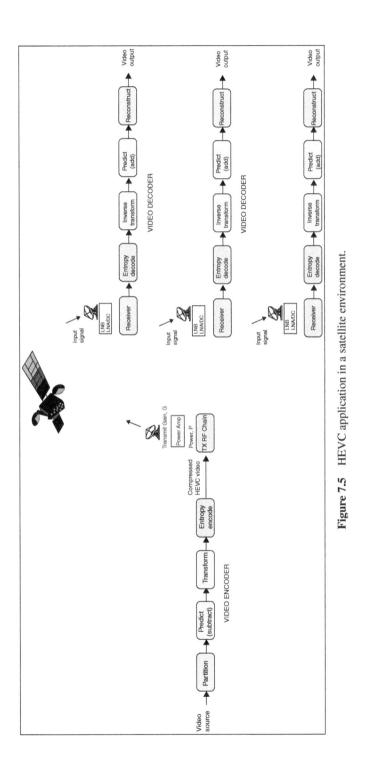

Figure 7.5 HEVC application in a satellite environment.

prediction from the unit itself to create a "delta" residual signal (the residual is the difference between the original video unit material and the predicted unit video). The next step entails transforming and then quantizing the residual signal. The bitstream thus generated is entropy-encoded, along with prediction information, mode information, and appropriate Moving Pictures Expert Group (MPEG) headers. The information is thus transmitted (this will entail adding the Forward Error Correction (FEC) and then modulating the underlying carrier). After the signal is received (the signal is demodulated and the FEC is consumed), the H.265 decoder handles the reverse process. The received stream is entropy decoded, and the elements of the coded sequence are extracted. The transformed signal is inverted and rescaled. The units are predicted, and the prediction is added to the output of the inverse transform. The final step is the reconstruction of the video image (combining the constituent units).

In HEVC (as in other ITU-T and ISO/IEC video coding standards), only the bitstream structure and syntax are standardized in addition to constraints on the bitstream and its mapping for the generation of decoded pictures [SUL201201]. Table 7.1 provides the definition of some key terms used in HEVC.

HEVC effectively doubles the data compression ratio compared to H.264/MPEG-4 AVC at the same level of video quality. The standard is ideally positioned for UltraHD TV displays and content capture systems that utilize progressive scanned frame rates; it efficiently supports display resolutions from Quarter Video Graphics Array (QVGA) resolution (320 × 240) all the way up to 4320p (8192 × 4320). In addition to more efficient data rates it provides improved picture quality as measured by noise level, color resolution, and dynamic range. Benchmarking firms were conservatively estimating at press time a 25–35% lower bit rates at a given Peak Signal to Noise Ratio (PSNR) [ANG201301].

To achieve these efficiencies, the HEVC specification provides a flexible frame representation mechanism by introducing the concepts of coding unit (CU), prediction unit (PU), and transform unit (TU). Individual video frames are partitioned into tiles (also known as slices), which are, in turn, partitioned again into units known as Coding Tree Units (CTUs), comparable to the macroblocks in MPEG-2/4 environments. Instead of H.264's 16 × 16-pixel macroblocks, HEVC employs a CTU that can be as large as 64 × 64, describing less complex areas in the picture more efficiently.

The CTUs are the basic 64 × 64 pixel elements that comprise the basic unit of coding. CTUs are, in turn, subdividable in smaller squares known as CUs, which themselves are partitionable into one or more PUs. The PUs are employed to make the "predictions" using intra(frame) or inter(frame) prediction. In intra(frame) prediction, each PU is predicted from the neighboring image data in the same frame using DC prediction[1] (this uses an average value for the PU), planar prediction (this fits a plane surface to the PU), or directional prediction (this extrapolates from neighboring image data). In inter(frame) prediction, each PU is predicted from image data

---

[1] An averaging sample mode. DC informally refers to "direct current," in the sense that a DC signal is a signal that has a constant value with no non-zero frequency content. So a DC intra prediction mode is a mode in which the prediction signal has a constant value.

TABLE 7.1 Definition of Key Terms Used in HEVC (from Joint Collaborative Team on Video Coding of ITU-T SG16 WP3 Documents)

Term	Definition
Access unit (AU)	A set of NAL units that are associated with each other according to a specified classification rule, are consecutive in decoding order, and contain exactly one coded picture.
Coded video sequence (CVS)	A sequence of access units
Coding block	An N×N block of samples for some value of N such that the division of a coding tree block into coding blocks is a partitioning.
Coding tree block	An N×N block of samples for some value of N such that the division of a component into coding tree blocks is a partitioning.
Coding tree unit (CTU)	A coding tree block of luma samples, two corresponding coding tree blocks of chroma samples of a picture that has three sample arrays, or a coding tree block of samples of a monochrome picture, or a picture that is coded using three separate color planes, and syntax structures used to code the samples.
Coding unit (CU)	A coding block of luma samples, two corresponding coding blocks of chroma samples of a picture that has three sample arrays, or a coding block of samples of a monochrome picture, or a picture that is coded using three separate color planes, and syntax structures used to code the samples.
Context-adaptive binary arithmetic coding (CABAC)	An entropy encoding methodology (a lossless compression technique) used in H.264/MPEG-4 AVC video encoding as well as in the HEVC. It combines an adaptive binary arithmetic coding technique with context modeling, to achieve a high degree of adaptation and redundancy reduction
DC transform coefficient	A transform coefficient for which the frequency index is zero in all dimensions.
encoding process	A process not specified in this in HEVC that produces a bit stream conforming to the H.265 specification
Inter coding	Coding of a coding block, slice, or picture that uses inter prediction.
inter prediction	A prediction derived in a manner that is dependent on data elements (e.g., sample values or motion vectors) of pictures other than the current picture.
Intra coding	Coding of a coding block, slice, or picture that uses intra prediction.
Intra prediction	A prediction derived from only data elements (e.g., sample values) of the same decoded slice.

(continued)

TABLE 7.1 (*Continued*)

Term	Definition
Network abstraction layer (NAL) unit	A syntax structure containing an indication of the type of data to follow and bytes containing that data
Prediction	An embodiment of the prediction process.
Prediction block	A rectangular M×N block of samples on which the same prediction is applied.
Prediction process	The use of a predictor to provide an estimate of the data element (e.g., sample value or motion vector) currently being decoded.
Prediction unit (PU)	A prediction block of luma samples, two corresponding prediction blocks of chroma samples of a picture that has three sample arrays, or a prediction block of samples of a monochrome picture, or a picture that is coded using three separate color planes and, syntax structures used to predict the prediction block samples.
Predictive (P) slice	A slice that may be decoded using intra prediction or inter prediction using at most one motion vector and reference index to predict the sample values of each block.
Residual	The decoded difference between a prediction of a sample or data element and its decoded value.
Transform unit (TU)	A transform block of luma samples of size 8 × 8, 16 × 16, or 32 × 32; or four transform blocks of luma samples of size 4 × 4, two corresponding transform blocks of chroma samples of a picture that has three sample arrays; or a transform block of luma samples of size 8 × 8, 16 × 16, or 32 × 32; or four transform blocks of luma samples of size 4 × 4 of a monochrome picture or a picture that is coded using three separate color planes; and syntax structures used to transform the transform block samples.

in one or two reference pictures (typically before and/or after the current picture in temporal order) using motion-compensated prediction.

The residual information that is present after the prediction difference is transformed into TUs utilizing a Discrete Sine Transform (DST) or a Discrete Cosine Transform (DCT). The block transforms applied against the residual information in each CU is 32×32, 16×16, 8×8, or 4×4 in scope. As noted, HEVC uses an enhancement to CABAC where the entropy coding of transform coefficients have been selected for a higher throughput than H.264/MPEG-4 AVC. A coded HEVC stream comprises quantized transform-resulting coefficients, prediction information (motion vectors and prediction modes), partitioning information, and header information. The entire bitstream comprising these data items is encoded using the CABAC.

In HEVC, intra coding supports 35 prediction modes: a planar mode, a DC mode, and 33 angular prediction modes. HEVC CTUs are an extension of the 16×16 luma samples macroblock in the AVC standard. A CTU encompasses $2N \times 2N$ luma samples, with $N = 4$, 8, 16, or 32 (luma is an adjective used in the standard represented by the symbol or subscript Y or L, specifying that a sample array or single sample represents the monochrome signal related to the primary colors). The larger CTUs support improved compression efficiency. Each CTU represents a top-level CU; each CU can be adaptively partitioned into four sub-CUs until a minimum size is obtained; this forms a recursive quadtree structure. The CU contains the PU that defines the prediction being made. With intra coding, the PU has the same size as a $2N \times 2N$ CU (a bottom-level CU with the minimum size contains four $N \times N$ PUs). The TU contains the transformation data. Adaptive transforms of 4×4, 8×8, 16×16, and 32×32 are supported [SHA201301].

Further to the discussion above, HEVC pictures are divided into coding tree blocks (CTBs); these appear in the picture in raster order. Depending on the stream parameters, the CBTs are 64×64, 32×32, or 16×16. Each CTB can be split recursively in a quadtree structure, all the way down to 8x8. For example, a 32×32 CTB can consist of three 16×16 and four 8×8 regions. These regions are called CUs. Thus, CUs can be 64×64, 32×32, 16×16, or 8×8. CUs are the basic unit of prediction in HEVC. The CUs in a CTB are traversed and coded in Z order. See Figure 7.6 for an example of ordering.

[BRO201301] explains that the samples are processed in units of CTBs. Each CTB is assigned a partition signaling to identify the block sizes for intra or inter prediction and for transform coding. The partitioning is a recursive quadtree partitioning. The root of the quadtree is associated with the CTB. The quadtree is split until a leaf is reached, which is referred to as the coding block. (When the component width is not an integer number of the CTB size, the CTBs at the right component boundary are incomplete; when the component height is not an integer multiple of the CTB size, the CTBs at the bottom component boundary are incomplete.) The coding block is the root node of two trees: the prediction tree and the transform tree. The prediction tree specifies the position and size of prediction blocks. The transform tree specifies the position and size of transform blocks. The splitting information for luma and chroma is identical for the prediction tree and may or may not be identical for

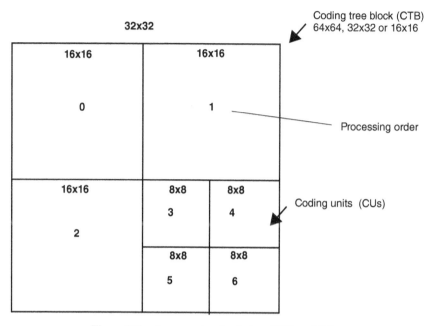

Figure 7.6   An example of ordering, CBT, and CUs.

the transform tree. The blocks and associated syntax structures are encapsulated in a "unit" as follows:

- One prediction block or three prediction blocks (luma and chroma) and associated prediction syntax structures units are encapsulated in a PU.
- One transform block or three transform blocks (luma and chroma) and associated transform syntax structures units are encapsulated in a TU.
- One coding block or three coding blocks (luma and chroma), the associated coding syntax structures, and the associated prediction and TUs are encapsulated in a CU.
- One CTB or three CTBs (luma and chroma), the associated coding tree syntax structures, and the associated CUs are encapsulated in a CTU.

The following divisions of processing elements of the H.265 specification form spatial or component-wise partitionings (also see Figure 7.7):

- The division of each picture into components,
- The division of each component into CTBs,
- The division of each picture into tile columns,
- The division of each picture into tile rows,
- The division of each tile column into tiles,

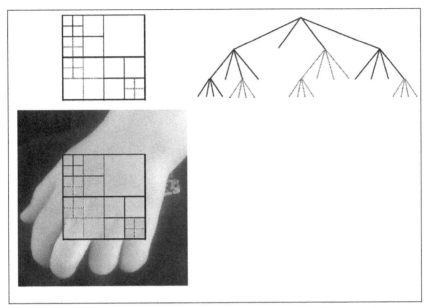

Subdivision of a CTB into CBs [and transform block (TBs)], along with corresponding quadtree
Solid lines indicate CB boundaries and dotted lines indicate TB boundaries.

Figure 7.7   Spatial or component-wise partitionings.

- The division of each tile row into tiles,
- The division of each tile into CTUs,
- The division of each picture into slices,
- The division of each slice into slice segments,
- The division of each slice segment into CTUs,
- The division of each CTU into CTBs,
- The division of each CTB into coding blocks, except that the CTBs are incomplete at the right component boundary when the component width is not an integer multiple of the CTB size, and the CTBs are incomplete at the bottom component boundary when the component height is not an integer multiple of the CTB size,
- The division of each CTU into CUs, except that the CTUs are incomplete at the right picture boundary when the picture width in luma samples is not an integer multiple of the luma CTB size, and the CTUs are incomplete at the bottom picture boundary when the picture height in luma samples is not an integer multiple of the luma CTB size,
- The division of each CU into PUs,
- The division of each CU into TUs,
- The division of each CU into coding blocks,

TABLE 7.2   A Comparison Between Coding Schemes

	HEVC	H.264/AVC
Partition size	(Large) Coding unit 8×8 to 64×64	Macroblock 16×16
Partitioning	Prediction Unit Quadtree down to 4×4 Square, symmetric, and asymmetric (only square for intra)	Sub-block down to 4×4
Transform	Transform unit square IDCT from 32×32 to 4×4 + DST Luma Intra 4×4	Integer DCT 8×8, 4×4
Intra prediction	35 predictors	Up to 9 predictors
Motion prediction	Advanced motion vector prediction AMVP (spatial + temporal)	Spatial median (3 blocks)
Motion precision	¼ Pixel 7 or 8-tap, 1/8 Pixel 4-tap chroma	½ Pixel 6-tap, ¼ Pixel bi-linear

- The division of each coding block into prediction blocks,
- The division of each coding block into transform blocks,
- The division of each PU into prediction blocks, and
- The division of each TU into transform blocks.

Table 7.2 (partially synthetized from reference [VIE201201] compares HEVC with AVC. In some configurations HEVC also may generate high data rates; Table 7.3, also based on [VIE201201], depicts what the stream output can be, which is based on various standard profiles (as defined by Levels and Tiers).

In summary, HEVC intra prediction mechanisms operates as follows: Frames are processed in $4 \times 4 - 64 \times 64$ blocks of pixels in (mostly) top-left to bottom-right order; the algorithm can use the (previously processed) upper and left neighboring pixels to estimate (predict) the current block of pixels. The video, as is normally the case, consists of 1 luma and 2 chroma streams (YC_rC_b colorspace); 4:2:0 subsampling implies that the luma is at $2 \times$ the x and y resolution. The prediction is done separately for all 3 streams. The H.265 algorithm allows one to predict a block of pixels as: (i) the average of its neighbors (DC mode; "direct current" analogy); (ii) a smooth gradient based on its neighbors (planar mode); and (iii) a linear extension of its neighbors in one of 33 directions (angular mode). A total of 35 total modes are available, increased from 8 in H.264. See Figure 7.8 for an illustration of the prediction modes. The DC mode predicts that all pixels in the block are the average of the

TABLE 7.3   Data Rates Generated by HEVC

Level	Max Luma Picture Size (samples)	Max Luma Sample Rate (samples/s)	Main Tier Max Bit Rate (Mbps)	High Tier Max Bit Rate (Mbps)	Min Comp. Ratio
1	36,864	552,960	0,128		2
2	122,880	3,686,400	1,5		2
2.1	245,760	7,372,800	3		2
3	552,960	16,588,800	6		2
3.1	983,040	33,177,600	10		2
4	2,228,224	66,846,720	12	30	4
4.1	2,228,224	133,693,440	20	50	4
5	8,912,896	267,386,880	25	100	6
5.1	8,912,896	534,773,760	40	160	8
5.2	8,912,896	1,069,547,520	60	240	8
6	33,423,360	1,069,547,520	60	240	8
6.1	33,423,360	2,005,401,600	120	480	8
6.2	33,423,360	4,010,803,200	240	800	6

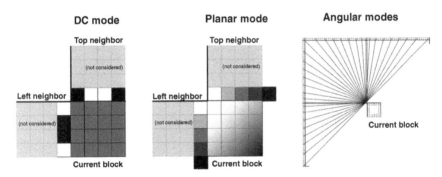

Figure 7.8   Prediction modes.

edge pixels of top and left neighbor blocks; this is ideal for the compression of flat (one color) frame regions. The planar mode predicts that the block forms a smooth gradient defined by its top and left neighbors; it is computed by the averaging of two linear interpolations; this is ideal for the compression of smoothly varying frame regions. The angular mode entails more coverage close to horizontal and vertical pixels (extend neighbor pixels into current block at specific angle); varying features along different directions (including areas with straight edges) are in fact common in actual video frames. Figure 7.9 depicts the positioning of HVEC functionality in the transmission protocol stack. Thus, HVEC affords higher coding efficiency than H.264 (as high as 50%); it has higher complexity than H.264, allowing encoding and decoding to be parallelizable, but thus requiring faster chipsets; and it supports higher resolutions for evolving video applications (e.g., UltraHD).

The interested reader should directly consult the standard for additional information and detailed protocol specifics.

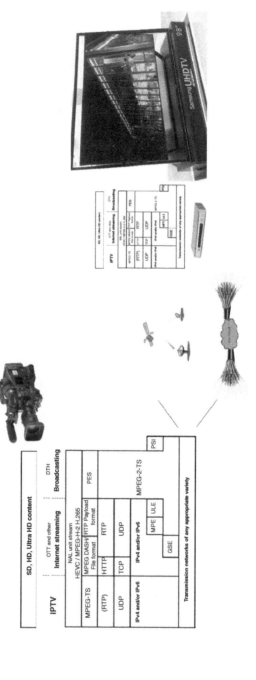

Figure 7.9 Positioning of HVEC functionality in the transmission protocol stack.

7.2 BANDWIDTH/TRANSMISSION REQUIREMENTS

The bit rates required to support UltraHD can be very high, especially if full quality is desired. Uncompressed UtraHD would require more than 50 Gbps of bandwidth. No residences have such access capacity today, and the network transport/transmission of multiple channels of this type of video would be quite taxing, if not impractical. Urban users in highly developed regions may have access bandwidth in the 100–500 Mbps range, but most users even in advanced countries have less than 100 Mbps "entrance" facilities. Current compression technology would put the delivery of UltraHD in the 30–120 Mbps range per channel. A household watching, say, four independent simultaneous channels would need 120–480 Mbps, or the equivalent of an OC-3–OC-12 or a GbE service. That is why new compression techniques are highly desirable; roughly, a 50% bandwidth saving is feasible with HVEC – however, we saw earlier that HEVC may still generate high data rates.

Proponents make the case that new capabilities and possibilities emerge as follows when afforded with a 50% video transmission rate reduction compared with the status quo:

- Improvements for existing applications
  - IP Television (IPTV) over Digital Subscriber Line (DSL): potentially greater IPTV penetration
  - Greater deployment of Over The Top (OTT) and multiscreen services
  - Improved archiving facilities, including cloud-based services
- Enablement of emerging services
  - Introduction of 1080p60/50 with bit rates comparable to 1080i
  - Support of immersive viewing experience (UltraHD 4K, 8K) both from stored media as well as from transmitted programming
  - Provision of premium services (sporting events, concerts, etc.) for home theaters, bars, electronic cinema

Regarding HEVC, two sets of performance results are listed next, on the basis of data from industry assessments, particularly [VIE201201].

1. Comparisons have been made with H.264/AVC. A Joint Model (JM) reference software has been developed by the MPEG and VCEG Joint Video Team for H.264/MPEG-4 AVC to assess performance results. Versions JM 18.3 and 18.4 (High Profile) have been compared with HM 7 and 8 (Main Profile), assessing performance with video ranging from WVGA (Wide Video Graphics Array) to Full HD video sequences. Note: WVGA is any display resolution comparable to VGA but somewhat wider, for example, 720×480 pixels (3:2 aspect), 800×480 pixels (aspect ratio 5:3), 848×480 pixels, 852×480 pixels, 853×480 pixels, or 854×480 pixels (aspect ratio approximately 16:9) – WVGA is a common resolution used by LCD projectors. VGA video refers to the video achievable with display hardware first introduced with the IBM PS/2 line of computers in

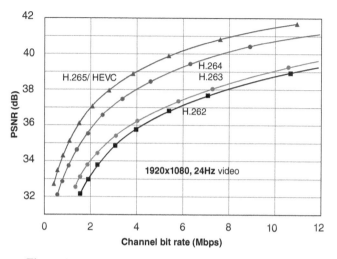

Figure 7.10   Comparison between various H.xxx schemes.

the late 1980s, later followed by widespread adoption as an analog computer display standard. Two indicators are assessed:

- Objective quality (PSNR-Bjontegaard benchmarking [BJO200301]): an average bit rate savings of 35% for entertainment applications has been validated. Coding performance increases with resolutions: >39% for HD and beyond. In summary, average bit rate savings of 40% for low delay applications (low delay) have been demonstrated (also see Figure 7.10 [UNT201201]; notice in the figure that to achieve a quality of a PSNR = 37 dB, for example, H.265 requires 2 Mbps, while H.264 requires 3 Mbps).

- Subjective quality (Mean Opinion Score): an average bit rate savings of 50% for equivalent perceived quality have been demonstrated. For some (not as-dynamic 1080p) content savings from 30% up to 67% have been demonstrated.

2. Performance assessments for 4K native UltraHD content have been undertaken. Both H.264/AVC and HEVC encoders were tested, with Constant Bit Rate (CBR) outputs and equivalent Group of Pictures (GOPs); 4:2:0 sampling and 8-bit material were used. A 38–50% bit rate savings were observed (PSNR-Bjontegaard benchmarking). The implications are that the observed gain would allow the broadcast of 4K UltraHD TV (50/60p) at video bit rates around 13 Mbps (as confirmed by subjective testing) and around 50 Mbps (SONET STS-1) for 8K UltraHD.

Table 7.4 summarizes and generalizes some of the results of this discussion (but rounds up the 4K/8K requirements to slightly more conservative targets). Note, as discussed earlier, that if content distributors just implement H.265 en bloc, without consideration to any UltraHD services, the number of transponders needed for a given

TABLE 7.4   Data Rates and System Capacity for Various Commercial TV Transmission Schemes

Data rate for commercial TV distribution in Mbps

	SD	HD	4K	8K
MPEG-2/H.263	4	20	80	320
MPEG-4/H.264	2	10	40	160
MPEG-4/H.265	1	5	20	80

Number of commercial TV channels per 36 MHz transponder with DVB-2SX

	SD	HD	4K	8K
MPEG-2/H.263	25	5	1	0
MPEG-4/H.264	50	10	2	0
MPEG-4/H.265	100	20	5	1

Number of commercial TV channels per a GbE (fiber) connection to the home (also for DOCSIS systems, e.g., 3.0)

	SD	HD	4K	8K
MPEG-2/H.263	250	50	10	3
MPEG-4/H.264	500	100	20	6
MPEG-4/H.265	1000	200	50	12

(assumes no Internet traffic on the connection)

Number of commercial TV channels per a 10 GbE (fiber) connection to the home (also for DOCSIS systems, e.g., 3.1)

	SD	HD	4K	8K
MPEG-2/H.263	2500	500	100	30
MPEG-4/H.264	5000	1000	200	60
MPEG-4/H.265	10000	2000	500	120

(assumes no Internet traffic on the connection)

bouquet of channels will be cut in half, thereby severely impacting the revenue stream to the satellite infrastructure providers (which is exactly what happened a few years ago when Germany converted its TV distribution from analog to digital: the number of satellite transponders needed went down by an order of magnitude).

7.3   TERRESTRIAL DISTRIBUTION

Naturally, for urban environments in developed countries, fiberoptic distribution of UltraHD can in principle be easily achieved, as seen in Table 7.4 when assessing typical fiber system capacities. However, both the actual plant upgrade to ascertain that all elements of the network (especially the last mile) support the higher speed(s), as well as the penetration per se throughout a given metropolitan area, may represent obstacles, or at least, retardant factors to the deployment of UltraHD TV. Suburban, exurban, and rural areas may find that satellite distribution is the only available choice.

In the context of terrestrial distribution, a short mention of Data Over Cable Service Interface Specification (DOCSIS) is made herewith. DOCSIS is a standard that supports the overlay of high-speed data transfer onto an existing hybrid fiber-coaxial (HFC) infrastructure used in a Cable TV system. DOCSIS was developed by Cable-Labs and other contributing companies in the 1990s. It has gone through a number of releases. DOCSIS 1.0 was released in 1997; it included basic elements from preceding proprietary cable modem products. DOCSIS 1.1 was released in 1999 incorporating additional standardization and quality of service (QoS) capabilities. DOCSIS 2.0 was released in 2001 with the objective of enhancing upstream transmission speeds. DOC-SIS 3.0 was released in 2006; the specification was revised to significantly increase transmission speeds (both upstream and downstream – in the 1 Gbps range). DOCSIS 3.1 was released in 2013, focusing on support capacities of at least 10 Gbps downstream and 1 Gbps upstream using 4096 Quadrature Amplitude Modulation (QAM); it does away with previous 6/8 MHz-wide channel spacing and uses smaller orthogonal frequency division multiplexing (OFDM) subcarriers that can be bonded inside a block spectrum (to about 200 MHz wide).

DOCSIS supports a number of Open Systems Interconnection (OSI) layers with 1 and 2 options, some of which are as follows:

- Physical layer channel width: All versions of DOCSIS up to 3.0 utilize either 6 MHz channels (e.g., North America) or 8 MHz channels (Europe, EuroDOC-SIS version) for downstream transmission over HFC plants. DOCSIS 2.0 is also used over microwave frequencies (10 GHz) in Ireland, and DOCSIS 1.x, 2.0, and 3.0 is also used for fixed wireless systems utilizing the 2.5–2.7 GHz MMDS microwave band in the United States.

- Physical layer modulation: All versions of DOCSIS specify the use of 64-level or 256-level QAM (64-QAM or 256-QAM) for the modulation of downstream data, utilizing the ITU-T J.83-Annex B standard for 6 MHz channel operation, and the DVB-C modulation standard for 8 MHz (EuroDOCSIS) operation. DOCSIS 2.0 and 3.0 also support 128-QAM with trellis coded modulation, and DOCSIS 3.1 adds 4096-QAM.

- Data link layer transmission: DOCSIS utilizes a mixture of deterministic access methods: TDMA for DOCSIS 1.0/1.1 and both TDMA and S-CDMA for DOC-SIS 2.0 and 3.0.

Terrestrial transmission of UltraHD would use DOCSIS in the "distribution" side of the content delivery (not on the "contribution side"), that is, in the "last mile."

7.4 SATELLITE DISTRIBUTION

Satellite operators are planning to position themselves in this market segment, with some generally available broadcast services planned for 2020, and with more limited, targeted transmission starting at press time. However, the projections for the deployment of UltraHD are modest, and mass deployment is not expected to occur before 2023. The number of satellite TV signals distributed globally reached approximately 30,000 in 2013; market researchers expect the addition of 11,000 new satellite TV

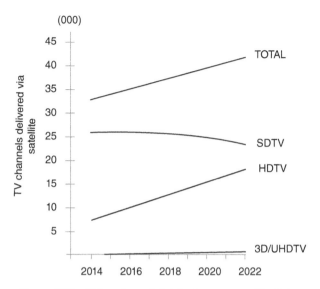

Figure 7.11 Video channel distribution over satellite links.

signals by 2023, with HDTV and emerging markets driving growth, but UltraHD still being "a drop in the bucket" at that time [BUC201401]. See Figure 7.11, which is based on industry data.

DTH techniques, in particular, with High Throughput Satellites (HTSs) operation in the Ku or Ka are available to UltraHD distribution, consistent with the parameters of Table 7.4 and expanded in Table 7.5.

As an illustrative example, Eutelsat Communications provided the world's first demo 4K channel in 2013 and launched in Europe in 2014 the first HEVC channel. As of press time, Eutelsat, working in partnership with ST Teleport, was planning to introduce its 4K TV channel on the EUTELSAT 70B satellite that reaches across Southeast Asia and Australia. The content displayed by Eutelsat's new 4K channel is encoded in HEVC by ATEME and Thomson Video Networks at 50 frames per second with 10-bit color depth (one billion colors – HEVC Main 10 profile); it will be displayed on the Eutelsat stand at CommunicAsia 2014 using a Samsung 65 inch panel (UE65HU7500) and can be received directly by the latest consumer 4K TV panels equipped with DVB-S2 demodulators and HEVC decoders. The 4K TV channel is broadcasting a growing library of documentary, cultural and sports content filmed by television channels, production companies and Eutelsat to showcase the viewing experience of UltraHD [SAT201401].

7.5 HYBRID DISTRIBUTION

Hybrid distribution, with satellite distribution complement by terrestrial transmission either over local cable/fiber facilities (e.g., a metropolitan area network) or by wireless

TABLE 7.5   Satellite Distribution of UltraHD TV Channels

Number of commercial TV channels per 24-transponder satellite with 36 MHz transponders operating with DVB-S2X

	SD	**HD**	**4K**	**8K**
MPEG-2/H.263	600	120	24	0 (*)
MPEG-4/H.264	1200	240	48	0 (*)
MPEG-4/H.265	2400	480	120	24

Note: tradeoffs will be needed between operating all 24 transponders at a fully-shared power level, or operating a smaller number of transponders at a higher power level (thereby improving reception quality)

*Channel bonding could be employed (as discussed in Chapter 2) to support UltraHD in these situations – an upgrade to H.265 is a better strategy.

Number of commercial TV Channels per 24-transponder satellite with H.265 and DVB-S2X and with a channel-type mix

Total channels		*250*	
	%		Nominal bandwidth
SD channels	20	50	50,000,000
HD channels	70	175	875,000,000
4K channels	4	10	20,000,000
8K channels	6	15	1,200,000,000
	100		2,325,000,000
		400	
	%		Nominal bandwidth
SD channels	40	160	160,000,000
HD channels	50	200	1,000,000,000
4K channels	8	32	640,000,000
8K channels	2	8	640,000,000
	100		2,440,000,000
		600	
	%		Nominal bandwidth
SD channels	75	450	450,000,000
HD channels	15	90	450,000,000
4K channels	9	54	1,080,000,000
8K channels	1	6	480,000,000
	100		2,460,000,000

technologies [WiMax, 3G/4G Long Term Evolution (LTE)], is always an option. One approach is for traditional distribution to receiving intermediate stations that offer caching (say a 4G/LTE or WiMax node), in turn using local distribution to the user community. This approach would work well, in particular, for nonlinear video.

7.6 DEPLOYMENT CHALLENGES, COSTS, ACCEPTANCE

As discussed above, the introduction of HEVC elements throughout the video chain (content capture, content encoding, content storage/media, content transmission, content reception, and content display) is needed at the practical level to make UltraHD a reality. This will certainly have a nontrivial cost implication, not only in the production side, but also in the transmission side (with new encoders, transcoders, modulators, perhaps new wide-band transponders/spacecraft), as well in the reception side (new receivers/set-top boxes and new home display systems).

Following the limited adoption of 3D channels as alluded to earlier in the chapter, UltraHD is being introduced progressively alongside the value chain: TV producers and broadcasters (NHK) and video distribution providers (e.g., Eutelsat, Intelsat) with prevalence in cinema production. UltraHD may initially be a branding tool for the operator; at a later stage the operators will focus on premium subscribers. The time and significant capital investments required for broadcasters, producers, TV manufacturers, infrastructure owners, and consumers to upgrade to 4K are limiting factors for the immediate development of UltraHD TV services. Some were expecting the 2016 Olympic Games to represent a milestone for the launch of UltraHD channels. As noted above, currently, two 4K UHD channels could be broadcast on a 36MHz transponder; additional encoding improvements may occur over the next few years. These efficiency gains may likely result in the deployment of more TV channels, reducing video storage and distribution costs while maintaining or increasing the quality of the experience for the viewer [BUC201401].

REFERENCES

[ANG201301] C. Angelini, "Next-Gen Video Encoding: x265 Tackles HEVC/H.265", Tom's Hardware online magazine, July 23, 2013.

[BJO200301] T. Wiegand, G. J. Sullivan, G. Bjontegaard, and A. Luthra, "Overview of the H.264/AVC Video Coding Standard", IEEE Transactions on Circuits and Systems for Video Technology, 13(7):560–576, 2003.

[BRO201301] B. Bross, et al., "High Efficiency Video Coding (HEVC) text specification draft 10 (for FDIS & Last Call)", Output Document of JCT-VC, 12th Meeting: Geneva, CH, 14–23 Jan. 2013, Joint Collaborative Team on Video Coding (JCT-VC) of ITU-T SG 16 WP 3 and ISO/IEC JTC 1/SC 29/WG 11.

[BUC201401] D. Buchs, M. Welinski, M. Bouzegaoui, *Video Content Management and Distribution, Key Figures, Concepts and Trends*", Euroconsult Report, March 2014, Euroconsult, 86 Blvd. Sebastopol, 75003 Paris, France. www.euroconsult-ec.com

[MAR200301] D. Marpe, H. Schwarz, and T. Wiegand, "Context-Based Adaptive Binary Arithmetic Coding in the H.264/AVC Video Compression Standard", IEEE Transactions On Circuits And Systems For Video Technology, VOL. 13, NO. 7, JULY 2003.

[RIC201301] I. Richardson, "HEVC: An introduction to High Efficiency Video Coding", White Paper. Vcodex Limited, 35 Regent Quay, Aberdeen AB11 5BE.

[SAT201401] Eutelsat Communications + ST Teleport—Technology Showcase @ Commu-
nicAsia2014 (SatBroadcasting™—4k), Satnews Daily, June 13th, 2014.

[SHA201301] M. S. Sharabayko, N. G. Markov, "Iterative Intra Prediction Search for
H.265/HEVC", 2013 International Siberian Conference on Control and Communication
(SIBCON), IEEE.

[SUL201201] G. J. Sullivan, J. -R. Ohm, W. -J. Han, and T. Wiegand, "Overview of the High
Efficiency Video Coding (HEVC) Standard", IEEE Transactions On Circuits And Systems
For Video Technology, VOL. 22, NO. 12, December 2012.

[UNT201201] A. Unterweger, "What is New in HEVC/H.265?" Department of Computer
Sciences University of Salzburg, October 17, 2012.

[VIE201201] J. Vieron, "HEVC: High-Efficiency Video Coding Next Generation Video Compression", WBU-ISOG FORUM, 27–28 November 2012. EBU headquarters in Geneva,
Switzerland.

[X26201401] http://x265.org/hevc.html. x265 is an open-source project and free application
library for encoding video streams into the H.265/ HEVC format.

8

SATELLITE TECHNOLOGY ADVANCES: ELECTRIC PROPULSION AND LAUNCH PLATFORMS

In this chapter we provide a basic survey of two developments related to the deployment of satellites: (i) electric propulsion (EP) and (ii) new launch platforms. Each of these advances can ultimately reduce the cost of commercial satellite communications, thus improving their competiveness compared with terrestrial transmission, in appropriate application segments.

It should be noted, however, that there are other ways also to improve the competiveness of satellite services, namely, direct price reduction. Price reduction can be achieved by further optimizing the operations (also known as run-the-engine) costs. For example, the satellite operators are, in a number of instances, swamped with superfluously redundant, zero-revenue, negative cash-flow ground infrastructure and staff in outdated, often overbuilt, "ungreen" teleports; one way to address costs is to consolidate these inefficient teleports and outsource the ground infrastructure function to companies with much lower overhead and personnel costs and with nimble management. Additionally, advantageously to the operators and their shareholders but disadvantageously to the user/consumer, the Earnings Before Insurance, Tax, Depreciation, and Amortization (EBITDA) of satellite operators is kept at around 75–85% by these operators. A commercial spacecraft on station may typically cost $150 M (including on-the-ground Ground Control System [GCS]); with a 15-year lifecycle, and, say 48 transponders (24 at C-band and 24 at Ku band), the yearly cost per transponder could thus be $208,333; adding a 20% operating

Innovations in Satellite Communications and Satellite Technology: The Industry Implications of DVB-S2X, High Throughput Satellites, Ultra HD, M2M, and IP, First Edition. Daniel Minoli.

cost[1] factor, adding a 22% Selling, General & Administrative Expense (SG&A),[2] and adding a 8% profit factor, brings this figure to around $312,500; even assuming a partial 75% transponder fill rate, the figures would be $278,000 and $416,666, respectively. However, the typical yearly transponder fee charged by satellite operators are, on the average, much higher, specifically[3] $1.6 M. Clearly, there are opportunities for price reductions by improved corporate management, besides various technical innovations.

8.1 BASIC TECHNOLOGY AND APPROACH FOR ELECTRIC PROPULSION

Work on EP for commercial satellites goes back to the 1980s. In principle, EP can be used for some of the propulsion function of a spacecraft, or for all propulsion functions – obviously the spacecraft design has to match the stated operator's propulsion policy. "All-electric propulsion" (all-EP) is a topic receiving a lot of attention of late. All-EP offers the opportunity to reduce the spacecraft mass and the launch costs; however, the technology is not yet fully tested for commercial applications, especially in terms of multiple missions with multiple operational spans. Clearly, there are competing tradeoffs in selecting a satellite configuration: larger satellites allow for larger payloads and higher power levels and, thus, support enhanced revenue generating capabilities; however, heavier satellites cost more to manufacture and require more powerful and more expensive launch vehicles.

We saw in Chapter 1 that the launch rocket places the spacecraft into an initial (or "transfer") orbit just a few 100 miles above the earth's surface; the satellite must then propel itself into its final operating orbit. This process normally takes from several days to several weeks; this process entails traversing the radiation belts mentioned in Chapter 3, which preferably is done in an expeditious manner. Once on-station, the spacecraft must be kept at the specified orbital location. During the lifecycle (say, 15 years) it may be desirable to relocate the spacecraft to some other orbital location to serve different markets, or to support a Bring Into Use (BIU) mission (this to secure orbital rights by actually being present/operating at a "granted" location). To be able

---

[1]The 15-year total operations cost per spacecraft for a large operator (where staff/infrastructure efficiencies ought to apply) was independently computed to be $5.75–11.5M in 2003, or in the range of $7.5–15M in 2014 (with a 3% annual inflation rate) [GRE200401]. Even taking the higher of these two numbers, that would equate to $1 M/year/satellite, or $20,000/transponder/year; our assumption of a 20% markup on the annualized equipment cost cited above is more than double this industry-documented figure for the operations cost. Looking at this issue from another perspective, an operator with, say 60 spacecraft would have, under our 20% ops markup heuristic, a yearly run-the-engine budget of $120 M; that would seem to be a fairly substantial generous figure to live on and at the same time still charge the user $417 K/year/transponder, rather than $1,600,000 – any discrepancy could be addressed by the operator making an explicit commitment to deprecate intrinsic inertia fostered by middle management and by proper reengineering of the operations function to be more cost-effective and less elephantine.

[2]See Appendix A at end of chapter.

[3]See Appendix A at end of chapter.

to support these activities, the spacecraft must be equipped with a number of (small) thrusters.

Focusing specifically on Geosynchronous Orbit (GSO) applications, a GSO spacecraft requires propulsion systems from its separation from the third stage of the launcher until the de-orbiting operation, years later. At this separation with the third stage, the satellite is injected on a transfer orbit with an apogee at 36,000 km from the earth surface; at the apogee of this orbit an applied thrust of 1–3 N gives an increment of velocity of ~2 m/s and moves the satellite to a quasi-circular and quasi-equatorial orbit. Additional maneuvering and In Orbit Testing (IoT) are required to make the spacecraft's orbit circular and equatorial, to open solar panels, to verify the instruments, to modify the satellite attitude and to set the satellite at its working longitude. This process usually takes a couple of months. The satellite has to be maintained at this working position throughout the mission (15–20 years). However, the satellite moves in 3-dimensional space in accordance with lunar and sun trajectories, inhomogeneity of the earth's gravitational pull, and radiative sun pressure. At the altitude of the GSO the predominant effect is the lunar-solar interactions, these effects being greater than the earth inhomogeneity and the sun radiation effects. Consequently, spacecraft thrusters are required to bring the satellite back to its nominal (initial) working place, and to perform daily north–south and east–west corrections. The thrusters deliver a thrust in the range 80–100 mN; the variation of velocity is 50 m/s (north–south) and 5 m/s (east–west) per year [DUD201001]. In the vacuum of space one achieves movement through reactive forces, ejecting matter in one direction and achieving movement in the opposite direction. Further to the discussion above, a spacecraft requires propulsion for the following reasons:

1. To transition from the Launch and Early Orbit Phase (LEOP), to the parking orbit, to the intermediate orbit, to the transfer orbit, to the target orbit. This activity (known in aggregate as "orbit raising") is seen by some as being comprised of a "orbit topping" phase where the satellite is brought up close to the target orbit and a "final GSO insertion" phase;

2. To maintain the satellite on-station within the permitted box (as granted by regulatory entities); this is done via the telemetry, tracking and command (TT&C) mechanism and the GCS;

3. To relocate the spacecraft to other GSO locations to support new traffic opportunities at new orbital slots;

4. To move the satellite during the mission to avoid a collision with an object (debris) moving in space; and,

5. To de-orbit the spacecraft at the end of its mission (the de-orbiting is achieved with an increment of velocity $\Delta V = 3$ m/s in order to send the satellite on a higher orbit of about 200 miles).

Currently there are two approaches to (spacecraft) propulsion. The traditional method is fully-chemical propulsion; it uses chemical reactions to produce a flow

TABLE 8.1   Possible Application of Various Propulsion Methods

Activity		All-Chemical	Hybrid Chemical	Electric	All-Electric (All-EP)
Orbit raising	Orbit topping	Used	Used		Used
	Final Insertion	Used		Used	Used
Station-keeping		Used		Used	Used
Relocations		Used	Used		Used
Emergency collision avoidance		Used	Used		Used
De-orbit		Used	Used	Perhaps used	Used

of fast-moving hot gas, thereby providing a push thrust. The newer method uses the electrical power that can be generated from sunlight with solar photovoltaic panels to propel the spacecraft using a number of atomic/subatomic particle ejection, thereby providing a push thrust. Electric thrusting of propellants is useful only for interorbital transportation, not for launch from the Earth's or Moon's surface. As seen in Table 8.1, and alluded to earlier, spacecraft may use a combination of both methods (a hybrid arrangement), with EP currently being used almost exclusively for station-keeping – this EP approach has been used on some commercial satellites since the 1980s. Or, the spacecraft could use all-EP, this being the latest innovation, now being commercially investigated. The adoption of EP technology has already brought a number of benefits, including longer operational life and increased satellite dry mass for a given launch mass (and conversely, decreased launch mass for a given payload mass). For commercial satellite operators, these factors directly translate into the ability to decrease costs for equivalent satellite capability and increase the amount of total life revenue generated by a given satellite [ALL201301]. Hence, the key technical and business question at this juncture is: "Should the operator chose to go all-EP on new procurements?"

At this time, spacecraft propulsion for commercial GSO satellites is typically supported by chemical propellant systems, especially for environments where long orbital drifts are contemplated at various points in the spacecraft's mission. Specifically, powerful chemical propulsion engines are used to quickly move satellites through Earth's hazardous radiation belt [PAT201301]. The issue is that chemical propellants add launch weight and eventually are exhausted (typically in 15 years). With chemical propulsion there needs to be one or two substances that can be burned in a controlled manner. The combustion takes place in a chamber and the resulting hot gases are ejected through an opening, the nozzle, to provide the desired thrust. Heating the gas in a chamber produces an increase in pressure and letting the gas escape through a small aperture produces a fast jet. By varying the amount of fuel burned and the shape of the nozzle, which controls the velocity of the exhaust gases coming out of the engine, or thruster, one can maneuver the satellite (note that in space thrusters must work without the oxygen present in the Earth's atmosphere). Chemical propulsion is a complex process and a network of tanks, pipes, valves and delicate control mechanisms is needed; hence the desire to look at

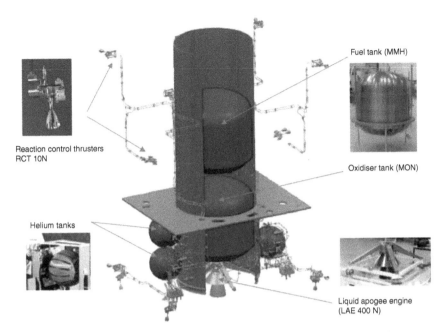

Fuel tank (MMH)

Reaction control thrusters
RCT 10N

Oxidiser tank (MON)

Helium tanks

Liquid apogee engine
(LAE 400 N)

Figure 8.1   Illustrative example of chemical propulsion.   Courtesy: ThalesAleina space.

alternatives. For illustrative purposes Figures 8.1 and 8.2 from Thales Alenia Space depict a chemical bi-propellant propulsion system. This kind of traditional thruster is usable for applications dealing with satellite orbit raising, east/west and north/south station-keeping, satellite attitude control, satellite attitude momentum control, and post-operational de-orbiting. Key features (of this specific model, discussed for illustrative purposes) include: same propellant supply for all apogee maneuvers and station keeping maneuvers resulting in consumption optimization; bi-propellant system, so that thrusters are not prone to chemical degradation; 2-tank configuration, implying lower disturbances than with a 4-tank configuration, lower operational complexity, and avoidance of non-parallel depletion errors; all thrusters in stand-by mode to be operated at any time without any valve switching for activating branches, implying high availability, flexibility and reliability; and, liquid apogee engine multiple-burn capability providing high accuracy of insertion in GSO than other systems.

In terms of the underlying flight dynamics principles, the concept of EP and traditional propulsion are similar: a force is a push or a pull that acts upon an object as a result of its interaction with another object; some forces result from physical contact interactions, while other forces are the result of action-at-a-distance interactions (e.g., gravitational, electrical, and magnetic forces). According to Newton, whenever objects A and B interact with each other, they exert forces upon each other. These two forces are called *action* and *reaction* forces. Newton's third law of motion states that for every action, there is an equal and opposite reaction. When gasses (or

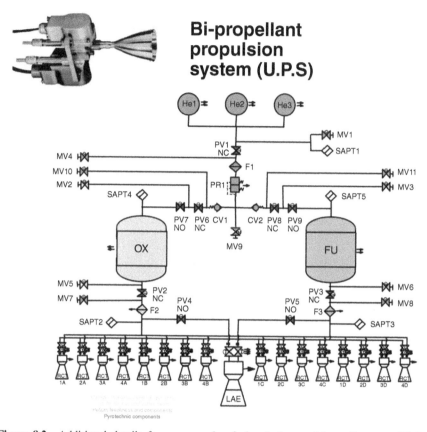

Figure 8.2 Additional details for an example of chemical propulsion. Courtesy: Thales-Aleina space.

ions) are expelled from a nozzle on a spacecraft (thing being the topic at hand), they exercise pressure against atoms in the surrounding space (or surrounding electromagnetic field), thus generating thrust.

Reference [JAH200201] provides a rather lucid characterization of EP when noting that the science and technology of EP encompass a broad variety of strategies for achieving very high exhaust velocities in order to reduce the total propellant burden and the corresponding launch mass of present and future space transportation systems; and, when noting that these techniques group broadly into three categories: (i) electrothermal propulsion, wherein the propellant is electrically heated, then expanded thermodynamically through a nozzle; (ii) electrostatic propulsion, wherein ionized propellant particles are accelerated through an electric field; and (iii) electromagnetic propulsion, wherein current driven through a propellant plasma interacts with an internal or external magnetic field to provide a stream-wise body force.

Such EP systems can produce a range of exhaust velocities and payload mass fractions an order of magnitude higher than that of the most advanced chemical rockets, which can thereby enable or substantially enhance many attractive space missions. The attainable thrust densities (thrust per unit exhaust area) of these systems are much lower, however, which, in turn, imply longer flight times and more complex mission trajectories. In addition, these systems require space-borne electric power supplies of low specific mass and high reliability, interfaced with suitable power processing equipment. Optimization of EP systems thus involves multidimensional trade-offs among mission objectives, propellant and power plant mass, trip time, internal and external environmental factors, and overall system reliability. A program of research and development of viable electric thrusters has been in progress for several decades, and over the past few years this has led to the increasing use of a number of EP systems on commercial and governmental spacecraft. Meanwhile, yet more advanced EP concepts have matured to high credibility for future mission applications. Two broad tradeoffs come into play: (i) the more energy available, the less propellant required, which in turn means less mass is required; (ii) the more time allowed for a space maneuver, the less power needed.

As noted, EP is an approach to accelerate a propellant (such as ions) through electro(magnetic) fields. The speed to which the propellant can be accelerated is a design consideration and the energy available on board is the other practical limitation. The simplest way to achieve EP is to replace the heat generated by combustion in chemical engines with electrical heating. Another way to heat a stream of gas is to use a controlled electrical discharge (arc). However, there are other, more sophisticated and more efficient ways of obtaining fast jets of gas. For example, ionized particles of propellant launched past a grid pushes the spacecraft in the opposite direction. Another possibility is to use the combined effect of an electric field to set the particles in motion and a magnetic field to accelerate particles of propellant. While with the use of an electric field alone, only charged particles of one polarity (opposite to that of the grid) provide propulsion, with the combined action of electric and magnetic fields both polarities are accelerated [ESA200201]. EP typically implies high ion exhaust speed, this being much greater than in conventional (chemical) rockets. Also, much less propellant consumption is involved (implying higher efficiency in the fuel utilization). Continuous propulsion means applying a smaller thrust for a longer time. The propellant employed varies with the propulsion type, and can be a rare gas (e.g., xenon or argon), a liquid metal (e.g., cesium or indium) or, a conventional chemical propellant (e.g., hydrazine, ammonia, or nitrogen).

There are two *commercialized* EP technologies already utilized in space today, as follows [PRA201401]: (i) Some American interorbital satellites today use electric "*ion drive*" for stationkeeping. Ion drive is a simple and fairly mature technology (the Deep Space 1 probe, launched in October 1998, was the first vehicle to depend upon electric ion drive for all of its propulsion needs, to perform a close flyby of an asteroid). (ii) The Russians have used extensively an EP technique called a "*plasma thruster*" for at least 10 years, which they have begun to market overseas.

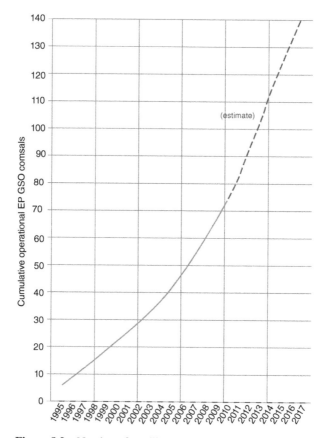

Figure 8.3   Number of satellites using EP for station-keeping.

This thruster has been used in approximately one hundred Russian military satellites, but is relatively unknown in the west.

EP has a relatively long and successful history in the context of station-keeping (e.g., Boeing's satellites have used ion thrusters for fine orbit control since 1997), but not with orbit raising, when satellites use their own propulsion system to reach orbit after separating from the rocket. Between 1995 and 2010 the cumulative number of operating GSO satellites that utilize EP for station-keeping increased more than ten-fold to over 70 spacecraft (see Figure 8.3).

8.2   EP ENGINES

There are a variety of potential approaches to accelerate ions or plasmas; most of these approaches utilize grids or electrodes (also see Table 8.2 based on [JAH200201]). Some examples discussed in the literature include:

- Ion engine.
- Hall thruster (hall current thrusters [HCT]).
- MagnetoPlasma dynamic (MPD) thrusters.
- RF plasma thrusters (ECR, VASIMR, helicon double layer).
- Plasmoid accelerated thrusters.

The different types of EP have different performances and optimal applications. Some types are more suitable for missions requiring higher thrust levels to reduce the trip time, some are better for high-precision positioning (flight dynamics) applications, while others are better for long mission-duration (minimizing the use of propellant). The field of electric thrusters is very large and the suggested plasma sources are numerous. However, only a few have been (fully) tested, validated and used in space. Other propulsion systems are futuristic concepts. Table 8.3 based directly on

TABLE 8.2 Key EP Concepts

Concept	Definition
Arcjet	A device that heats a propellant stream by passing a high-current electrical arc through it, before the propellant is expanded through a downstream nozzle.
Electromagnetic propulsion	Propulsion wherein the propellant is accelerated under the combined action of electric and magnetic fields.
Electrostatic propulsion	Propulsion wherein the propellant is accelerated by direct application of electrostatic forces to ionized particles.
Electrothermal propulsion	Propulsion wherein the propellant is heated by some electrical process, then expanded through a suitable nozzle.
Hall effect	Conduction of electric current perpendicular to an applied electric field in a superimposed magnetic field.
Inductive thruster	Device that heats a propellant stream by means of an inductive discharge before the propellant is expanded through a downstream nozzle.
Ion thruster	Device that accelerates propellant ions by an electrostatic field.
Magnetoplasmadynamic thruster	Device that accelerates a propellant plasma by an internal or external magnetic field acting on an internal arc current.
Plasma	Heavily ionized state of matter, usually gaseous, composed of ions, electrons, and neutral atoms or molecules, that has sufficient electrical conductivity to carry substantial current and to react to electric and magnetic body forces.
Resistojet	Device that heats a propellant stream by passing it through a resistively heated chamber before the propellant is expanded through a downstream nozzle.
Thrust	Unbalanced internal force exerted on a rocket during expulsion of its propellant mass.

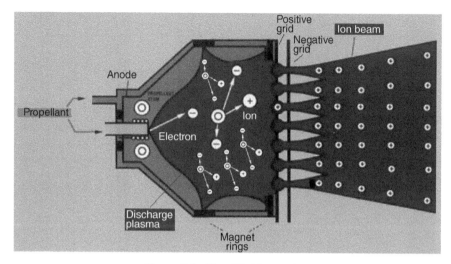

Figure 8.4   Simplified ion engine.

information from reference [DUD201001] provides a summary of some of the more well-known technologies (refer to reference for additional information).

8.2.1   Ion Engines

Ion Engines were first developed by Russian researchers. This technology is also known as the Stationary Plasma Thrusters (SPT) technology. The expectation is that ion engines will be commonly used on commercial satellites in the near future, especially for north–south station-keeping [GOE200801]. A simplified ion engine is shown in Figures 8.4 and 8.5. Electrons emitted by the (hollow) cathode traverse the discharge plasma and are collected by the anode; along the way they impact atoms and create ions. The magnetic field inside the chamber enhances the ionization process and its efficiency. As they exit the engine the ions are electrostatically accelerated. In an ion thruster, the propellant (Xenon) atoms are ionized in a discharge chamber (anode). An electrostatic field is then used to accelerate the positive ions to produce the required thrust. To prevent the spacecraft from charging, the positive ion beam must be neutralized by an equivalent negative charge [BRO200101]. NASA has developed a Deep Space One Ion Engine, the NASA's Evolutionary Xenon Thruster (NEXT).

8.2.2   Hall Effect Thrusters

The Hall Effect Thruster (HET) makes use of the Hall Effect (discovered in the 1870s by Edwin Hall). The Hall Effect is the production of a voltage difference across an electrical conductor, orthogonal to an electric current in the conductor and a magnetic field perpendicular to the current; an electric force F_e is generated

TABLE 8.3   Electric Thruster Technologies

Type	Description	Examples
Ablative Pulsed Plasma Thrusters (PPT)	Ions are produced by successive sparks with a frequency of a few Hz between two high voltage electrodes set in front of a solid propellant (Teflon). PPT delivers a high specific impulse (800–1200 s), a thrust around 1 mN and operates with an electric power lower than 100 W. Used for attitude control, micro-satellites and low thrust maneuvers.	Zond2 (Soviet Union – 1964)
Field Emission Electric Plasma thruster (FEEP)	Uses the flow of a liquid metal (cesium or indium) through a slit limited by two metal surfaces. The extracted positive ions Cs+ are accelerated by an electric field (potential ∼10 kV). Provide very low thrust (micro-N to a few milli-N). A high specific impulse is obtained by this micro-propulsion engine: more than 10,000 s. Used to compensate the drag effect, for attitude control or for formation flying.	ESA/NASA LISA Pathfinder and Microscope missions. Also under development at Alta SpA (Italy) and Space Propulsion (Austria)
Gridded Ion Engine (GIE)	(also called Gridded Ion Thruster [GIT]). Positive ions (Xenon) are obtained by electron impacts (Kaufman or radiofrequency ion thrusters), then extracted and accelerated by a set of 2 or 3 multi-aperture grids. This electrostatic ion engine delivers a thrust up to 670 mN, and an Isp up to 9,620 s for a power of 39.3 kW. Used for satellite station keeping and also for deep space trips.	Exploration of the Borrelly Comet (NASA Deep Space 1 mission; 1998–2001), used the NEXT thrusters (90 mN at a power 2.3 kW) from the Jet Propulsion Laboratory. Others include the Hayabusa spacecraft (JAXA, Japan), and Astra 2A launched in 1998 (Hughes Space and Boeing Satellites Systems)
Arc-jet thrusters	These are electro-thermal engines. The gas (hydrogen, ammonia, hydrazine) enters in a chamber and crosses the throat of a nozzle used as anode. An arc is sustained between this anode and a cathode set in the chamber. The gas is heated and expanded through the nozzle. Arc-jets generate a thrust in the range 0.01–0.5 N with a specific impulse between 500 and 1000 s.	A 750 W ammonia arc-jet thruster has been manufactured at the Institut für Raumfahrtsystem (IRS – Stuttgart university) (AMSAT P3-D satellite, 1994 Germany)

(*continued*)

TABLE 8.3 *(Continued)*

Type	Description	Examples
Hall Effect Thrusters (HET)	(also named "closed-drift thrusters," Stationary Plasma Thrusters [SPT] or Propulseurs par Plasma pour Satellites [PPS].) Hall effect thrusters are advanced electro-magnetic propulsion devices. They use a partially magnetized plasma discharge in a cross-electromagnetic field. The plasma discharge is sustained between two coaxial dielectric cylinders and between an external hollow cathode emitting electrons and an anode set at the bottom of the annular chamber. The discharge voltage is around 300 V (high voltage up to 1000 V have been tested in order to increase the specific impulse). The propellant is generally injected through the anode. Xenon is most suitable thanks to its low first ionization level and its high mass. Due to the low pressure of the channel it is necessary to trap the electrons by a radial magnetic field. This magnetic field is created by external magnetization coils (inner and outer coils). Enjoys economy in propellant mass (saving about 400 kg for a 4 t-class satellite for a mission of 15 years). Used to maintain the satellites in geostationary orbit with north–south and east–west corrections.	The first HET was used on the Meteor meteorological satellite (two SPT-60, USSR, 1972). A few hundreds of SPTs were used on-board Russian satellites. Several HETs with different powers are available: Fakel EDB, Russia: SPT 25, 35, 50, 60, 70, 100, 140, 200, 290, Snecma-Safran Group, France: PPS-1350 and PPS-5000 since 1999 for future satellites, Busek Co., USA: BHT200, BHT400, BHT1000, BHT1500, BHT8000, BHT20k, Pratt & Whitney Space Propulsion, USA: T-40, T-140, T-220 Keldysh Research Institute in Russia: KM-32 and in collaboration with Astrium (now Airbus Defence and Space) the ROS-99 and the ROS- 2000 Rafael Space Systems, Israel: IHET-300. Two IHET-300 operate on the Rafael Venus Satellite.

orthogonal to the direction of conventional electric current from charge buildup; and a magnetic force F_m is generated on the negative charge buildup. Electromagnets are used to generate magnetic fields. The Hall Effect confines electrons: these electrons impact the propelled fuel's atoms to create ions. Hall thrusters provide high specific impulse (compared to chemical thrusters) and high thrust-to-power ratio (compared to ion thrusters). One example is NASA's High Voltage Hall Accelerator (HiVHAC) Thruster. See Figure 8.6.

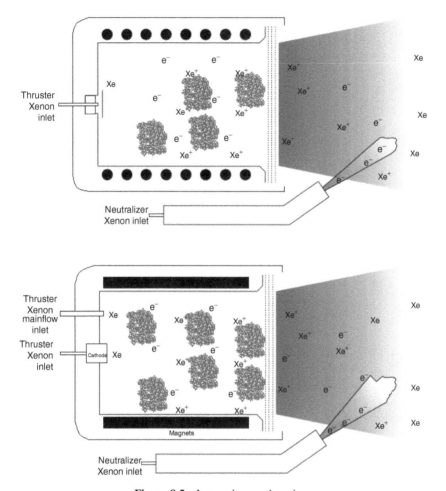

Figure 8.5   Ion engine – other view.

8.2.3   MagnetoPlasma Dynamic Thruster

The MagnetoPlasma Dynamic Thruster (MPDT) is proposed by the EP and Plasma Dynamics Lab of Princeton University. MPDT entails an ionization step, an energizing step, an acceleration step, and a detachment step. Superconducting magnets generate a magnetic field where the fuel gas is transformed into a cold plasma (by a plasma source antenna); a Radio Frequency (RF) booster antenna energizes the plasma and accelerates it, then, it is ejected. The actual working of magneto-electric thrusters is relatively complex. Charged particles are set in motion – in opposite directions – by the electric field. A magnetic field has an effect on a moving charge

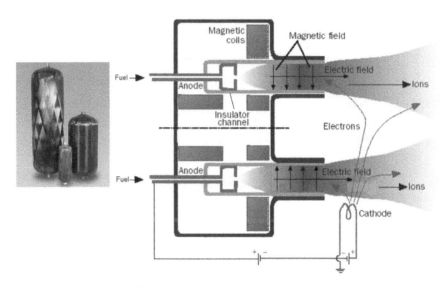

Figure 8.6   Hall thruster engine figure.

known as the "Lorentz force," similar to that which an electric field has on a stationary charge: it pulls the charged and moving particle sideways. By combining the effects of the electric and magnetic fields, the flow of charged particles is accelerated and ejected by the thrusters, producing the desired push [ESA200201]. See Figure 8.7.

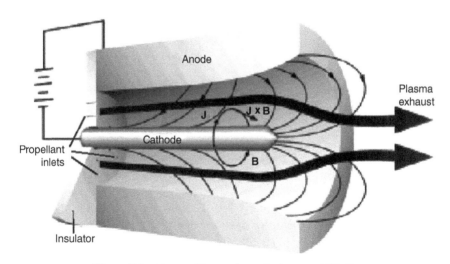

Figure 8.7   MagnetoPlasma dynamic thruster (MPDT).

8.3   ADVANTAGES AND DISADVANTAGES OF ALL-EP

All-EP (or even hybrid EP) offers several advantages, including [ESA200101], [PAT201301], [CLA201401], [ALL201301]:

- Electric-propulsion engines are more efficient than chemical ones: they require significantly less propellant to produce the same overall effect, for example a specific increase in spacecraft velocity. The propellant is ejected up to 20 times faster than from chemically-based thrusters and, thus, the same propelling force is obtained with a lot less propellant. An all-EP spacecraft utilizes fuel very efficiently: a bi-propellant system may weigh four metric tons to do the same type of activity that a two metric ton EP would do (the xenon ion propulsion system is 10 times more efficient than chemical thrusters).
- The total mass of the satellite is approximately half of what a traditional chemical propulsion satellite system would be. For many missions, launch costs range between one-fourth and one-half of total mission costs; hence, reducing these costs can have a major impact on the overall return on investment. EP's ability to reduce or eliminate the weight of a satellite's chemical propulsion fuel and hardware allows operators to select a less-costly launch vehicle or make more efficient use of a given launch vehicle's capability.
- Because of lower mass operators may employ a dual launch configuration (e.g., on the newer SpaceX Falcon 9 rocket), reducing launch cost; without the mass burden of large fuel tanks, satellite operators have the option of tailoring their spacecraft to be more lightweight, allowing the satellites to be launched on smaller rockets or in tandem to spread launch costs between two missions. Launching two all-electric telecom payloads on a Falcon 9 rocket puts launch costs for each satellite at about $30 million (as of press time).
- All-EP satellites will improve capital investment efficiencies and are likely to expand business opportunities.
- While all-electric spacecraft will require longer orbit raising intervals, operators may be able offset the longer time frames with shorter production cycles.
- All-electric satellites can also carry larger payloads: because of the efficiencies of EP, larger payloads can fit on a smaller platform.
- Electric thrusters provide the ability to regulate the force applied to the spacecraft very accurately, making it possible to control the spacecraft's position and orientation along its orbit with improved precision.
- Once the satellite is in orbit, during the execution of a maneuver in the stationary orbit, one will have a better pointing accuracy toward the earth; signal variations will be reduced; instead of doing one station-keeping maneuver every two weeks, operators will be doing one maneuver every day.
- By offering a dual launch configuration, more easily achieved in all-EP-based spacecraft, the financial cost and risk is cut in half for operators, which lowers barrier-to-entry into the space market for small providers and entrepreneurs.

- EP is safer, as it involves smaller amounts of flammable fuels during the space-craft preparation for launch.
- The force EP produces can be applied continuously for very long periods – months or even years.
- Further improvements in satellites' propulsion and power systems may support an increase their service life to 20–30 years from the current 10–15 years.
- While the benefits of using EP for station-keeping have been significant, traditional chemical propulsion systems are still the main workhorses for more heavy-duty in-space propulsion tasks such as orbit-insertion. As early as 1999, satellite manufacturers saw success with EP "orbit-topping." In orbit-topping, chemical propulsion provides the first series of orbit-raising maneuvers and afterward the orbit-raising campaign is completed using EP. Satellites that use EP for orbit-topping are said to have "hybrid" chemical/EP systems. The benefits of using EP for orbit-topping are the same as those for station-keeping, but with even greater financial and operational payoff.

EP also has disadvantages, including, but not limited to the following:

As we have seen, chemical engines can eject large amounts of propellant, while electric thrusters produce very small flows; in turn, this implies that it takes much longer to achieve a particular speed and when high acceleration is a requirement, electrical propulsion cannot be employed with the current-generation technologies. It takes up to eight months to move a satellites from temporary transfer orbit to their final operating GSO slot. (The U.S. Air Force and Lockheed Martin achieved an inadvertent demonstration of all-electric orbit-raising in 2010 and 2011 following a mishap with the primary chemical propulsion system on an Advanced EHF strategic communications satellite – the AEHF 1 spacecraft reached its operational orbit nine months after launch [CLA201401].) See Figure 8.8.

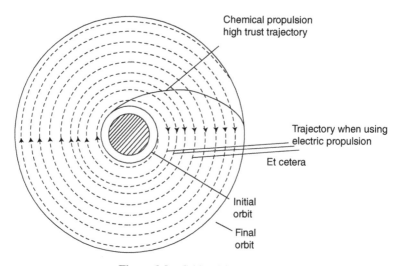

Figure 8.8   Orbit raising with EP.

Because the thrust of EP systems is lower than with chemical propulsion systems, as noted, it will take satellites up to 6 months longer to reach their final GSO orbit compared with systems with chemical or hybrid engines. This not only exposes the spacecraft to increased exposure risk, but also reduces the amount of time the spacecraft is able to generate revenues (in a true sense delaying market entry). Typically operators want to raise through the Earth's radiation belt in a rapid manner. The Van Allen Belt contains trapped high-energy particles captured by the earth's magnetic field; the Van Allen Belt exists in a middle-level zone located below the GSO and above the Low Earth Orbit (LEO). The radiation degrades the solar cells a little bit each time they pass through. Spacecraft with all-EP may ultimately require adding more shields, which in turn adds mass and reduces the size of payload, minimizing one of the system's key advantages.

Some practitioners are still reluctant to endorse all-electric satellites because these designs introduce new failure modes on the spacecraft: if one eliminates the chemical system completely, one has some new risks; for example, as the satellite separates from the launch vehicle, the spacecraft can also tumble out of control due to an anomaly – in this situation EP will not easily be able to correct the spacecraft if that occurs – chemical thrusters would be able to handle that situation better [PAT201301]. Thus, according to these practitioners, an "all-EP" spacecraft would use newer, not fully-space-tested technology, as an aggregate system, where the technology risk is not fully retired.

Station-keeping maneuvers can be supported, but orbit raising, drifts, and BIUs may not be as efficient as with traditional propulsion. An orbital drift maneuver with chemical propulsion achieves a movement of $1.25 - 1.75°$ a day – a drift of $180°$ may take 3-to-4 months; with EP this drift may take longer, also impacting revenues.

Some have raised issues related to the interaction between the plasma plume of an ion engine and the spacecraft. Generally, the phenomenon of surface degradation (erosion and sputtered particles re- deposition) is clearly identified as potential interaction that can degrade the spacecraft's structure. [BRO200101].

8.4   BASICS ABOUT STATION-KEEPING

Spacecraft in GSO (and other orbits) gradually drift north–south on a daily basis due to the influence of the sun and moon. There is a gradual increase in the inclination of the orbit. If left alone, a satellite that has initial zero inclination will have its inclination increase at the rate of $0.8°$ per year; a GSO spacecraft moves through space at 6,878 mph. It is usual not to let the maximum north–south movements exceed $\pm 0.05°$ to allow the use of large numbers of fixed pointing moderate sized antennas. Typically, a GSO box is around $0.1°$ of longitude which is approximately 70 km (43.5 miles) in length.[4,5] The "operational box" is $\pm 0.05°$ horizontally and vertically

---

[4] Slightly different boxes are used by some operators.
[5] Rec. ITU-R S.484-3 1 Recommendation, "Station-keeping in longitude of geostationary satellites in the fixed-satellite service (1974-1978-1982-1992)" states that " ... space stations on geostationary satellites

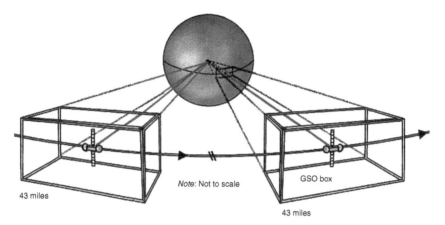

Figure 8.9   GSO box.

from the location the spacecraft is allowed to operate in (also known as east/west and north/south travel or station-keeping). See Figure 8.9. In some situations, a satellite operator might be locating multiple satellites within the same slot, this grouping being called a cluster. Clustering allows more satellites to be located at an orbital slot, always within the established box corresponding with that orbital location, but they give rise to challenges for keeping the individual satellites separated. Clusters are used such that a Direct to Home (DTH) antenna pointing at a given orbital location will see multiple satellites, each perhaps supporting (say) four transponders, so that each spacecraft can use more of its available power just for those transponders (a cluster of 6 spacecraft would cover the entire 24 transponders).

using frequency bands allocated to the fixed-satellite service should maintain their positions within $\pm 0.1°$ of longitude of their nominal positions irrespective of the cause of variation … ". The recommendation also makes these observations: A satellite is ideally geostationary if its orbital elements satisfy the following conditions: (i) semi-major axis a = synchronous radius = 42 165 km; (ii) eccentricity e = 0; (iii) inclination i = 0. The orbit gradually departs from the geostationary state, however, due to the perturbation forces acting on the satellite. The distortion of the earth's gravitational field due to the non-sphericity of the Earth causes either a steady increase or decrease in the semi-major axis, according to the stationary longitude of the satellite. No change however occurs at four special longitudes called the equilibrium points. When the semi-major axis deviates from the synchronous radius by Δa (km) the satellite longitude drifts at the rate of $-0.013 \Delta a$ degrees per day. The solar radiation pressure, acting in proportion to the ratio of satellite cross section to mass, changes the eccentricity with time. There is also a minor effect due to the Moons gravity. A non-zero eccentricity e yields a diurnal libration motion in the satellite longitude of as much as $\pm 2e \times 180/\pi$ degrees. The gravity of the moon and the sun acting as a tidal force on the satellite changes the inclination at a rate that varies from year to year between 0.75 and 0.95 per year. A non-zero inclination i (degrees) yields a half-diurnal longitudinal libration motion of as much as $\pm (i2/4) (\pi/180)$ degrees. This becomes significant for longitudinal station- keeping when the inclination is greater than a few degrees.

Spacecraft have a specified amount of rocket fuel that is used periodically (every few weeks) to correct the trend toward orbit inclination increase. Typically a brief burst firing of the north–south thrusters will be made as the satellite crosses up across the equator, so that instead of continuing gradually upwards it instead goes gradually downwards, effectively into negative inclination. The inclination then reduces over the next few weeks to zero and then increases again, by which time another thruster firing is needed. The net result is that over a period of 10–15 years the station-keeping fuel is gradually used up, but for this entire period the satellite is maintained within ±0.05° north–south of the equator. A small proportion of the total fuel is used for east–west orbit adjustments, since there is a tendency for satellite to very slowly drift sideways due to the triaxial nature of the earth (gravity is slightly stronger at three points around the equator). Once a satellite is getting near the end if its normal north–south station-keeping the satellite operators decide to stop and concentrate the remaining fuel on the much more economic east–west station-keeping, so as to extend the life by several more years. During this period the satellite is kept in its east–west position so that interference to adjacent satellites is avoided, but its inclination is allowed to increase to, say, ±5° over a half-a-dozen years. The communications payload continues to operate, with some loss of performance at the edges of the coverage beams since they no longer always point accurately at the countries on the ground all of the time. During the inclined orbit years earth stations must have tracking antennas so that their pointing is adjusted to aim at the satellite all during the day; the beam pointing movement is a lissajous "figure of 8", but at times it might be a tall slim elliptical shape or tilted elliptical shape [SAT201301].

The thrust required for the different space missions falls into the ranges: 200–400 N for elliptic-circular orbit transfer, 80–100 mN for station keeping of a GSO satellite and a few micro-N for precise positioning of scientific probes. The thrust T is defined as the product of the mass flow rate by the exhaust velocity of the propellant. Thrust is not the only parameter that needs to be taken into account when choosing the best thruster: another significant parameter is the specific impulse (Isp) defined as the ratio of the thrust to the mass flow to the intensity of gravity at the surface of earth. The specific impulse is related to the mass consumption for a defined mission associated to a required velocity variation of the spacecraft. A high specific impulse allows for a decrease in the mass of propellant and, consequently, a decrease in the mass of the satellite or, for the same mass, having more on-board electronics; as a tradeoff, a higher thrust allows a faster change of orbit [DUD201001]. The main parameters of concern are:

- Axial thrust.
- Specific impulse.
- Mass consumption.
- Power per mN.
- Efficiency.

Station-keeping maneuvers include

- East–west maneuvers.
- Drift and eccentricity maneuvers.
- Inclination maneuvers.
- Momentum adjustments.

East–west maneuvers are achieved with a single-part thruster firing, using thrusters on either the east or west side of spacecraft; the maneuver adjusts satellite longitudinal drift to remain within defined limits. During these maneuvers one should plan to minimize eccentricity magnitude to the extent possible and to minimize fuel use. Both the earth and the spacecraft are rotating/revolving in an easterly direction. Since the spacecraft is drifting west, we can think of the earth moving (rotating) underneath it, but faster than the satellite is revolving in its orbit above. Therefore, the spacecraft needs to move faster (more eastward motion) in its orbit. The spacecraft's velocity can be increased by reducing the size of the semi-major axis (SMA) of its orbit. The SMA is reduced by firing east. This imparts a negative delta-V. Doing so at perigee yields a negative delta-eccentricity, which also reduces its daily longitude excursions.

Drift and eccentricity (D&E) maneuvers entail two large east–west maneuvers having additive effect on eccentricity in inertial space, while effectively canceling their longitudinal and drift effects in an earth reference. These are used to correct for solar radiation pressure (this being greater on larger spacecraft and on those with larger surface areas), and triaxiality.

Inclination maneuvers are needed routinely, as noted above. Out-of-plane gravitational torque effects from sun and moon "pull" the angular momentum vector toward the Vernal Equinox. Viewed from the equinoctal plane, the inclination vector moves toward Summer Solstice. Lunar gravitation causes cyclic variation and solar gravitation causes regression of moon's line of nodes (the Kamel/Tibbitts cycle).

Momentum adjustment maneuvers (momentum management) are performed entirely by thrusters. Inertial attitude is updated using earth sensor and sun sensor data.

8.5 INDUSTRY APPROACHES

EP-equipped satellites (all-EP, and station-keeping EP) are likely to be increasingly utilized because of the benefits of saving fuel, launch mass reduction, extended life, and larger payloads. Hybrid spacecraft will continue to see deployment and "all-EP" systems are now being deployed. Observers argue that depending upon how expensive or complicated the mission is, large operators plan on taking advantage of all technologies (all-electric, all-chemical, and hybrids); while no operator might be willing to test an all-electric system on a large $250 million satellite, some would take a chance with a smaller 10-kilowatt satellite [PAT201301]. All-EP proponents make the following assertions [ALL201301]:

- EP technology offers potentially disruptive cost savings that have implications across the satellite value chain;
- Over the next 3–5 years (2019), use of EP for satellite orbit-insertion is likely to see rapid growth;
- Expanded use of EP will offer satellite operators the opportunity to reduce mission costs and increase revenue, but will also present challenging strategic questions

In 2012 Satmex (recently acquired by Eutelsat) and Asia Broadcast Satellite (ABS) each ordered satellites with all-EP from Boeing. Also in 2013, an unnamed US government customer ordered three light-class geostationary satellites from Boeing. Boeing's 702SP (small platform) design uses EP to perform all of its orbit-raising maneuvers; the spacecraft has a dry mass percentage of 80%, far higher than typical industry ranges of 40–60%. That high dry mass ratio will enable the lightweight 702 SP to dual-launch onboard a Falcon 9 v1.1 without sacrificing payload capability (thereby cutting launch costs nearly in half). The Boeing 702SP platform is a smaller version of Boeing's 702 satellite bus used by commercial and military operators. With the introduction of the Boeing 702 SP, Boeing has done away with chemical propulsion altogether. When launched in 2015, the 702 SP will be the first commercial spacecraft designed to use EP for the full orbit-raising campaign. According to observers, the mission and financial parameters of the Satmex/ABS spacecraft – which would be impossible without EP technology – "*sent shockwaves through the global commercial space industry… several major operators [now] indicated they expected to be purchasing their own all-electric satellites in the near future … likewise several satellite manufacturers stated that they would soon introduce their own designs for an all-EP satellite.*" [ALL201301]. The first two Boeing 702SP satellites, ABS 3A and Satmex 7, will launch in early 2015 in a dual-launch aboard a SpaceX Falcon 9 rocket; ABS 2A and Satmex 9 were expected to launch on another Falcon 9 booster later in 2015. Each of the ABS and Satmex satellites will have a launch mass of less than 4,000 lb but still offer communications throughput and power comparable to larger classic satellites with conventional fuel [CLA201401].

The Boeing 702 thruster is a xenon ion propulsion system (XIPS) and is the culmination of nearly four decades of research into the use of EP at Boeing. Boeing's 702SP satellite has a capability of 7.5 kilowatts of payload but the mass of the satellite is effectively half of what a traditional satellite system because it does not carry chemical liquid fuel tanks (many other satellites with that power range are up to 2,000 lb heavier at launch than the Boeing 702SP). The propulsion system works by accelerating electrically-ionized xenon gas through a thrust chamber at more than 60,000 mph; the thrust of an ion engine is lower than chemical propulsion, but the engines can fire for thousands of hours and consume less propellant. Boeing also redesigned its 702HP (high power), which uses both electric and bipropellant fuel to provide a configuration compatible with the Falcon 9, and introduced the 702MP (medium power) in 2009, which uses bipropellant but was designed to also accommodate other, more efficient EP [PAT201301].

As of press time, only three major commercial satellite manufacturers were offering GSO communications satellites that use EP for the full orbit-raising campaign, while at least five manufacturers were offering hybrid chemical/EP satellites. Lockheed Martin Commercial Space Systems had not decided to produce an all-electric satellite as of press time, despite its success with the AEHF 1 satellite's orbit-raising. The firm has electrical propulsion technology, but not at a commercial-operator price point. Lockheed's standard offering has been its A2100 product since 1996, which uses arc jets that enhance the capability of chemical propulsion systems; the company reportedly expects to continue HCTs as standard equipment. (HCTs provide more thrust than most other EP systems – as noted they utilize magnetic fields to focus and accelerate ions while traditional ion engines use electric fields to accelerate ions.) Space Systems/Loral was reportedly developing an ion thruster system to shorten the orbit-raising period to as low as three months for availability in 2015 [CLA201402].

Approximately one quarter of the spacecraft in the Intelsat fleet use a mixture of electrical and chemical propulsion systems, while the rest rely on chemical propulsion systems; the EP technology uses can be XIPS, arc jets and, since 2000, a Russian-developed ion propulsion system called SPT. None of the spacecraft use an all-EP system.

As an illustrative example of engine subcomponents, Space Power, Inc. part of Pratt & Whitney Space Propulsion, offers fully integrated propulsion systems that include HTCs, Power Processing Units, propellant management systems, tanks and other components. The press time family of HTCs is shown in Figure 8.10.

8.6   NEW APPROACHES AND PLAYERS FOR LAUNCH PLATFORMS

Delivery of payload in space, including into the GSO, still remains an expensive and often delay-prone proposition. Depending on the size and weight of payload, the cost of a single launch ranges from $100 M to 250M, with commercial GSO-class satellites being at the lower end of that range. Until early 2014, most commercial operations used Russia's space agency in Kazakhstan or Europe's Space Agency, which launches from French Guiana.

8.6.1   Space Exploration Technologies Corporation (SpaceX)

SpaceX is a new commercial entrant that recently charged only $55 M for the launch of a GSO-class satellite. They use some new technologies to reduce costs. Whether these figures are the target point or whether this was an incentivized first-GSO launch price listing remains to be fully determined. SpaceX officials have reportedly targeted a $60 million typical price tag for such a launch; industry observers, however, expect SpaceX's prices eventually will climb to about $100 million per launch. Nonetheless, the entry of Space X will bring competition to the effective duopoly for commercial launches that existed. SpaceX's Falcon 9 rocket has a modular design

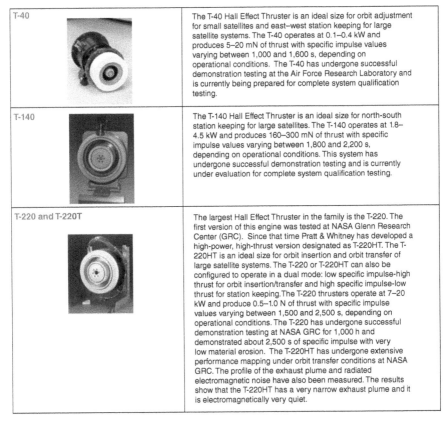

T-40	The T-40 Hall Effect Thruster is an ideal size for orbit adjustment for small satellites and east–west station keeping for large satellite systems. The T-40 operates at 0.1–0.4 kW and produces 5–20 mN of thrust with specific impulse values varying between 1,000 and 1,600 s, depending on operational conditions. The T-40 has undergone successful demonstration testing at the Air Force Research Laboratory and is currently being prepared for complete system qualification testing.
T-140	The T-140 Hall Effect Thruster is an ideal size for north-south station keeping for large satellites. The T-140 operates at 1.8–4.5 kW and produces 160–300 mN of thrust with specific impulse values varying between 1,800 and 2,200 s, depending on operational conditions. This system has undergone successful demonstration testing and is currently under evaluation for complete system qualification testing.
T-220 and T-220T	The largest Hall Effect Thruster in the family is the T-220. The first version of this engine was tested at NASA Glenn Research Center (GRC). Since that time Pratt & Whitney has developed a high-power, high-thrust version designated as T-220HT. The T-220HT is an ideal size for orbit insertion and orbit transfer of large satellite systems. The T-220 or T-220HT can also be configured to operate in a dual mode: low specific impulse-high thrust for orbit insertion/transfer and high specific impulse-low thrust for station keeping. The T-220 thrusters operate at 7–20 kW and produce 0.5–1.0 N of thrust with specific impulse values varying between 1,500 and 2,500 s, depending on operational conditions. The T-220 has undergone successful demonstration testing at NASA GRC for 1,000 h and demonstrated about 2,500 s of specific impulse with very low material erosion. The T-220HT has undergone extensive performance mapping under orbit transfer conditions at NASA GRC. The profile of the exhaust plume and radiated electromagnetic noise have also been measured. The results show that the T-220HT has a very narrow exhaust plume and it is electromagnetically very quiet.

Figure 8.10   Examples of hall effect thrusters.   Courtesy: Space Power, Inc. / Pratt & Whitney Space Propulsion.

and reusable elements. The upgraded version of the Falcon 9 rocket can launch up to 10,000 lb (4,500 kg) to geostationary transfer orbit; the Falcon 9 rocket lifts less mass than of its major competitors, but it is also less expensive. SpaceX's strategy is to attract customers with lower prices and with a more expedited launch schedule than the incumbents. The company already had a $4 billion book of business at press time.

One of the most expensive factors impacting launch economics is that the rocket typically burns up in the atmosphere when it returns to earth or is lost to the ocean. That is the reason Grasshopper, the company's prototype for a reusable Falcon 9 rocket, is very promising. The cost of the Falcon 9 rocket is $55 million and it utilizes $200,000 worth of fuel. Hence, in principle, one wants to get to a situation where the rocket is actually reusable, and the only costs to launch a satellite are then the fuel, the refurbishing rocket costs, and the ground support services; that would reduce the launch costs considerably.

8.6.2   Sea Launch

Sea Launch provides heavy lift launch services based on the Zenit-3SL launch system. The company has gone through several reorganizations and some lauch failures. In May 2014 they successfully launched Eutelsat 3B. The Sea Launch Zenit rocket blasts off from a mobile platform on the equator in the Pacific Ocean. Sea Launch's Zenit rocket is available for missions with single payloads of more than 6 metric tons. Under leadership from Energia, the organization has completely restructured the contracting for the Zenit-3SL supply chain, streamlined the oversight of all Russian/Ukrainian contracts for hardware and services, introduced a new level of transparency in conducting customer audits and obtained performance guarantees from key suppliers.

8.6.3   Traditional Launchers

The key traditional satellite launchers include:

- *Arianespace*. The firm utilizes the Ariane 5 rocket; the rocket is tailored for tandem launches of satellites weighing six metric tons total or individually three metric tons. The Ariane 5 launch base is in French Guiana, just north of the equator. Equatorial launches are beneficial for geostationary orbiting satellites because they require less powerful rockets and less liquid propellant to reach a spacecraft's final station (there is a mass penalty for satellites launching from higher inclinations: the spacecraft must fly aboard a larger launcher or carry more fuel in its tanks to make up the difference and reach its ultimate destination in space) [CLA201402].
- *International Launch Services (ILS)*. The Proton rocket and Breeze M upper stage can support single or dual launches with a total payload up more than six metric tons. In 1995, International Launch Services was established, upon the merger of Lockheed and Martin Marietta companies, to market Proton and Atlas launch services to the commercial satellite telecommunications marketplace worldwide. Prior to the merger, each of these companies were competing in the commercial launch services market with the Proton and Atlas rockets. With a history of demonstrated performance, the Proton launcher provides proven on-time reliability and expanded commercial capability, launched from the dedicated world-class facilities in Baikonur, Kazakhstan. Built by the Khrunichev State Research and Production Center, the Proton rocket is the largest Russian launch vehicle in operational service. Proton launches both geostationary and interplanetary missions, and is the principal workhorse of the Russian space program. Up to press time the partnership accounts for launching 30% of the global commercial space market. Some recent launches include:
  - KazSat-3/Luch-5 V – April 28, 2014.
  - Express AT1/AT2 – March 16, 2014.
  - TURKSAT-4A – February 14, 2014.
  - Express-AM5 – December 26, 2013.

- Inmarsat-5 F1 – December 8, 2013.
- Federal – November 12, 2013.
- Sirius FM-6 – October 26, 2013.
- ASTRA 2E – September 30, 2013.
- GLONASS – July 2, 2013.
- SES-6 – June 3, 2013.
- Eutelsat 3D – May 14, 2013.
- Anik G1 – April 16, 2013.
- Satmex_8 – March 27, 2013.
- Yamal 402 – December 8, 2012.
- EchoStar XVI – November 21, 2012.
- Yamal 300 K/Luch 5B – November 3, 2012.
- Intelsat 23 – October 14, 2012 (ILS).
- Telkom 3/ Express MD2 – August 7, 2012.
- SES-5 – July 10, 2012.
- Nimiq 6 – May 18, 2012.
- China's Long March 3C launcher has a lift capacity of 5,500 kg to geostationary transfer orbit.

REFERENCES

[ALL201301] G. Allen, R. Dalby, *et al.,* "Satellite Electric Propulsion: Key Questions for Satellite Operators and their Suppliers", Avascent White Paper, March 21, 2013. Avascent, 1615 L Street, NW, Suite 120, Washington, DC 20036. www.avascent.com.

[BRO200101] S. Brosse, O. Chanrion, V. Perrin, "Electric Effects OF Plasma Propulsion On Satellites", Spacecraft Charging Technology, Proceedings of the Seventh International Conference held 23–27 April, 2001 at ESTEC, Noordwijk, the Netherlands. Edited by R.A. Harris, European Space Agency, ESA SP-476, 2001. p. 139.

[CLA201401] S. Clark, "Boeing Reveals Government's All-Electric Satellite Purchase" Spaceflight Now, March 12, 2014.

[CLA201402] S. Clark, "Electric Propulsion Could Launch New Commercial Trend" Spaceflight Now, March 19, 2014.

[DEL201301] Telecom Update, Deloitte Corporate Finance LLC, Jun 30, 2013 (also, referencing Capital IQ/Thomson Financial). Retrieved at www.deloitte.com.

[DUD201001] M. Dudeck, F. Doveil, N. Arcis, S. Zurbach, "Plasma Propulsion For Geostationary Satellites And Interplanetary Spacecraft", 15th International Conference on Plasma Physics and Applications, 1–4 July 2010, Iasi, Romania. Romanian Journal of Physics, Vol. 56, Supplement, P. 3–14, Bucharest, 2011.

[ESA200201] Electric Propulsion White Paper, July 2002, European Space Agency (ESA).

[GRE200401] D. Green, The 'total cost of ownership', Satellite Manufacturing Special – LMCSS (Lockheed Martin Commercial Space System), July/August 2004, pages 36–39. www.satellite-evolution.com. Reporting on a 2003 Futron study entitled 'GEO Commercial Satellite Bus Operations: A Comparative Analysis, 2003'.

[GOE200801] D. M. Goebel, I. Katz, *Fundamentals of Electric Propulsion: Ion and Hall Thrusters*, Wiley, 2008.

[JAH200201] R. G. Jahn, E. Y. Choueiri, *"Electric Propulsion"*, Encyclopedia of Physical Science and Technology, Third Edition, Volume 5, Academic Press, 2002.

[PAT201301] C. Patton, "All Electric Satellites: Revolution or Evolution?" Via Satellite, May 1, 2013.

[PRA201401] M. Prado, "Electric Propulsion for Inter-Orbital Vehicles", The Permanent Website, http://www.permanent.com/space-transportation-electric.html.

[SAT201301] Inclined orbit operation of geostationary satellites, November 20, 2013, Retrieved at http://www.satsig.net/satellite/inclined-orbit-operation.htm.

[SEL201201] P. B. de Selding, "Rising Transponder Prices Mask Regional Disparity", Space News, August 23, 2012

[STE201401] A. Damodaran, "Operating and Net Margins", New York University Stern School of Business. Retrieved at http://pages.stern.nyu.edu/~adamodar/New_ Home_Page/datafile/margin.html (also see http://www.stern.nyu.edu/~adamodar/pc/data sets/margin.xls).

APPENDIX 8A

TRANSPONDER COSTS

This appendix has two sections; the first assesses EBITDA for various industries; the second looks at transponder costs.

8A.1   TYPICAL SG&A AND EBITDA FOR THE GENERAL COMMERCIAL WORLD AND SATELLITE FIRMS

New York University Stern School of Business published the EBITDA as a percentage of sales for 7,766 firms, such figure being 14.7% [STE201401]. See Table 8A.1.

Table 8A.2 focuses more closely to the wireless/satellite industry, based on data from Deloitte [DEL201301].

Table 8A.3 depicts the Selling, General & Administrative Expense (SG&A) costs of major industries. The average SG&A is 17%; for the aerospace industry it is 8% and for the wireless industry is 33% (if we used an average between these two industries, the average is 21%, right on-mark with a telecom industry SG&A of 22%).

Innovations in Satellite Communications and Satellite Technology: The Industry Implications of DVB-S2X, High Throughput Satellites, Ultra HD, M2M, and IP, First Edition. Daniel Minoli.
© 2015 John Wiley & Sons, Inc. Published 2015 by John Wiley & Sons, Inc.

TABLE 8A.1   EBITDA By Industry (New York University Stern School of Business Data)

Industry Name	Number of Firms in Study	EBITDA/Sales (%)
Utility (water)	20	42.50
Precious metals	166	42.36
Oil/Gas (production and exploration)	411	42.09
Tobacco	12	41.50
Railroad	10	37.71
Pharma and drugs	138	34.02
Computer software	273	30.75
Cable TV	16	28.96
Semiconductor	104	27.61
Broadcasting	30	27.45
Information services	71	27.44
Utility (general)	20	27.32
Power	106	26.90
Healthcare equipment	193	26.43
Telecommunications services	82	25.68
Internet software and services	330	23.50
Telecommunications equipment	131	23.36
Entertainment	85	22.99
Metals and mining	134	22.90
Computers/Peripherals	66	22.64
Beverage (alcoholic)	19	22.38
Biotechnology	349	22.31
Hotel/Gaming	89	22.13
Real estate (general/diversified)	11	21.96
Beverage	47	21.54
Oil/Gas (integrated)	8	21.10
Diversified	20	20.30
Coal and related energy	45	20.03
Chemical (specialty)	100	19.60
Restaurant	84	19.60
Telecom (wireless)	28	19.28
Household products	139	19.23
Insurance (general)	26	19.21
Environmental and waste services	108	18.71
Healthcare information and technology	125	18.58
Recreation	70	18.15
Healthcare products	58	17.58
Semiconductor equipment	51	16.15
Shipbuilding and marine	14	15.85
Healthcare facilities	47	15.79
Publishing and newspapers	52	15.78
Electrical equipment	135	15.63
Advertising	65	15.52

TABLE 8A.1 *(Continued)*

Industry Name	Number of Firms in Study	EBITDA/Sales (%)
Educational services	40	15.38
Machinery	141	15.26
Heavy construction	46	14.38
Apparel	70	14.23
Chemical (diversified)	10	14.22
Chemical (basic)	47	14.11
Packaging and container	24	13.61
Construction	18	13.60
Trucking	28	13.29
(see reference for rest of table)		
Total	77,663	14.71

Source: NYU Stern.

TABLE 8A.2 EBITDA for Telecom/Wireless/Satellite Industry

$US in Millions	Reported Date	Market Cap	Enterprise Value (EV)	EBITDA Margin (%)
Multiline Telecom Service Providers				
Nippon Telegraph and Telephone Corporation	Mar 31, 2013	$61,392	$93,822	29.2
AT&T, Incorporated	Mar 31, 2013	$190,452	$261,492	22.7
Verizon Communications Incorporated	Jun 30, 2013	$144,023	$191,369	27.7
Telefonica, S.A.	Jun 30, 2013	$58,785	$127,638	31.4
Deutsche Telekom AG	Mar 31, 2013	$49,977	$99,587	26.6
Vodafone Group Public Limited Company	Mar 31, 2013	$138,112	$182,010	44.2
Cegedim SA	Mar 31, 2013	$427	$1,069	17.2
America Movil S.A.B. de C.V.	Mar 31, 2013	$79,255	$108,812	33.0
Orange	Jun 30, 2013	$24,824	$65,394	27.5
KDDI Corporation	Mar 31, 2013	$39,744	$49,011	25.7
China Telecom Corporation Limited	Dec 31, 2012	$38,608	$49,372	24.6
SoftBank Corporation	Mar 31, 2013	$69,513	$84,940	33.0
China Unicom (Hong Kong) Limited	Dec 31, 2012	$31,394	$51,256	29.3
Telecom Italia S.p.A.	Mar 31, 2013	$13,384	$54,943	39.1
Sprint Corporation	Mar 31, 2013	N.M.	N.M.	12.7

(continued)

TABLE 8A.2   *(Continued)*

$US in Millions	Reported Date	Market Cap	Enterprise Value (EV)	EBITDA Margin (%)
BT Group plc	Jun 30, 2013	$37,039	$51,264	29.6
Telstra Corporation Limited	Dec 31, 2012	$54,167	$66,532	39.1
Oi SA	Mar 31, 2013	$2,901	$17,190	30.1
KT Corporation	Mar 31, 2013	$7,617	$16,448	17.9
BCE, Incorporated	Mar 31, 2013	$31,747	$50,789	40.2
Century Link, Incorporated	Mar 31, 2013	$21,530	$41,842	41.6
Level 3 Communications, Incorporated	Mar 31, 2013	$4,668	$12,649	22.6
VimpelCom Limited.	Dec 31, 2012	$17,582	$41,444	42.0
China Mobile Limited	Dec 31, 2012	$209,922	$149,887	46.4
T-Mobile US, Incorporated	Mar 31, 2013	$17,995	$34,872	24.8
Leap Wireless International Incorporated	Mar 31, 2013	$532	$3,168	16.3
United States Cellular Corporation	Mar 31, 2013	$3,071	$3,420	19.7
Mobile Telesystems OJSC	Dec 31, 2012	$18,834	$25,644	43.4
Average Multiline Telecom Service Providers				29.9
Median Multiline Telecom Service Providers				29.3
Telecom Equipment Manufacturers				
Ericsson	Jun 30, 2013	$36,335	$30,646	7.7
Nokia Corporation	Jun 30, 2013	$13,734	$8,040	8.8
Alcatel-Lucent, S.A.	Mar 31, 2013	$4,142	$4,713	4.3
QUALCOMM Incorporated	Jun 30, 2013	$105,553	$94,112	34.8
Juniper Networks, Incorporated	Mar 31, 2013	$9,776	$8,118	15.9
Novatel Wireless Incorporated	Mar 31, 2013	$134	$95	8.8
Sierra Wireless Incorporated	Mar 31, 2013	$393	$337	1.2

TABLE 8A.2 *(Continued)*

$US in Millions	Reported Date	Market Cap	Enterprise Value (EV)	EBITDA Margin (%)
F5 Networks, Incorporated	Mar 31, 2013	$5,421	$4,899	31.5
ZTE Corporation	Mar 31, 2013	$7,136	$9,750	0.9
Cisco Systems, Incorporated	Apr 27, 2013	$130,054	$98,914	27.3
Fujitsu Limited	Mar 31, 2013	$8,549	$11,427	6.7
American Tower Corporation	Mar 31, 2013	$28,938	$37,351	63.1
SBA Communications Corporation	Mar 31, 2013	$9,458	$14,695	62.5
EchoStar Corporation	Mar 31, 2013	$3,465	$4,401	18.7
Average Telecom Equipment Manufacturers				19.3
Median Telecom Equipment Manufacturers				12.3
Cable & Satellite Service Providers				
Comcast Corporation	Mar 31, 2013	$109,945	$152,491	33.8
Time Warner Cable Incorporated	Mar 31, 2013	$32,728	$56,262	37.2
Liberty Global plc	Mar 31, 2013	$27,919	$57,829	45.8
DIRECTV	Mar 31, 2013	$34,429	$51,115	24.9
Dish Network Corporation	Mar 31, 2013	$19,332	$24,117	20.1
British Sky Broadcasting Group plc	Mar 31, 2013	$18,990	$21,165	23.7
Rogers Communications Incorporated	Jun 30, 2013	$20,145	$31,372	37.9
Charter Communications, Incorporated	Mar 31, 2013	$12,540	$25,344	35.1
Cablevision Systems Corporation	Mar 31, 2013	$4,492	$14,296	26.6
Shaw Communications, Incorporated	May 31, 2013	$10,785	$15,487	41.2
Kabel Deutschland Holding AG	Mar 31, 2013	$9,640	$13,253	43.5

(*continued*)

TABLE 8A.2 *(Continued)*

$US in Millions	Reported Date	Market Cap	Enterprise Value (EV)	EBITDA Margin (%)
Telenet Group Holding NV	Dec 31, 2012	$5,253	$9,125	48.1
SKY Perfect JSAT Holdings Incorporated	Mar 31, 2013	$1,536	$1,270	27.0
Eutelsat Communications S.A.	Dec 31, 2012	$6,214	$9,697	77.6
Average Cable & Satellite Service Providers				37.3
Median Cable & Satellite Service Providers				36.1

Source: Deloitte/IQ capital.

TABLE 8A.3   SG&A By Industry (New York University Stern School of Business Data)

Industry Name	SG&A as % of Revenues
Advertising	35.93
Aerospace/Defense	8.01
Air transport	14.50
Apparel	33.39
Auto parts	9.63
Automotive	11.85
Beverage	34.06
Biotechnology	55.46
Building materials	16.23
Cable TV	10.11
Chemical (basic)	9.83
Chemical (diversified)	14.80
Chemical (specialty)	13.73
Coal	4.83
Computer software	44.94
Computers/Peripherals	19.43
Diversified company	13.46
Drug	42.46
E-commerce	38.51
Educational services	38.39

TABLE 8A.3 *(Continued)*

Industry Name	SG&A as % of Revenues
Electric utility (central)	3.01
Electric utility (east)	0.00
Electric utility (west)	1.27
Electrical equipment	21.10
Electronics	11.05
Engineering and construction	4.96
Entertainment	14.42
Entertainment technology	44.84
Environmental	12.23
Financial Svcs. (division)	41.83
Food processing	18.51
Foreign electronics	27.03
Funeral services	13.13
Furn/Home furnishings	23.14
Healthcare information	44.24
Heavy truck and equipment	12.47
Homebuilding	11.30
Hotel/Gaming	16.14
Household products	29.43
Human resources	16.97
Industrial services	11.58
Information services	21.46
Internet	34.35
Investment Company	38.25
IT services	17.10
Machinery	20.58
Maritime	7.29
Med supp invasive	39.77
Med supp non-invasive	12.22
Medical services	13.27
Metal fabricating	15.51
Metals and mining (division)	3.05
Natural gas (division)	9.84
Natural gas utility	18.53
Newspaper	32.14
Office equipment/supplies	23.11
Oil/Gas distribution	7.76
Oilfield Svcs/Equipment	7.18
Packaging and container	10.31
Paper/Forest products	7.93
Petroleum (integrated)	4.62
Petroleum (producing)	4.31
Pharmacy services	13.80

(continued)

TABLE 8A.3   *(Continued)*

Industry Name	SG&A as % of Revenues
Pipeline MLPs	3.09
Power	6.83
Precious metals	8.63
Precision instrument	29.23
Public/Private equity	17.28
Publishing	26.49
R.E.I.T.	6.71
Railroad	0.55
Recreation	16.13
Restaurant	11.04
Retail (hardlines)	24.88
Retail (softlines)	26.50
Retail automotive	17.89
Retail building supply	23.44
Retail store	18.24
Retail/Wholesale food	15.15
Securities brokerage	4.32
Semiconductor	30.19
Semiconductor equipment	27.67
Shoe	32.35
Steel	5.38%
Telecommunications equipment	31.36
Telecommunications services	22.07
Telecommunications utility	30.27
Tobacco	9.83
Toiletries/Cosmetics	51.85
Trucking	4.26
Water utility	6.39
Wireless networking	33.37
Total market	17.10

Source: NYU Stern.

8A.2   TRANSPONDER COSTS

The article that follows, quoted nearly verbatim from [SEL201201], intrinsically makes important observations.

"The average price of leasing a telecommunications satellite transponder increased just about everywhere in 2011 except in North America, the Middle East and North Africa, according to a market assessment released August 23, 2012. The report by Euroconsult of Paris said the average transponder price of $1.62 million per year for 36 megahertz of capacity hides a wide regional price disparity. The report says there is a "risk of price erosion" in the coming years as some markets soften with new satellites entering service. As has been the case for years, Western Europe continues to be the highest-price market for leasing satellite bandwidth, at an average of $3.2 million per transponder in 2011, Euroconsult said.

Northeast Asia featured the second-most-costly transponders, at $2.6 million per transponder per year, followed by the region around Australia and New Zealand, where prices averaged $1.7 million per year. The least-expensive transponders were to be found in South Asia, where prices averaged only slightly more than $1 million per transponder per year. Satellite operators have said that absent revolution in satellite or launch costs, $1 million is a kind of threshold price below which it is difficult to make a profit.

For some smaller satellite operators financed by their governments, making a profit may not be a primary concern. As these operators proliferate, their approach to the market may affect the business models of the established operators for which profitability is the principal motivation. Not surprisingly, North America appears in the Euroconsult report as a place where overall market size and prices, at around $1.4 million per transponder per year, have been flat in recent years and are likely to remain so for some time …

Euroconsult said that for the operators of fixed satellite services that report financial results, the average EBITDA – earnings before interest, taxes, depreciation and amortization – was 75 percent of revenue in 2011. But as with transponder prices, this figure masked wide variations, with the most profitable operators reporting 80 percent EBITDA margins, while others reported margins of 50 percent. Television broadcasts remain the core of the market's profitability as demand for high-definition television programming, which requires more bandwidth, has outpaced advances in video compression that puts downward pressure on bandwidth demand.

Global military demand for commercial fixed satellite services bandwidth, which has been led by the U.S. Defense Department, is likely to decrease in the coming years with the winding down of U.S.-led military coalition activity in Afghanistan and Iraq, Euroconsult said. But countering this trend will be growth in U.S. military use of Ka-band satellite broadcast frequencies. The U.S. military spent more than $450 million purchasing conventional commercial fixed satellite service capacity in 2011, according to Euroconsult. To this figure is added L-band capacity purchases from mobile satellite services operators such as Iridium of McClean, Va., and Inmarsat of London.

One of the highest-growth segments of the telecommunications satellite industry is in delivering broadband Internet access to unserved or underserved areas. A new generation of platforms, called high-throughput satellites, is being fielded in North America, Europe and Australia, with more to come in other regions, according to most market assessments. Euroconsult forecasts that high-throughput satellites, which provided 35 gigabits per second of capacity worldwide, will be beaming around 850 gigabits per second in a decade's time. More than half that capacity will be used to provide consumers with broadband access from their homes. North America, which has been the biggest early adopter of satellite broadband despite insignificant government support, will remain the biggest market for this technology through the next decade, Euroconsult said. ViaSat Inc. of Carlsbad, Calif., and Englewood, Colo.-based EchoStar Corp.'s Hughes division are both deploying high-throughput satellites for their established consumer broadband businesses."

REFERENCES

[AER200801] The Aerospace Corporation, Satellite Communications Glossary, In *Crosslink* (ISSN 1527–5264 [print], ISSN 1527–5272 [Web]). Corporate Communications, P.O. Box 92957, M1-447, Los Angeles, CA 90009–2957. The Aerospace Press, P.O. Box 92957, Los Angeles, CA 90009–2957.

[AMS200001] T. S. Glickman, Editor, Glossary of Meteorology, American Meteorological Society, Cambridge, Massachusetts, June 2000.

[ANS200001] ANS T1.523-2001, *Telecom Glossary 2000*, American National Standard (ANS), an outgrowth of the Federal Standard 1037 series, *Glossary of Telecommunication Terms*, 1996.

[BAR200101] M. Bartlett, Satellite FAQ, S+AS Limited, 6 The Walled Garden, Wallhouse, Torphichen, West Lothian, EH48 4NQ, SCOTLAND, July 2001.

[BLO200701] Staff, "QAM Defined", Blonder Tongue Laboratories, Inc., One Jake Brown Road, Old Bridge, NJ 08857, 2007.

[DEL201301] Telecom Update, Deloitte Corporate Finance LLC, June 30, 2013 (also, referencing Capital IQ/Thomson Financial). Retrieved at www.deloitte.com.

[EUT200701] Eutelsat, Glossary, 70, rue Balard, F-75502 Paris Cedex 15, France, http://www.eutelsat.com/business/2_6_2.html.

[GEO200101] Staff, "Geostationary, LEO, MEO, HEO Orbits Including Polar and Sun-Synchronous Orbits with Example Systems and a brief section on Satellite History", 2001, http://www.geo-orbit.org.

[HEN200201] H. Hendrix, "Viterbi Decoding Techniques for the TMS320C54x DSP Generation", Texas Instruments, Application Report SPRA071A – January 2002. Texas Instruments, Post Office Box 655303, Dallas, Texas 75265.

[JAH200201] R. G. Jahn and E. Y. Choueiri, *"Electric Propulsion"*, Encyclopedia of Physical Science and Technology, Third Edition, Academic Press, 2002. Volume 5.

[JOU199901] M. K. Juonolainen, Forward Error Correction in INSTANCE, Cand Scient Thesis, 1/2/1999, University of Oslo, Department of Informatics.

[LAU200701] J. E. Laube, HughesNet, "Introduction to the Satellite Mobility Support Network, HughesNet User Guide" 2005–2007.

[ORT200801] ORTEL/EMCORE, "RF and Microwave Fiber-Optic Design Guide", Application Note, March 7, 2003, 10420 Research Road, SE, Albuquerque, NM 87123.

[SEL201201] P. B. de Selding, "Rising Transponder Prices Mask Regional Disparity", Space News, August 23, 2012.

[STE201401] A. Damodaran, "Operating and Net Margins", New York University Stern School of Business. Retrieved at http://pages.stern.nyu.edu/~adamodar/New_Home _Page/datafile/margin.html (also see http://www.stern.nyu.edu/~adamodar/pc/datasets/ margin.xls).

[SAT200501] Satellite Internet Inc., Satellite Physical Units & Definitions, http://www. satellite-internet.ro/satellite-internet-terminology-definitions.htm.

[SAT200801] http://www.satsig.net.

[USA199801] U.S. Army Information Systems Engineering Command Fort Huachuca, Arizona, "Automated Information Systems, Design Guidance, Commercial Satellite Transmission", August 1998.

[ZYR199801] J. Zyren, A. Petrick, "Tutorial on Basic Link Budget Analysis", Application Note AN9804.1, June 1998, Intersil Corporation, 1001 Murphy Ranch Road, Milpitas, CA 95035.

APPENDIX A

PARTIAL LISTING OF SYSTEM-LEVEL US PATENTS FOR SPOT-BEAM/ MULTI-BEAM SATELLITES

Patent	Filing Date	Publication Date	Applicant	Title
US3810255	Jun 10, 1971	May 7, 1974	Communications Satellite Corporation	Frequency translation routing communications transponder
US4228401	Dec 22, 1977	Oct 14, 1980	Communications Satellite Corporation	Communication satellite transponder interconnection utilizing variable bandpass filter

Innovations in Satellite Communications and Satellite Technology: The Industry Implications of DVB-S2X, High Throughput Satellites, Ultra HD, M2M, and IP, First Edition. Daniel Minoli.
© 2015 John Wiley & Sons, Inc. Published 2015 by John Wiley & Sons, Inc.

Patent	Filing Date	Publication Date	Applicant	Title
US4689625	Nov 6, 1984	Aug 25, 1987	Martin Marietta Corporation	Satellite communications system and method therefor
US4813036	Nov 27, 1985	Mar 14, 1989	National Exchange, Incorporated	Fully interconnected spot beam satellite communication system
US4706239	Dec 18, 1985	Nov 10, 1987	Kokusai Denshin Denwa Company, Limited	Communications satellite repeater
US4858225	Nov 5, 1987	Aug 15, 1989	International Telecommunications Satellite	Variable bandwidth variable center-frequency multibeam satellite-switched router
US4931802	Mar 11, 1988	Jun 5, 1990	Communications Satellite Corporation	Multiple spot-beam systems for satellite communications
US5119225	Jan 18, 1989	Jun 2, 1992	British Aerospace Public Limited Company	Multiple access communication system
US4903126	Feb 10, 1989	Feb 20, 1990	Kassatly Salim A	Method and apparatus for TV broadcasting
US5394560	Sep 30, 1992	Feb 28, 1995	Motorola, Incorporated	Nationwide satellite message delivery system

Patent	Filing Date	Publication Date	Applicant	Title
US5428814	Aug 13, 1993	Jun 27, 1995	Alcatel Espace	Space communications apparatus employing switchable band filters for transparently switching signals on board a communications satellite, payload architectures using such apparatus, and methods of implementing the apparatus and the architectures
US5412660	Sep 10, 1993	May 2, 1995	Trimble Navigation Limited	ISDN-to-ISDN communication via satellite microwave radio frequency communications link
US5424862	Apr 28, 1994	Jun 13, 1995	Glynn; Thomas W.	High capacity communications satellite
US5594780	Jun 2, 1995	Jan 14, 1997	Space Systems/Loral, Incorporated	Satellite communication system that is coupled to a terrestrial communication network and method
US5640386	Jun 6, 1995	Jun 17, 1997	Globalstar L.P.	Two-system protocol conversion transceiver repeater

(*continued*)

Patent	Filing Date	Publication Date	Applicant	Title
US5552920	Jun 7, 1995	Sep 3, 1996	Glynn; Thomas W.	Optically crosslinked communication system (OCCS)
US5680240	Jun 12, 1995	Oct 21, 1997	Glynn; Thomas W.	High capacity communications satellite
EP07480 64A2	Apr 3, 1996	Dec 11, 1996	Globalstar L.P.	Two-system protocol conversion transceiver repeater
US5822312	Apr 17, 1996	Oct 13, 1998	Com Dev Limited	Repeaters for multibeam satellites
US5809141	Jul 30, 1996	Sep 15, 1998	Ericsson Incorporated	Method and apparatus for enabling mobile-to-mobile calls in a communication system
US5898681	Sep 30, 1996	Apr 27, 1999	Amse Subsidiary Corporation	Methods of load balancing and controlling congestion in a combined frequency division and time division multiple access communication system using intelligent login procedures and mobile terminal move commands

Patent	Filing Date	Publication Date	Applicant	Title
US5963862	Oct 25, 1996	Oct 5, 1999	Pt Pasifik Satelit Nusantara	Integrated telecommunications system providing fixed and mobile satellite-based services
US5864546	Nov 5, 1996	Jan 26, 1999	Worldspace International Network, Incorporated	System for formatting broadcast data for satellite transmission and radio reception
US5697050	Dec 12, 1996	Dec 9, 1997	Globalstar L.P.	Satellite beam steering reference using terrestrial beam steering terminals
US6256496	Mar 10, 1997	Jul 3, 2001	Deutsche Telekom Ag	Digital radio communication apparatus and method for communication in a satellite-supported VSAT network
US5884142	Apr 15, 1997	Mar 16, 1999	Globalstar L.P.	Low earth orbit distributed gateway communication system
EP08832 52A2	May 30, 1998	Dec 9, 1998	Hughes Electronics Corporation	Method and system for providing wideband communications to mobile users in a satellite-based network

(*continued*)

Patent	Filing Date	Publication Date	Applicant	Title
US6278876	Jul 13, 1998	Aug 21, 2001	Hughes Electronics Corporation	System and method for implementing terminal to terminal connections via a geosynchronous earth orbit satellite
US6496682	Sep 14, 1998	Dec 17, 2002	Space Systems/Loral, Incorporated	Satellite communication system employing unique spot beam antenna design
US6415329	Oct 30, 1998	Jul 2, 2002	Massachusetts Institute of Technology	Method and apparatus for improving efficiency of TCP/IP protocol over high delay-bandwidth network
EP0967 745A2	Jun 23, 1999	Dec 29, 1999	Matsushita Electric Industrial Co., Limited	Receiving apparatus for receiving a plurality of digital transmissions comprising a plurality of receivers
US6240124	Nov 2, 1999	May 29, 2001	Globalstar L.P.	Closed loop power control for low earth orbit satellite communications system
US7245933	Mar 1, 2000	Jul 17, 2007	Motorola, Incorporated	Method and apparatus for multicasting in a wireless communication system

Patent	Filing Date	Publication Date	Applicant	Title
US7068616	Apr 20, 2001	Jun 27, 2006	The Directv Group, Incorporated	Multiple dynamic connectivity for satellite communications systems
US7809403	May 15, 2001	Oct 5, 2010	The Directv Group, Incorporated	Stratospheric platforms communication system using adaptive antennas
US20030 109220	Dec 12, 2001	Jun 12, 2003	Hadinger Peter J.	Communication satellite adaptable links with a ground-based wideband network
US6965755	Nov 8, 2002	Nov 15, 2005	The Directv Group, Incorporated	Comprehensive network monitoring at broadcast satellite sites located outside of the broadcast service area
US7069036	Feb 13, 2003	Jun 27, 2006	The Boeing Company	System and method for minimizing interference in a spot beam communication system
US7715838	Apr 2, 2003	May 11, 2010	The Boeing Company	Aircraft based cellular system
US7653349	Jun 18, 2003	Jan 26, 2010	The DirecTV Group, Incorporated	Adaptive return link for two-way satellite communication systems
US8358971	Jun 23, 2003	Jan 22, 2013	Qualcomm Incorporated	Satellite-based programmable allocation of bandwidth for forward and return links

(*continued*)

Patent	Filing Date	Publication Date	Applicant	Title
US7525934	Sep 13, 2004	Apr 28, 2009	Qualcomm Incorporated	Mixed reuse of feeder link and user link bandwidth
US7580708	Sep 8, 2005	Aug 25, 2009	The DirecTV Group, Incorporated	Comprehensive network monitoring at broadcast satellite sites located outside of the broadcast service area
US7636546	Dec 20, 2005	Dec 22, 2009	Atc Technologies, Llc	Satellite communications systems and methods using diverse polarizations
WO200612 3064A1	May 19, 2006	Nov 23, 2006	Centre Nat Etd Spatiales	Method for assigning frequency subbands to upstream radiofrequency connections and a network for carrying out said method
US7962134	Jan 17, 2007	Jun 14, 2011	M.N.C. Microsat Networks (Cyprus) Limited	Systems and methods for communicating with satellites via non-compliant antennas
US8078141	Jan 17, 2007	Dec 13, 2011	Overhorizon (Cyprus) Plc	Systems and methods for collecting and processing satellite communications network usage information

Patent	Filing Date	Publication Date	Applicant	Title
US8326217	Jan 17, 2007	Dec 4, 2012	Overhorizon (Cyprus) Plc	Systems and methods for satellite communications with mobile terrestrial terminals
US2009 0191810	May 15, 2007	Jul 30, 2009	Mitsubishi Electric Corporation	Satellite communication system
US8346161	May 15, 2007	Jan 1, 2013	Mitsubishi Electric Corporation	Satellite communication system
US8050628	Jul 17, 2007	Nov 1, 2011	M.N.C. Microsat Networks (Cyprus) Limited	Systems and methods for mitigating radio relay link interference in mobile satellite communications
US7869759	Dec 12, 2007	Jan 11, 2011	Viasat, Incorporated	Satellite communication system and method with asymmetric feeder and service frequency bands
US8311498	Sep 29, 2008	Nov 13, 2012	Broadcom Corporation	Multiband communication device for use with a mesh network and methods for use therewith
US2009 0291633	Mar 18, 2009	Nov 26, 2009	Viasat, Incorporated	Frequency re-use for service and gateway beams
US2009 0298416	Mar 18, 2009	Dec 3, 2009	Viasat, Incorporated	Satellite Architecture

(*continued*)

Patent	Filing Date	Publication Date	Applicant	Title
US20120 276840	Mar 18, 2009	Nov 1, 2012	Viasat, Incorporated	Satellite Architecture
US8254832	Mar 18, 2009	Aug 28, 2012	Viasat, Incorportated	Frequency re-use for service and gateway beams
US8315199	Mar 18, 2009	Nov 20, 2012	Viasat, Incorporated	Adaptive use of satellite uplink bands
US8538323	Mar 18, 2009	Sep 17, 2013	Viasat, Incorporated	Satellite architecture
US8339309	Sep 24, 2010	Dec 25, 2012	Mccandliss Brian	Global communication system
US2012 0164941	Dec 16, 2011	Jun 28, 2012	Electronics and Telecommuni-cations Research Institute	Beam bandwidth allocation apparatus and method for use in multi-spot beam satellite system
US2012 0244798	May 30, 2012	Sep 27, 2012	Viasat, Incorporated	Frequency re-use for service and gateway beams
US8548377	May 30, 2012	Oct 1, 2013	Viasat, Incorporated	Frequency re-use for service and gateway beams

Note: table sorted by filling date

APPENDIX B

GLOSSARY OF KEY SATELLITE CONCEPTS AND TERMS

Absorption[a]	In the transmission of electrical/electromagnetic (or acoustic) signals, absorption is the conversion of the transmitted energy into another form, usually thermal energy. The conversion takes place as a result of interaction between the incident energy and the material medium, at the molecular or atomic level. Absorption is one cause of signal attenuation.
Additive Gaussian noise	Noise that disturbs the digitally-modulated signal during analog transmission, for instance in the analog channel. Additive superimposed noise normally has a constant power density and a Gaussian amplitude distribution throughout the bandwidth of a channel. If no other error is present at the same time, the points representing the ideal signal status are expanded to form circular "clouds" [BLO200701].

[a]This basic satellite glossary is synthetized from a variety of industry sources (including but limited to [SAT200501], [EUT200701], [ANS20001], and [GEO200101]).

Innovations in Satellite Communications and Satellite Technology: The Industry Implications of DVB-S2X, High Throughput Satellites, Ultra HD, M2M, and IP, First Edition. Daniel Minoli.
© 2015 John Wiley & Sons, Inc. Published 2015 by John Wiley & Sons, Inc.

Adjacent channel interference (ACI)	Extraneous power from a signal in an adjacent channel. ACI may be caused by (i) inadequate filtering of unwanted modulation products in frequency modulation (FM) systems, (ii) improper tuning, or (iii) poor frequency control, in either the reference channel or the interfering channel, or both.
Adjacent satellite interference (ASI)	For a ground station, extraneous power from a signal in an adjacent satellite (2° or 4° away) operating at the same frequency band. It may be due to the fact that the antenna is too small and is not able to focus properly on the specific satellite of interest, or due to the fact that the antenna is mispointed [ANS200001].
Amplifier	A (highly-linear) device used to increase the amplitude (power) of a signal. In an earth station the main amplifier is known as the High Power Amplifier (HPA). For a Very Small Aperture Terminal (VSAT) antenna, the amplifier may be included in the Block UpConverter (BUC). While a variety of power values are available, HPAs typically range from 500 to 3000 W; BUCs range from 5–25 W for commercial applications. Other amplifiers are used throughout the earth station to amplify various stages of the signal.
Amplitude imbalance	The different gains of the I and Q components of a modulated signal. In a constellation diagram, amplitude imbalance shows by one signal component being expanded and the other one being compressed [BLO200701].
Antenna	A device for transmitting or receiving radio waves. In commercial satellite communication, the antenna typically consists of a parabolic reflector and a feed horn. On the receiving link the reflector focuses radio waves onto the feed horn; the feed horn detects the signal and converts it into an electrical signal. On the transmitting side the reflector concentrates the radio waves emitted by the feed horn into a narrow beam that is aimed at the satellite in question.
Antenna alignment	The proper positioning and pointing of the antenna, so that it can receive the signal only from the intended satellite (thereby minimizing the ASI). The antenna uses a three axis pointing system,

which one needs to adjust when pointing it to a satellite: azimuth (the magnetic compass direction at which one points the dish); elevation (the angle above the horizon, at which one point the dish); and polarization (the horizontal/vertical rotation to align with the signal's polarization).

Antenna aperture — The effective area of an antenna that is capable of radiating or receiving radio frequency energy. In a parabolic antenna this dimension is equivalent to the diameter of the main reflector.

Antenna efficiency — (Also known as radiation efficiency.) The ratio of power applied to an antenna to the power actually radiated by the antenna, stated a percentage. The ratio of the signal strength transmitted towards (or received from) a particular direction in space by a real antenna, to the signal strength that would be obtained with a theoretical reference antenna of the same size.

Antenna gain — A measure of the "amplifying" or focusing power of an antenna when transmitting to, or receiving from, a particular direction in space. The gain of an antenna is the ratio of (i) the power radiated (or received) per unit solid angle by the antenna in a given direction to (ii) the power radiated (or received) per unit solid angle by an isotropic antenna fed with the same power. The gain is usually expressed in dBi.

Antenna illumination — On the transmit side, the radiation of electromagnetic energy from the feed horn (feed) to the surface of the parabolic reflector of a transmit antenna; on the receive side, the focusing of electromagnetic energy captured by the reflector of a receiving antenna towards the feed horn. With perfect illumination no signal energy is lost to the surrounding terrain; in practice there will be losses.

Antenna noise temperature — The temperature of a hypothetical resistor at the input of an ideal noise-free receiver that would generate the same output noise power per unit bandwidth as that at the antenna output at a specified frequency. The antenna noise temperature depends on antenna coupling to all noise sources in its environment, as well as on noise generated within the antenna.

	The antenna noise temperature is a measure of noise whose value is equal to the actual temperature of a passive device [ANS200001].
Aperture-medium coupling loss	Coupling loss at the antenna-LNA/LNB (Low Noise Amplifier/Low Noise Block Downconverter [LNB]) interface.
Apogee	Point in a satellite orbit (especially for highly-elliptical ones) which is farthest from the earth.
Arcjet	A device that heats a propellant stream by passing a high-current electrical arc through it, before the propellant is expanded through a downstream nozzle [JAH200201].
Atmospheric absorption	The attenuation of signal due to absorption; conversion of the transmitted RF energy into thermal energy. The absorption coefficient is a measure of the attenuation caused by absorption of energy that results from its passage through a given medium.
Atmospheric noise	Radio noise caused by natural atmospheric processes, primarily lightning discharges in thunderstorms.
Atmospheric signal scintillation	Atmospheric losses for radio signals passing through the atmosphere. In general, scintillation is a random fluctuation of the received field strength caused by irregular changes in the transmission path over time. Scintillation is produced by turbulent air with variations in the refractive index. These losses are dependent on, among other factors, elevation angle, and antenna size. Attenuation due to scintillation rapidly increases with increasing frequency (e.g., Ka-band signals that pass through a turbulent region are scattered by the turbulent cells and rapid amplitude changes around the average occurs) [AER200801].
Attenuation	The decrease in intensity of a signal, beam, or wave as a result of absorption of energy and of scattering of the path to the detector, but excluding the reduction due to geometric spreading. Attenuation is expressed in dB.
Attenuator	In electrical systems, a resistive network that reduces the amplitude of a signal without appreciably distorting its waveform (a comparable process is

	applicable to optical systems). Attenuators are usually passive devices. The degree of attenuation may be fixed, continuously adjustable, or incrementally adjustable. Fixed attenuators are often called pads, especially in telephony. The amount by which the signal power is reduced is usually expressed in dB [ANS200001].
Automatic Repeat reQuest (ARQ)	An error control mechanism for data transmissions. Lost data is detected with sequence numbers and timeouts. The communication about which Protocol Data Units (PDUs) are received or lost is accomplished with acknowledgments. If the acknowledgment back to the sender is negative, lost PDUs are retransmitted. The two most common strategies for retransmission are [JOU199901]:

- Go-Back N scheme goes back to the lost PDU and restarts transmission from the lost PDU.
- In selective repeat, only the lost packets are retransmitted.

Availability	The ratio of (i) the total time a functional unit is capable of being used during a given interval to (ii) the length of the given interval. Typical availability objectives are specified in decimal fractions, for example as 0.995.
Band pass filter	A filter that ideally passes all frequencies between two non-zero finite limits and blocks all frequencies not within the limits. Note: The cutoff frequencies are usually taken to be the 3-dB points. Thus, the band pass filter allows only a specified range of frequencies to pass from input to output, rejecting all signals at lower or higher frequencies.
Beam	A unidirectional flow of radio waves concentrated in a particular direction. A term commonly used to refer to an antenna's radiation pattern, often used to describe the radiation pattern of satellite antennas. The intersection of a satellite beam with the earth's surface is referred to as the (beam's) footprint.
Beamforming	Beamforming techniques entail techniques where multiple feed horns are energized with signals constructed on a "network," called beam-forming

	network (BFN), to alter the phase of the distinct feeds such that constructive interference can be used either to generate a large shaped beam (shaped to a specific geography), or to generate a pattern of spot beams. If a dynamically-reconfigurable pattern is desired, then BFNs can be used as a core component of the multiple beams antennas; however, BFNs add weight and complexity.
Beamforming network (BFN)	A beamforming approach used in the context of multiple feed horns (on a satellite). The BFN takes the signal to be transmitted and divides it in such a way that each feed horn in the array is excited by an appropriate amount of the original signal. Each individual feed horn illuminates the entire reflector, not just a portion. Since each of the feed horns in the field array form one single-component beam in the far field, there will be a need for as many feed horns as there are component beams. The amount of excitation for each feed horn is adjusted in both the amplitude and the phase to maximize the power within the coverage area while at the same time minimizing the sidelobes (radiation) outside the coverage area.
Beamwidth	A measure of the ability of an antenna to focus signal energy towards a particular direction in space (e.g., towards the satellite for a ground-based transmitting antenna), or to collect signal energy from a particular direction in space (e.g., from the satellite for a ground-based receiving antenna). The beamwidth is measured in a plane containing the direction of maximum signal strength. It is typically expressed as the angular separation between the two directions in which the signal strength is reduced to one-half of the maximum value (the $-3\,\text{dB}$ half-power points).
Block codes	Encoding algorithms where the encoder processes a block of message symbols and then outputs a block of codeword symbols. Reed–Solomon (RS) codes are one example.
Block up-converter (BUC)	Transmitter device that combines signal up-conversion and power amplification in a single unit. The BUC is typically located directly at the antenna output, or relatively close to it.

Boresight	The direction of maximum antenna gain. For a receiving antenna, the boresight is aligned with the satellite as accurately as possible in order to achieve maximum received signal strength.
Broadcast satellite service (BSS)	A satellite service that supports the transmission of signals that are intended for direct reception by consumers. BSS systems employ satellites capable of transmitting high power, enabling reception by small antennas (e.g., 60 cm to 1.2 m).
$C/(N + I)$ (carrier-to-noise-plus-interference-ratio)	A measure of the quality of a signal at the receiver input. It is the ratio of the power of the carrier to the combined power of noise and man-made interference, measured within a specified bandwidth (usually the modulated carrier's bandwidth). It is typically expressed in dB. The higher the ratio, the better quality of the received signal.
C/I (carrier-to-interference-ratio)	A measure of the quality of a signal at the receiver input. It is the ratio of the power of the carrier to the power of interference arising from man-made sources, measured within a specified bandwidth (usually the modulated carrier's bandwidth). It is typically expressed in dB. The higher the ratio, the better quality of the received signal.
C/N (carrier-to-noise-ratio)	An important measure of the quality of a modulated carrier at the receiver input. It is the ratio of the power of the carrier to the power of the noise introduced in the transmission medium, measured within a specified bandwidth (usually the modulated carrier's bandwidth). It is typically expressed in dB. The higher the ratio, the better quality of the received carrier.
C/N_o	Link performance metric, where C is the received carrier power in Watts and N_o is the noise power spectral density (C/N_o is measured in Hz – Note that C/N_o and E_b/N_o are relatable). This term is traditionally used in analog radios, where N is the receiver or channel noise equivalent bandwidth. Note that $C/N_0 = C/kT_e$ with T_e is Equivalent Noise Temperature (degree Kelvin) k (1.39×10^{-23} J/K) is Boltzmann's constant,

	B is the system bandwidth (Hz),
	$N = N_o \times B$ (linear expression) (B = Bandwidth),
	$N = N_o + 10 \log B$ (decibel expression).
Carrier ID	The inclusion of Carrier Identifier (CID) in a user-transmitted satellite signal, as specified in the DVB Carrier ID standard (DVB-CID – 2013). With this standard the DVB-CID signal is an overlay to original carrier; this approach is agnostic to traffic carrier or transport mechanism (TS video and IP data). It provides high robustness: it can be decoded even if the main carrier is jammed. The RF carrier ID standard was released by DVB as Bluebook A164 in February 2013 and it was adapted as ETSI TS 103 129 v1.1.1 in May 2013. The CID is injected by modulator into carrier with fixed source ID such as MAC address, and user configurable data such as GPS coordinates, carrier name, contact information.
Carrier suppression/leakage	A special type of interference in which its frequency equals the carrier frequency in the RF channel. Carrier leakage can be superimposed on the QAM signal in the I/Q modulator. In the constellation diagram, carrier leakage shows up as a shifting of the signal states corresponding to the DC components of the I- and Q-components [BLO200701].
Carrier-to-receiver noise density (C/kT)	The ratio of the received carrier power to the receiver noise power density. The carrier-to-receiver noise density ratio is usually expressed in dB. C is the received carrier power in watts. k is Boltzmann's constant in joules per kelvin. T is the receiver system noise temperature in kelvins. The receiver noise power density, kT, is the receiver noise power per hertz.
C-band	Original operating frequency band for communications satellites. It makes use of the following frequency ranges: 3.7–4.2 GHz for downlink frequencies and 5.925–6.425 GHz for uplink frequencies. This band is most resistant to rain fade of the commercial satellite bands.

Channel Noise	All objects that have heat emit RF energy in the form of random (Gaussian) noise [ZYR199801]. The amount of radiation emitted is $N = kTB$ with N = noise power (Watts), k = Boltzman's constant (1.38×10^{-23} J/K), T = system temperature, usually assumed to be 290 K, B = channel bandwidth (Hz).
Circular orbit	A satellite orbit where the distance between the center of mass of the satellite and of the earth is constant.
Circular polarization	A circularly-polarized wave in which the electric field vector, observed in any fixed plane normal to the direction of propagation, rotates with time and traces a circle in the plane of observation. Unlike linear polarization, circular polarization does not require alignment of earth station and satellite antennas with the polarization of the radio waves (namely, the feed does not need to be rotated to/by a specific angular rotation to receive the signal at full strength).
Clarke belt	The circular orbit (Geostationary Orbit) at approximately 35,786 km above the equator, where the satellites travel at the same speed as the earth's rotation and, thus, appear to be stationary to an observer on earth (named after Arthur C. Clarke who was the first to describe the concept of geostationary communication satellites).
Clear air attenuation	Attenuation of a signal passing through air (causes of attenuation include rain, scintillation, oxygen and water vapor). The amplitude, phase, and angle-of-arrival of a microwave signal passing through the atmosphere vary due to inhomogeneities in the refractivity of the atmosphere. The clear air attenuation for a satellite system is about 2 dB, and varies to some degree with the location and time.
Clear sky	A term describing the weather conditions encountered at the terrestrial end of an earth-space path of a satellite communication link. It is used to describe the condition where the attenuation of radio waves caused by precipitation (rain, snow, sleet, dew, etc.) is lowest (i.e., cloud-free sky and good visibility).

Coax(ial) cable	A cable used principally as a transmission line for RF signals that has relatively low loss at mid-to-high frequencies (3–20 GHz). It supports mid-range power levels (waveguide is used at high-power high-frequency). A coax cable consists of an inner conductor, which may be solid or stranded wire, and dialectic filler, covered by a flexible wire braid. (Metal foil may also be used.) All of these are covered by an insulating material.
Coding rate	The ratio of input data bits to bits transmitted, for example, 1/2, 7/8, etc.
Collocated satellites	Arrangement where two or more satellites occupying approximately the same geostationary orbital position such that the angular separation between them is effectively zero when viewed from the ground. To a small receiving antenna the satellites appear to be exactly collocated; in reality, the satellites are kept several km apart in space to avoid collisions. Different operating frequencies and/or polarizations are used.
Convolution channel coding gain	Channel coding is the application of processing algorithms, prior to transmission over a channel (and reverse algorithms at the receiver), used to improve data reliability of the received information. Coding Gain is the increase in efficiency that a coded signal provides over an uncoded signal. Expressed in decibels, the coding gain indicates a level of power reduction that can be achieved. Specifically, convolution channel coding gain relates to gain achieved with convolutional coding; convolutional coding is a Forward Error Correction (FEC) scheme, where the coded sequence is algorithmically achieved through the use of current data bits plus some of the previous data bits from the incoming stream [AER200801]. Note: the older Reed Solomon Viterbi (RSV) FEC methodology is inefficient when compared to modem Turbo Codes (TC) or Turbo Product Codes (TPC); this inefficiency results in higher bandwidth and power required for the same BER as TC or TPC. FEC codes are also referred to as channel codes.

Convolution codes	Encoding algorithms where, instead of processing message symbols in discrete blocks, the encoder works on a continuous stream of message symbols and simultaneously generates a continuous encoded output stream.
Co-polar(ized) (CP)	Of the same polarization. In contrast to XP (cross-polar).
Coupling loss	The loss that occurs when energy is transferred from one circuit, circuit element, or medium to another. Coupling loss is usually expressed in the same units – such as watts or dB – as in the originating circuit element or medium.
Cross polar(ized) (XP)	Term to describe signals of the opposite polarity to another desired signal being transmitted and received. For example, cross-polarization discrimination refers to the ability of a feed to detect one polarity and reject the signals having the opposite sense of polarity.
Cross-pol interference	Interference caused by a received signal that has opposite polarization to the signal of interest.
Cross-polar discrimination (XPD)	The ratio of the signal power received (or transmitted) by an antenna on one polarization (the polarization of the desired signal) to the signal power received (transmitted) on the opposite polarization. This ratio is usually expressed in decibels. XPD is a measure of the ability of the antenna to detect (emit) signals on one polarization and to reject signals at the same frequency having the opposite polarization.
Cross-polar isolation (XPI)	The ratio of the signal power received (or transmitted) by an earth station on one polarization (the desired signal) to the signal power received (transmitted) on the same polarization but originating from a cross-polar signal. This ratio is usually expressed in decibels. XPI is a measure of interference from cross-polar signals into the desired signal which occurs in all practical systems that exploit both orthogonal polarization. The terms "cross-polar isolation" and "cross-polar discrimination" have different meanings but are often used interchangeably.

Crosstalk (XT)	Undesired coupling of a signal from one circuit, part of a circuit, or channel, to another. Any phenomenon by which a signal transmitted on one circuit or channel of a transmission system creates an undesired effect in another circuit or channel.
Cumulative transit delay	The total transit delay applicable for a data call obtained by summing the individual transit delays of all component portions of the data connection.
Cyclic redundancy code (CRC)	A subclass of linear code used for detection of errors in computer data. An error-detection scheme that uses redundant encoded bits (parity bits), and appends them to the digital signal. The received signal is decoded and PDUs with errors are discarded. For example, an ARQ system can be combined with CRC, if error correction is required [JOU199901].
Data coding	Operating on data with an algorithm to accomplish encryption, error correction, compression, or some other feature.
Dawn-to-dusk orbit	(Also called a heliosynchronous orbit) An orbit where the satellite is in perpetual sunlight, allowing it to rely totally on solar panels to generate power. This orbit is used for earth observation, imaging and research satellites.
Demodulation	The process of retrieving information impressed upon a carrier wave during the process of modulation. Demodulation along with Forward Error Correction (FEC) aims at recovery of the original signal or data being transmitted with a high level of fidelity (extremely low Bit Error Rate).
Demodulator	A device that creates a facsimile of the original signal or data transmitted by a modulator.
Descrambler	A device that recovers the original signal from one that has been scrambled (encrypted).
Diffraction	The deviation of an electromagnetic wavefront from the path predicted by geometric optics when the wavefront interacts with, *that is*, is restricted by, a physical object such as an opening (aperture) or an edge. *Note:* Diffraction is usually most noticeable for openings of the

	order of a wavelength; however, diffraction may still be important for apertures many orders of magnitude larger than the wavelength.
Digital-to-analog converter	A device that converts a digital signal (a series of numbers or characters) into its equivalent analog waveform (a continuously-varying signal voltage).
Dish	A "non-technical term" for a parabolic antenna used for transmitting and/or receiving satellite signals. The term typically describes the entire antenna system, including the feed horn reflectors and all associated antenna structures.
Dispersion	Any phenomenon in which the velocity of propagation of an electromagnetic wave is wavelength-dependent. Processes by which an electromagnetic signal propagating in a physical medium is degraded because the various wave components (i.e., frequencies) of the signal have different propagation velocities within the physical medium.
Distortion	Any departure of the output signal waveform from what would be theoretically expected from input signal waveform's being operated on by the system's ideal function. Distortion may result from nonlinearities in the transfer function of an active device, such as a vacuum tube or transistor amplifier; it may also be caused by a passive component such as a coaxial cable or optical fiber, or by inhomogeneities, reflections, etc., in the propagation path.
Doppler effect	The change in the observed frequency (or wavelength) of a wave, caused by a time rate of change in the effective path length between the source and the point of observation.
Downconverter	A device for converting the frequency of a signal to a lower frequency; transceivers that take a C-, Ku-, or Ka-band signal and frequency-converts it to either 70 MHz, 140 MHz, or L-band 950–1450 MHz.
Downlink Doppler frequency shift	A phenomenon that results in a change in the received frequency (here the downlink frequency) when the transmitter and/or receiver are in motion. If the source is moving away

	(positive velocity) the observed frequency is lower and the observed wavelength is greater; if the source is moving towards (negative velocity) the observed frequency is higher and the wavelength is shorter.
Downlink frequency	Frequency of the downlink transmission for the appropriate band (e.g., C-band, Ku-band, BSS-band, Ka-band), also in reference to the specific transponder being used.
Downlink signal power	Power of the signal transmitted by the satellite.
Downlink thermal noise	The downlink portion of the satellite link adds noise. The downlink thermal noise is defined as the ratio of the power in the carrier that is being received to the noise added to the system by the downlink, and is designated C/T. This noise is dominated by the G/T of the satellite earth station.
Dropouts	Momentary loss of signal. Dropouts are usually caused by noise, propagation anomalies, or system malfunctions; also due to rain fade.
Dual feed	A term referring to an antenna system capable of simultaneously receiving from two different satellites at different orbital positions (the angular separation between satellite positions is typically 4–6°). This system consists of a reflector, a support structure and two Low Noise Block Downconverters (LNBs), each equipped with a separate feed horn or sharing an integrated feed assembly.
Dual-band feed	A feed horn that can simultaneously receive signals in two different frequency bands, typically the C-band and the Ku-band, often employed at earth stations dealing with hybrid satellites.
Dual-feed antenna	An antenna system consisting of a reflector, a support structure and two Low Noise Block Downconverters (LNBs), each equipped with a separate feed horn (feed) or sharing an integrated feed assembly. The focal point of each feed is set so that the antenna system can receive from two different satellite orbital positions simultaneously. The angular separation between satellite positions is usually around 6 degrees, although other angles are possible.

DVB-S2	Standardized modulation scheme used in satellite transmission. DVB-S2 is a second generation framing structure, channel coding and modulation system for broadcasting, interactive services, news gathering and other broadband satellite applications. DVB-S2 is a specification developed by the Digital Video Broadcasting Project (DVB) adopted by European Telecommunications Standards Institute (ETSI) standards now used worldwide. The specification was developed in 2003.
DVB-S2X	DVB-S2 extensions. DVB-S2X is a superset of DVB-S2 and provides increased availability. While DVB-S2 is optimized for non-linear operations, DVB-S2X handles both environments equally well. DVB-S2X allows increased granularity in MODCODs and higher order modulation schemes (e.g., 64APSK). The DVB steering committee approved the specifications for the new standard in February 2014.
Earth terminal antenna azimuth	Azimuth is the angular distance from true north along the horizon to a satellite, measured in degrees. It is the magnetic compass direction (angle of sighting) at which an operator points the antenna [LAU200701]. It is a side-to-side adjustment of the antenna to peak it to a specific satellite.
Earth terminal antenna elevation	Elevation is the angle above the horizon, at which the operator points the antenna; this is an up-and-down adjustment of the antenna.
Earth terminal to satellite slant range	The length of a line drawn from the antenna to the satellite.
Earth Station antenna	A highly-directional, high-gain antenna designed to transmit and receive radio signals to and from an orbiting spacecraft. Typically earth station antennas are parabolic; however, new type antennas are also being deployed including flat panel (phased array antennas) and spherical antennas (Luneberg Lens).
Earth station latitude	Latitude of the earth station.
Earth station longitude	Longitude of the earth station.

E_b/N_o	Commonly-used figure-of-merit ratio for systems making use of digital modulation. The ratio of bit energy per symbol to noise power spectral density, in decibels. Measure utilized in digital modulation/transmission environments. 0 dB means the signal and noise power levels are equal and a 3 dB increment doubles the signal relative to the noise. Digital modulation schemes are typically specified in terms of theoretical BER vs E_b/N_o performance. Both E_b and N_o are expressed in units of dBm/Hz (or J/s), hence the ratio is dimensionless and is expressed in dB. $E_b = C/$(data rate) (linear expression). $E_b = C - 10*\log$(data rate) (decibel expression). E_b/N_o is a good measure to compare digital systems using different modulation or encoding schemes and data rates. $E_b = P_t \times T_b$ or P_t/f_b where P_t is transmit power in W (or J/s), T_b is bit duration in seconds, f_b is the bit rate in bits per second (bps)
Effective isotropic radiated power (EIRP)	A measure of the signal strength that a satellite transmits towards the earth, or an earth station transmits towards a satellite, expressed in dBW. Also, the arithmetic product of (i) the power supplied to an antenna and (ii) its gain [ANS200001]. EIRP = $P_T \times G_T/L_T$, where G_T is the transmit antenna gain and L_T is transmission line losses including back-off loss, feeder loss, etc. EIRP(dBW) = P_T(dB) − L_{bo}(dB) − L_{bf}(dB) + G_T(dB), where P_T = actual power output of the transmitter (dBW). L_{bo} = back-off losses of HPA (dB). L_{bf} = total branching and feeder loss (dB). G_T = transmit antenna gain (dB).
Electromagnetic propulsion	Propulsion wherein the propellant is accelerated under the combined action of electric and magnetic fields [JAH200201].
Electrostatic propulsion	Propulsion wherein the propellant is accelerated by direct application of electrostatic forces to ionized particles [JAH200201].

Electrothermal propulsion	Propulsion wherein the propellant is heated by some electrical process, then expanded through a suitable nozzle [JAH200201].
Elevation	The angle measured in the local vertical plane between the satellite and the local horizon. It is the vertical co-ordinate that is used to align a satellite antenna.
End of life (EOL)	In this context, the point in the life of a communications satellite when it is no longer capable of carrying out its mission and must be retired.
Error correction code (ECC)	Added information that allows data that is being read or transmitted to be checked for errors and, when necessary, corrected "on the fly."
External interference	Interference caused by other sources, such as terrestrial microwave links, cosmic rays, intentional jamming, etc.
F/D Ratio	The ratio of an antenna's focal length F to its diameter D. This ratio describes the geometric architecture of the antenna, which impacts its physical size, design, and electrical performance.
Fade	In a received signal, the variation (with time) of the amplitude or relative phase, or both, of one or more of the frequency components of the signal.
Fade margin	A design allowance that provides for sufficient system gain or sensitivity to accommodate expected fading, for the purpose of ensuring that the required quality of service is maintained. The amount by which a received signal level may be reduced without causing system performance to fall below a specified threshold value
Feed horn (feed)	A device that (i) in a transmitting system directs the signals onto the reflector surface for focusing into a narrow beam aimed at the satellite, and/or (ii) in a receiving system collects signals reflected from the surface of the antenna. In a receiving system it collects microwave signals reflected from the surface of the antenna. In a transmitting system it directs microwave signals onto the reflector surface for focusing into a narrow beam aimed at the

	satellite. The feed is designed to match a particular antenna's geometry (*F/D* ratio) and is mounted at the focus of the parabolic reflector.
Filter	A passive electrical device that blocks signals or radiation of certain frequencies while allowing others to pass unaltered.
Fixed satellite service (FSS)	A satellite service between satellite terminals at specific fixed points using one or more satellites. Typically, FSS is used for the transmission of video, voice, and IP data over long distances. FSS makes use of geostationary satellites with fixed ground stations.
Flux density	The fundamental magnetic force field. "Flux" means to flow (around a current-carrying conductor, for example); "density" refers to the intensity within an enclosed area. Faraday's Law is used to determine the induced voltage. Also called the "induction field." The unit of flux density is a Tesla, volt-second per square meter per turn (the unit of magnetic flux density is the Gauss; there are 10,000 Gauss per Tesla).
Focal length	The distance F from the reflective surface of an antenna to its focal point, usually measured in the horizontal plane. Incoming satellite signals are directed to the feed horn (feed), which is normally located at the focal point.
Forward error correction (FEC)	A technique where extra (redundant) information is added to a data stream, so that errors that may occur during the transmission can be corrected by the receiver without interaction from the transmitter. If FEC techniques were not used and an error occurred, the receiver would have to request for re-transmission, which increases latency (or is not practical in some applications, especially for one-way applications).
Free-space attenuation	See geometric spreading.
Free-space loss	The signal attenuation caused by beam divergence, that is, signal energy spreading over larger areas at increased distances from the source, thereby decreasing the overall energy that is received at a point of reference. It is the attenuation that results when all absorbing, diffracting,

	obstructing, refracting, scattering, and reflecting influences are removed so as to have no effect on propagation.
Frequency colors	High Throughput Satellites (HTSs) have a large number of beams making heavy use of frequency re-use; they employ frequency-reuse concepts partially similar to small-cell cellular telephony. The beams can be arranged in a honeycomb to provide coverage of a service region. A number of distinct frequencies is employed; in fact, both frequency differentiation and polarization differentiation can be utilized among neighboring beams. The term "color" is used to discuss different non-overlapping (or non-interfering, by using different polarizations) frequency bands, clearly with a reference to the mathematical concept of the Four Color Theorem (4CT). Adjacent spots are of different "colors," differing in frequency or polarization; hence, they can support the transmission of different information without mutual interferences. Spots beams with the same color use the same frequency and the same polarization, but they are spatially isolated from each other, no spot has a neighbor with the same color: because of the spatial separation, spot beams with the same color can support the transfer of different information. In most cases, the four color approach is the best compromise between system capacity and performance; however, other frequency re-use schemes can also be utilized.
Frequency deviation	The amount by which a frequency differs from a prescribed value, such as the amount an oscillator frequency drifts from its nominal frequency. In frequency modulation, this is the maximum absolute difference, during a specified period, between the instantaneous frequency of the modulated wave and the carrier frequency.
Frequency drift	An undesired progressive change in frequency with time. Causes of frequency drift include component aging and environmental changes.

Frequency reuse	A technique for utilizing a specified range of frequencies more than once within the same satellite system so that the total capacity of the system is increased without increasing its allocated bandwidth. Frequency reuse schemes require sufficient isolation between the signals that use the same frequencies, so that mutual interference between them is controlled to an acceptable level. Frequency reuse can be achieved by using orthogonal polarization states (horizontal/vertical for linear, or LHC/RHC for circular) for transmission and/or by using satellite antenna (spot) beams that serve separate, non-overlapping geographic regions.
Frequency shift	Any change in frequency. Any change in the frequency of a radio transmitter or oscillator.
Fresnel diffraction pattern	See near-field diffraction pattern.
Fresnel zone	In radio communications, one of a (theoretically infinite) number of a concentric ellipsoids of revolution which define volumes in the radiation pattern of a (usually) circular aperture. Fresnel zones result from diffraction by the circular aperture. The cross section of the first Fresnel zone is circular; subsequent Fresnel zones are annular in cross section, and concentric with the first. Odd-numbered Fresnel zones have relatively intense field strengths, whereas even-numbered Fresnel zones are nulls.
G/T figure of merit	The figure of merit for an earth station, where G = Gain (of the antenna system in dB) and T = system noise temperature (in Kelvins). The ratio of the maximum gain G of a receiving antenna to the receiving system's equivalent noise temperature T. This value is usually expressed in dB/K. It is a measure of the ability of an earth station to receive a satellite signal with good quality (high carrier-to-noise ratio). In general, the G/T increases with increasing antenna diameter.
Geometric spreading (aka inverse-square law)	The physical law stating that irradiance, that is, the power per unit area in the direction of propagation, of a spherical wavefront varies

	inversely as the square of the distance from the source (assuming there are no losses caused by absorption or scattering). The power radiated from a point source, an omnidirectional isotropic antenna, or from any source at very large distances from the source compared to the size of the source, must spread itself over larger and larger spherical surfaces as the distance from the source increases.
Geostationary orbit (GEO)	(Also known as the Clarke Orbit, after Sir Arthur C. Clarke, its inventor.) A circular orbit orientated in the plane of the earth's equator allowing the satellite to appear stationary in relation to an observer on the earth's surface. This allows a satellite to provide communications services 24 h a day to an entire hemisphere and without the need for a tracking antenna (a fine-grain tracking antenna may still be used, however, for high-end large aperture applications).
Global positioning (service/) system (GPS)	A satellite service that uses a constellation of Low Earth Orbiting (LEO) satellites to provide global positioning information to earth stations small enough to be hand-held.
Ground-based beam forming (GBBF) systems	A GBBF system is a large signal processor that is designed to coordinate and process up to several hundred beams at once (up to 500). The GBBF system creates hundreds of small, flexible, adaptive "spot" beams on the earth that allow mobile handsets to communicate directly with the satellite using small antennas, and achieve high speeds. While the beams are projected to the earth by the satellite, the ground system performs the beam-shaping signal processing.
Group delay	The rate of change of the total phase shift with respect to angular frequency, $d\Theta/d\omega$, through a device or transmission medium, where Θ is the total phase shift, and ω is the angular frequency equal to $2\pi f$, where f is the frequency.

H.265	High Efficiency Video Coding (HEVC) is a newly-developed standard (first stage approval in January 2013) for video compression developed by the ITU-T and ISO; it is a joint publication of ISO/IEC and ITU-T, formally known as ISO/IEC 23008–2 (ISO-MPEG H Part 2) and ITU-T Recommendation H.265. H.265 is seen as a successor to H.264/MPEG-4 AVC. HEVC has the same general structure as MPEG-2 and H.264/AVC, but it affords more efficient compression; namely, it can provide the same picture quality as the predecessor (H.264) but with a lower datarate (and storage requirement), or it can provide better picture quality than its predecessor but at the same datarate (and storage requirement).
Hall effect	Conduction of electric current perpendicular to an applied electric field in a superimposed magnetic field.
High efficiency video coding (HEVC)	See H.265.
High power amplifier (HPA)	High-reliability high-power (500–3000 W) amplifier used in large earth station environments. Can be a Solid State Power Amplifier (SSPA), a Travelling Wave Tube (TWT), or klystron.
High throughput satellite (HTS)	A satellite system that makes use of a large number of geographically-confined spot beams distributed over a specified service area; offering a contiguous (or non-contiguous) covering of the that service area; and, providing high system capacity and user throughput at a lower net cost per bit. While most satellites in orbit operating at the Ku-band currently provide large contoured coverages for video broadcast applications, HTS satellites aim at providing broadband data services utilizing spot beams technology typically using Ka-band frequencies.
Highly elliptical orbits (HEOs)	Orbits that have a perigee (point in each orbit which is closest to the earth) at about 500 km above the surface of the earth and an apogee (the point in its orbit which is farthest from the earth) as high as 50,000 km [GEO200101].

Hit	A transient disturbance to, or momentary interruption of, a communication channel.
Horizontal polarization	Type of linear polarization where the electric field is approximately aligned with the local horizontal plane at an on-ground transmission or reception point.
Hybrid satellite	A satellite capable of operating in two or more frequency bands simultaneously, for example, C-band and Ku-band.
Inclination	The angle between the plane of the orbit of a satellite and the earth's equatorial plane. An orbit of a perfectly-geostationary satellite has an inclination of 0.
Inclined orbit	An orbit that approximates the geostationary orbit but whose plane is tilted slightly with respect to the equatorial plane. The satellite appears to move about its nominal position in a daily "figure-of-eight" motion when viewed from the ground. Spacecrafts (satellites) are often allowed to drift into an inclined orbit near the end of their nominal lifetime in order to conserve on-board fuel, which would otherwise be used to correct this natural drift caused by the gravitational pull of the sun and moon. North–south maneuvers are not conducted, allowing the orbit to become highly inclined.
Inductive thruster	Device that heats a propellant stream by means of an inductive discharge before the propellant is expanded through a downstream nozzle [JAH200201].
Interference	In general, extraneous energy, from natural or man-made sources, that impedes the reception of desired for example, a radio emission from another transmitter at approximately the same frequency, or having a harmonic frequency approximately the same as another emission of interest to a given recipient, and which impedes reception of the desired signal by the intended recipient.
	Note: Interference may be constructive or destructive, that is, it may result in increased amplitude or decreased amplitude, respectively. Two waves equal in frequency and amplitude, and out of phase by $180°$, will completely cancel one another; in phase, they create a resultant wave having twice the amplitude of either interfering beam.

Interferers	Spurious sinusoidal signals occurring in the transmission frequency range and superimposed on the modulated signal at some point in the transmission path. After demodulation, the interferer is contained in the baseband form of low-frequency sinusoidal spurious signals. The frequency of these signals corresponds to the difference between the frequency of the original sinusoidal interference and the carrier frequency in the RF band. In the constellation diagram, an interferer shows in the form of a rotating pointer superimposed on each signal status. The constellation diagram shows the path of the pointer as a circle around each ideal signal status [BLO200701].
Intermediate frequency (IF)	A frequency to which a carrier frequency is translated to, in the process of reception or transmission of the signal. IFs are found between a modulator and an upconverter, or between a downconverter and a demodulator. Typical IFs are: 70 MHz, 140 MHz and L-band 950–1450 MHz.
Intermodulation	The production, in a nonlinear element of a system, of frequencies corresponding to the sum and difference frequencies of the fundamentals and harmonics thereof that are transmitted through the element.
Intermodulation distortion	Nonlinear distortion characterized by the appearance, in the output of a device, of frequencies that are linear combinations of the fundamental frequencies and all harmonics present in the input signals. Harmonic components themselves are not usually considered to characterize intermodulation distortion.
Inter-symbol interference (ISI)	ISI is the interference between adjacent symbols often caused by system filtering, dispersion in optical fibers, or multipath propagation in radio system. This occurs when the signal is filtered enough so that the symbols blur together and each symbol affects those around it. This is determined by the time-domain response or impulse response of the filter.
Inverse-square law	See geometric spreading.
Ion thruster	Device that accelerates propellant ions by an electrostatic field.

Isotropic antenna	A theoretical point source of radiation used as a reference antenna when calculating antenna gain.
Ka-band	Third, and most-recently utilized, band of frequencies authorized for commercial satellite communications. It occupies approximately from 18 to 30 GHz in the radio spectrum. This band is the most susceptible to rain fade of all three satellite bands.
Kelvin (kelvin scale)	The Kelvin scale is a measure of temperature. The lowest value is Absolute Zero, the point where all molecular motion stops. Since all electrical devices contribute noise to a system based upon their temperature, noise contribution in satellite systems is expressed in Kelvins.
Klystron	A power amplifier tube used to amplify microwave energy (provided by a radio- frequency exciter) to a high power level. A klystron is characterized by high power, large size, high stability, high gain, *relatively narrow bandwidth* and high operating voltages. Electrons are formed into a beam that is velocity modulated by the input waveform to produce microwave energy. A klystron is sometimes referred to as a linear beam tube because the direction of the electric field that accelerates the electron beam coincides with the axis of the magnetic field [AMS200001].
Ku-band	Secondly-utilized band of frequencies authorized for satellite communications. It occupies approximately from 10 to 14.5 GHz in the radio spectrum. This band is more susceptible to rain fade than the C-band.
L-band	An Intermediate Frequency (IF) typically employed at an earth station to route traffic between various points over coaxial/waveguide facilities. The frequency range coves the 950–1450 MHz spectrum. Note that over-the-air L-band ranges are (slightly) different and are defined by various regulatory agencies. Satellite signals (at C-band and Ku-band frequencies) are converted down to L-band in the focal point of many dish antennas by the Low Noise Block Downconverter (LNB) for further distribution within the electronics subsystem or the earth station. At C-band the downconversion is typically as follows: 4200 to 950 MHz; 4180 to 970 MHz; 4160 to 990 MHz, and so on to 3700 to 1450 MHz. At the Ku-band, the

	downconversion is typically as follows: 11700 to 950 MHz; 11720 to 970 MHz; 11740 to 990 MHz, and so on to 12200 to 1450 MHz. The upconverter handles the opposite function.
Left-hand circularly polarized (LHC(P)) wave	An elliptically- or circularly-polarized wave in which the electric field vector, observed in any fixed plane normal to the direction of propagation, while looking in the direction of propagation, rotates with time in a left-hand or anticlockwise direction.
Line amplifier	An amplifier in a transmission line that boosts the strength of a signal level. Usually utilized in an InterFacility Link (IFL) between the antenna and the in-door electronics.
Line splitter	An active or passive device that divides a signal into two or more signals containing all the original information; this is done to feed the signal to two or more electronic distinct components or transmission lines. A passive splitter feeds an attenuated version of the input signal to the output ports; an active splitter amplifies the input signal to overcome the splitter's loss.
Linear polarization	Propagation of a wave in which the electric field vector, observed in any fixed plane normal to the direction of propagation, maintains a constant direction with time. With linear polarization, the earth station and satellite antennas of a particular earth-space link must be precisely aligned so that their reference polarization directions coincide, in order to obtain maximum reception signal and signal quality.
Link budget	A collection of the various system parameters of a satellite link and is used to either determine: (i) the link performance from a fixed set of system parameters, or, (ii) some aspect of the system parameters given particular link performance criteria.
Link budget analysis	Link Budget is a generic term used to describe a series of engineering calculations designed to model the performance of a communications link. In a typical simplex (one-way) satellite link, there are two link budget calculations: one link from the transmitting ground station to the satellite, and one link from the spacecraft to the receiving ground station.

Link budget analysis tools include the following issues [USA199801]:

a. Link Budget Model
   - Uplink and downlink carrier-to-receiver noise density (C/kT)
   - Satellite receive power flux density

b. Positional Data Model
   - Earth Terminal to satellite slant range
   - Earth terminal antenna elevation and azimuth
   - Uplink and downlink Doppler frequency shift

c. Benign Atmosphere Attenuation
   - Clear air attenuation
   - Rain fall Attenuation
   - Atmospheric signal scintillation

d. Modulation and Channel Encoding
   - Noise equivalent bandwidth
   - Modulation spectral efficiency
   - Demodulator implementation loss
   - Probability of detection of error
   - Convolution channel coding gain

e. Earth Terminal Model
   - Antenna model
   - Receive system model
   - Transmit system model

f. Satellite Model
   - Uplink signal power
   - Downlink signal power
   - Transponder signal and noise power sharing
   - Transponder uplink power flux

The following input items are needed to produce a link budget calculation for a spacecraft to ground station link:

- Earth station latitude
- Earth station longitude
- Spacecraft longitude

- Downlink frequency
- Antenna gain
- Antenna noise temperature
- Low noise amplifier
- Ortho mode transfer (OMT) loss
- Effective isotropic radiated power (EIRP)
- Intermediate frequency (IF) receive bandwidth
- Transmit data rate
- Link margin

Local oscillator (LO)	A single-frequency reference signal that is used by a mixer to convert a communications signal to a higher or lower frequency band. For example, used with a Low Noise Block Downconverter (LNB).
Low density parity check codes (LDPC)	A Forward Error Correction (FEC) mechanism; used in DVB-S2. LDPCs are block codes defined by a sparse parity check matrix; this sparseness admits a low complexity iterative decoding algorithm. The generator matrix corresponding to this parity check matrix can be determined, and used to encode a block of information bits.
Low earth orbits (LEO)	Either elliptical or (more commonly) circular orbits that are at a height of approximately 2,000 km or less above the surface of the earth. The orbit period at these altitudes varies between ninety minutes and two hours and the maximum time during which a satellite in LEO orbit is above the local horizon for an observer on the earth is up to 20 min [GEO200101].
Low noise amplifier (LNA)	A low-noise device that receives and amplifies satellite signals at the output of the antenna feed horn; an LNA does not change the frequency of the received signal. LNAs are designed to contribute a minimum amount of noise the signal received from the satellite in order to minimize the overall system noise temperature.
Low noise block downconverter (LNB)	A low-noise device that receives and amplifies satellite signals at the output of a feed horn, while also performing other functions as follows: signal detection, high-gain low-noise amplification, and *frequency conversion*. The frequency conversion

	downconverts a block of frequencies to a lower intermediate frequency range (typically in the L-band). For low-end applications (e.g., DBS consumer reception) the feed horn is integrated with LNB in a single mechanical unit.
Machine-to-Machine (M2M) communications	M2M a subset of the Internet of things (IoT), where standards have evolved and published by industry organizations such as European Telecommunications Standards Institute (ETSI). *M2M services* aim at automating decision and communication processes and support consistent, cost-effective interaction for ubiquitous applications. *M2M communications* per se is the communication between two or more entities that do not necessarily need direct human intervention: it is communication between remotely-deployed devices with specific roles and requiring little or no human intervention. M2M communication modules are usually integrated directly into target devices, such as Automated Meter Readers (AMRs), vending machines, alarm systems, surveillance cameras, and automotive/aeronautical/maritime equipment, to list a few. These devices span an array of domains including (among others) industrial, trucking/transportation, financial, retail Point of Sales (POS), energy/utilities, smart appliances, and healthcare.
Magnetoplasma dynamic thruster	Device that accelerates a propellant plasma by an internal or external magnetic field acting on an internal arc current [JAH200201].
Maritime mobile satellite service (MMSS)	A satellite service between mobile-satellite earth stations and one or more satellites.
Medium earth orbits /intermediate circular orbits (MEO/ICOs)	Circular orbits at an altitude of around 10,000 km. Their orbit period is in the range of 6 h.
Mixer	A device in which two or more input signals are combined to generate a single output signal.
Mobile satellite service (MSS)	A satellite service intended to provide wireless communication to any point on the globe.

Modulation spectral efficiency	The ratio of the transmitted bit rate R_c to the bandwidth occupied by the carrier. The bandwidth occupied by the carrier depends on the spectrum of the modulated carrier and the filtering it undergoes. Filtering is employed at the transmitter and receiver side to limit the interference to adjacent out-of-band carriers; however, filtering can generate intersymbol interference which in turn impacts the Bit Error Rate (BER) of the received signal.
Modulator	A device that superimposes a signal (intelligence) onto a wave or signal (called a carrier), which is then used to convey the original signal via a transmission medium. Modulation techniques include amplitude modulation, frequency modulation, or phase modulation. For satellite applications Phase Shift Keying (PSK) is fairly common.
Multibeam	Refers to the use of multiple antenna beams on board the satellite to cover a contiguous geographical area, instead of a single wide-area beam. Multibeam architectures are often utilized in satellites operating in the Ka-band, which is characterized by narrower beamwidths with respect to the Ku-band. Single, wide-area beams predominate in the latter.
Multipath	The propagation phenomenon that results in radio signals' reaching the receiving antenna by two or more paths. Causes of multipath include atmospheric ducting, ionospheric reflection and refraction, and reflection from terrestrial objects, such as mountains and buildings. The effects of multipath include constructive and destructive interference, and phase shifting of the signal.
Near-field diffraction pattern	The diffraction pattern of an electromagnetic wave, which pattern is observed close to a source or aperture, as distinguished from a far-field diffraction pattern. The pattern in the output plane is called the near-field radiation pattern.
Noise	(Also known as Thermal Noise.) Any undesired electrical disturbance in a circuit or communication channel. When combined with a received signal, it affects the receiver's ability to correctly reproduce the original signal.

Noise equivalent bandwidth (NEB)	Given the amplitude frequency response of an RF device, the noise-equivalent bandwidth is defined as the width of a rectangle whose area is equal to the total area under the response curve and whose height is that of the maximum amplitude of the response. This is the bandwidth used to compute the total noise power passed by a device (generally not the same as the 3 dB bandwidth) [ORT200801].
Noise figure	A method for quantifying the electrical noise generated by a practical device. The noise figure is the ratio of the noise power at the output of a device to the noise power at the input to the device, where the input noise temperature is equal to the reference temperature (290 K). The noise figure is usually expressed in dB.
Noise temperature	A mathematical construct for predicting the influence of noise in a communications system. It is a measure of the noise power generated by a practical device, expressed as the equivalent temperature of a resistor which, when placed at the input of a perfect noise-free device, generates the same amount of output noise. The noise temperature is usually expressed in kelvin (K).
Off-axis	Any direction in space that does not correspond to an antenna's boresight direction.
Offset (feed) antenna	An antenna having a feed horn that is offset from the center of the reflector. It generally offers better performance than a symmetrically-fed antenna because the feed system does not block the main reflector aperture. Offset antennas can be easily modified to accept dual or multiple feeds allowing them to receive signals from more than one satellite.
Offset feed	A Low Noise Block Downconverter (LNB) that is slightly displaced with respect to the focal point of the reflector so that it receives signals originating from a different direction to that obtained with an LNB placed at the focal point. A technique used in dual-feed reception systems, which receive signals from satellites located at two different orbital locations.

Orbit	The path described by the center of mass of a satellite in space.
Orbital plane	The plane containing the center of mass of the earth and the velocity vector (direction of motion) of a satellite.
Ortho mode transducer (OMT)	Physical element that is directly behind the feed horn and supports functions relating to reception and transmission of satellite signals. The main function of the OMT is to transfer RF to individual ports (e.g., transmit and receive) and to provide an isolation between them; for example, to isolate 90° orthogonal (vertical and horizontal) signals.
Ortho mode transfer loss	The Ortho Mode Transducer is used at the antenna for splitting the signal into vertical and horizontal polarization channels. There are losses associated with this device/function, these being known as Ortho Mode Transfer Loss.
Overall noise performance	An overall noise specification for the entire link used in the link analysis to determine how much noise is present in the received signal. If the overall combined noise it too high the performance of the receiving modems will degrade. For example, if a system has most of the noise in the uplink, one may not be able to get beyond a specific noise level for the entire system. Using a larger antenna will increase the signal level and decrease the noise level only for the downlink portion of the system only.
Parabolic	A geometric shape formed by the intersection of a cone by a plane parallel to its side.
Parabolic antenna	An antenna having a main reflector surface that is a paraboloid or is shaped like a paraboloid. It has the property of reflecting parallel incoming signals to a single focal point. A paraboloid is a geometric surface whose sections parallel to two co-ordinate planes are parabolic and whose sections parallel to the third plane are either elliptical or hyperbolic.
Parallel concatenated convolutional codes (PCCC)	The original Turbo Convolutional Code introduced in the early 1990s.

Path loss	In a communication system, the attenuation undergone by an electromagnetic wave in transit between a transmitter and a receiver. Path loss may be due to many effects such as free-space loss, refraction, reflection, aperture-medium coupling loss, and absorption. Path loss is usually expressed in dB.
Perigee	Point in a satellite orbit (especially for highly-elliptical ones) which is closest to the earth.
Phase error	The difference between the phase angles of the I and Q components of a modulated signal referred to 90°. A phase error is caused by an error of the phase shifter of the I/Q modulator. The I and Q components are in this case not orthogonal to each other after demodulation [BLO200701].
Phase Jitter or phase noise	Noise caused by transponders in the transmission path or by the I/Q modulator. It may be produced in carrier recovery, a possibility that is to be excluded here. In contrast to the phase error, phase jitter is a statistical quantity that affects the I and Q path equally. In the constellation diagram, phase jitter shows up by the signal states being shifted about their coordinate origin [BLO200701].
Phase shift	The change in phase of a periodic signal with respect to a reference.
Plasma	Heavily ionized state of matter, usually gaseous, composed of ions, electrons, and neutral atoms or molecules, that has sufficient electrical conductivity to carry substantial current and to react to electric and magnetic body forces [JAH200201].
Pointing angles	The elevation and azimuth angles which specify the direction of a satellite from a point on the earth's surface.
Pointing error (antenna)	A value that quantifies the amount by which an antenna is misaligned with the satellite's position in space. This is either expressed as an angular error, or as a loss in signal strength with respect to the maximum that would be achieved with a perfectly aligned antenna.

Polar mount	A mechanical support structure for an earth station antenna that permits all satellites to be tracked with movement of only one axis.
Polar satellites	Satellites in LEO orbits that are in a plane of the two poles. Their applications include the ability to view only the poles (e.g., to fill in gaps of GEO coverage), or to view the same place on earth at the same time each 24-h day. Polar orbits are typically used by LEO communications satellites as well as research, weather, and surveillance satellites.
Polarization	Transmission approach where radio waves are restricted to certain directions of electrical and magnetic field variations, where these directions are perpendicular to the direction of wave travel. By convention, the polarization of a radio wave is defined by the direction of the electric field vector. Four senses of polarization are used in satellite transmissions: horizontal linear polarization, vertical linear polarization, right-hand circular polarization and left-hand circular polarization.
Polarization alignment	The process of aligning the reference polarization plane of a linearly-polarized antenna with a particular reference direction. For individual and collective systems receiving linearly-polarized signals, this consists of rotating the Low Noise Block Downconverter (LNB) about the feed axis so that its radio wave detector is aligned with the electric field vector of the incoming signal (to achieve detected signal strength).
Polarization switching	The process of selecting one of two orthogonal polarizations (e.g., linear horizontal or linear vertical) for reception of satellite signals. Polarization switching is implemented in the Low Noise Block Downconverter (LNB) or, more rarely, in a separate device inserted between the feed horn and the LNA/LNB or integrated with the feed horn.
Power flux density (PFD) (aka power density)	The power crossing unit area normal to the direction of wave propagation. Measured in units of watts per square meter (W/m^2).

Power spectral density (PSD)	Amount of power per unit (density) of frequency (spectral) as a function of the frequency. A power spectral density function is a real-valued continuous function of frequency, shown as frequency on the horizontal axis and density on the vertical axis where the signal is compared with itself.
Precipitation loss	Signal loss due to rain attenuating the signal, particularly in the Ku-band, and further dispersing the signal as it passes through the drops of water.
Radiation pattern	A three-dimensional representation of the gain of a transmit or receive antenna as a function of the direction or radiation or reception.
Radio frequency interference (RFI)	(*Synonym* of electromagnetic interference (EMI).) RFI is any electromagnetic disturbance that interrupts, obstructs, limits or otherwise degrades the effective performance of electronics/electrical equipment. RFI can be induced intentionally, as in some forms of electronic warfare, or unintentionally, as a result of spurious emissions and responses, intermodulation products, and so on.
Rain	Atmospheric precipitation that falls to earth in drops more than 0.5 mm in diameter. Rainfall is the amount of precipitation of any type, primarily liquid. It is usually the amount that is measured by a rain gauge.
Rain fade	Fade cause by heavy downpours of rain. Impacts Ku- and Ka-bands more so than C-band operation.
Rain fall attenuation	Atmospheric losses for passing radio signals through (moderate-to-heavy) rain.
Rain outage	Loss of signal due to absorption and increased sky-noise temperature caused by heavy rainfall.
Rain zone (aka precipitation zone)	To aid in calculating the effect of precipitation loss, the world is divided into precipitation zones or rain climatic zones, each of which has a numerical value defined by the International Telecommunications Union (ITU), used in the calculation of a link budget.
Receiver noise temperature	The equivalent noise temperature of a complete receiving system, excluding contributions from the antenna and the physical connection to the antenna, referred to the receiver input.

Reed–Solomon (RS)	A type of block code for forward error correction (FEC). Developed in 1960 by Irving S. Reed and Gustave Solomon. It is the most widely used algorithm for error correcting. The Reed–Solomon code is relatively easy to implement and provides a good tolerance to error bursts. These codes correct many bit errors within large message blocks.
Reed–Solomon/ Viterbi	Reed–Solomon is a block-oriented coding system that is generally applied on top of standard Viterbi coding to correct the bulk of the data errors that are not detected by the other coding systems, thereby further reducing the bit error rates.
Reflection	The change in direction of a wave front at an interface between two dissimilar media so that the wave front returns into the medium from which it originated. Reflection may be specular (i.e., mirror-like) or diffuse (i.e., not retaining the image, only the energy) according to the nature of the interface. Depending on the nature of the interface, that is, dielectric-conductor or dielectric-dielectric, the phase of the reflected wave may or may not be inverted.
Refraction	Retardation, and possibly redirection, of a wavefront passing through (i) a boundary between two dissimilar media, or (ii) a medium having a refractive index that is a continuous function of position, for example, a graded-index optical fiber. For two media of different refractive indices, the angle of refraction is closely approximated by Snell's Law.
Resistojet	Device that heats a propellant stream by passing it through a resistively heated chamber before the propellant is expanded through a downstream nozzle [JAH200201].
Right-hand circularly polarized (RHC(P)) wave	An elliptically- or circularly-polarized wave, in which the electric field vector, observed in any fixed plane normal to the direction of propagation while looking in the direction of propagation, rotates with time in a right-hand or clockwise direction.
S/N (signal-to-noise ratio, aka SNR)	A measure of the quality of an electrical signal, usually at the receiver output. It is the ratio of the signal level to the noise level, measured within a specified bandwidth (typically the bandwidth of the signal). It is typically expressed in dB. The higher the ratio, the better quality of the signal.

Satellite bus	Key components that most satellites are equipped with, such as the control system, battery, solar panel, communication system, and frame structure of the spacecraft itself.
Satellite payload	Components of a satellite which serve the mission function such as a TWT amplifiers to serve as a signal relaying function, camera system for remote sensing, and antenna/receivers for information collection.
Satellite receive power flux density (PFD)	Satellite uplink power flux density is a calculated parameter. The figure is derived from the power flux density to saturate the transponder (PFD/sat) – this figure is available from the uplink coverage map and the satellite transponder specification. A user needs to adjust the uplink earth station transmit power (and possibly uplink antenna size) to obtain the figure one requires at the satellite. For example, the PFD/sat at beam center might be -83 dBW/m^2; the PFD/sat on the -4 dB contour would then be -79 dBW/m^2. (Note that the PFD/sat figures can be altered by commanding changes in the gain setting of the satellite transponder; it may be possible to negotiate changes in gain setting, particularly if one leases an entire transponder.) The value just described refers to an entire transponder: if one transmits an uplink signal with sufficient power to produce the PFD/sat just described, one will saturate the transponder [SAT200801].
Satellite receiver	A receiver designed for satellite reception; it receives modulated signals from a Low Noise Amplifier (LNA) or Low Noise Block Downconverter (LNB) and converts them into their original form.
Satellite transmit power flux density (PFD)	Parameter defined as EIRP/$4\pi d^2$.
Scattering	The scattering of a wave propagating in a material medium. A phenomenon in which the direction, frequency, or polarization of the wave is changed when the wave encounters discontinuities in the medium, or interacts with the material at the atomic or molecular level. Scattering results in a disordered or random change in the incident energy distribution.

Scrambler	A device that renders a signal unintelligible; of interest for modern communication is the process of encryption that has high cryptographic strength.
Serially concatenated convolutional codes (SCCC)	A type of Turbo Convolutional Code.
Shaped beam	The radiation pattern of a satellite antenna that has been designed so that its footprint follows the boundary of a specified geographical area (the area of service provision) as closely as possible. Shaped beams maximize the antenna gain over the service area and reduce the likelihood of interference into systems serving other geographical areas.
Sidelobe	Part of an antenna's radiation pattern that can detect or radiate signals in an unwanted direction (i.e., off-axis). This radiation can produce interference into other systems or susceptibility to interference from other systems. The larger the side lobes, the more noise and interference an antenna can detect. Sidelobe levels are determined by the design of the antenna.
Slant range	The length of a line drawn from the antenna to the satellite.
Solid state power amplifier (SSPA)	A HPA using solid-state technology (i.e., transistors). Originally used for low and medium power applications; however reliable medium-power to high-power SSPA technology has emerged and is used routinely at earth stations and on spacecraft.
Spacecraft longitude	Orbital location of (geosynchronous) satellite.
Spectral power flux density (SPFD)	SPFD is PFD per unit bandwidth. Its units are $W/m^2/Hz$ or Jansky (1 Jansky = 10^{-26} $W/m^2/Hz$ implying $-260\,dB/Wm^2/Hz$
Splitter	A device that takes an input signal and splits it into two or more identical output signals, each a replica of the input signal (typically with reduced amplitude, e.g., $-3\,dB$, but active devices can also operate at $0\,dB$ loss).
Spot beams	An antenna radiation pattern designed to serve a relatively small or isolated geographic area, usually with high gain. Spot beams are areas of discrete signal reception on the ground and discrete

transmission reception in the spacecraft as implemented by supporting antenna structures. In spot beam satellite communication environments, the satellite provides coverage of only a portion of the earth, usually a nation or subcontinent, by using shaped narrow beams pointed to different geographic areas. The advantage of this approach is a higher satellite antenna gain due to a reduction in the aperture angle of the antenna beam, which in turn implies that the user can employ a small aperture antenna. Additionally, the multi-beam technique supports the reuse of frequencies for different beams, thereby effectively increasing the total system capacity.

Spread spectrum	1. Telecommunications techniques in which a signal is transmitted in a bandwidth considerably greater than the frequency content of the original information. Frequency hopping, direct sequence spreading, time scrambling, and combinations of these techniques are forms of spread spectrum.
	2. A signal structuring technique that employs direct sequence, frequency hopping or a hybrid of these, which can be used for multiple access and/or multiple functions. This technique decreases the potential interference to other receivers while achieving privacy and increasing the immunity of spread spectrum receivers to noise and interference. Spread spectrum generally makes use of a sequential noise-like signal structure to spread the normally narrowband information signal over a relatively wide band of frequencies. The receiver correlates the signals to retrieve the original information signal [ANS200001].
Steerable beam	An antenna beam that can be re-pointed by mechanical and/or electrical means. Usually used to refer to relatively narrow satellite beams that can be steered over a part or the whole of the portion of the earth's surface that is visible from the satellite's orbital position.

Sun outage	A natural phenomenon that occurs twice a year (in the spring and fall), when the sun appears to be passing directly behind the satellite, as seen from a receiving earth station. Since the sun is a strong source of radio frequency energy, the earth station's receivers may become overwhelmed by the sun's "noise" output, and reception becomes degraded and eventually impossible for a brief period of time, usually less than 15 min. An observer at the earth station will notice that the antenna feed's shadow will fall exactly in the center of the reflector during the peak of the sun-outage period; this indicates that the antenna, the satellite and the sun are in direct alignment. At this point in time, the sun's radio signals are being focused directly into the antenna's receive feed. This results in a temporary degradation in the signal-to-noise ratio of the signal received from the satellite.
Sun-satellite conjunction	The alignment of the sun with the satellite as seen from an earth station, which takes place twice a year for several minutes around local midday. This event can affect the performance of receiving earth stations.
Sun-synchronous orbit (SS)	A special polar orbit that crosses the equator and each latitude at the same time each day; this orbit can make data collection a convenient task. Satellites in polar orbits are typically used for earth-sensing applications. Typically such a satellite moves at an altitude of 1,000 km [GEO200101].
Symbol error rate (SER)	The probability of receiving a symbol in error.
Syndrome	In Reed-Solomon ECC chips, typically 32 syndromes make up the Syndrome Polynomial that contains the information necessary to find and correct errors in a given message. The syndrome is a unique collection of bits that identifies error.
System noise temperature	A value, expressed in Kelvins, that accounts for the noise contribution of all components in the earth station's receive chain. The equivalent noise temperature of a complete receiving system, taking into account contributions from the antenna, the receiver and the transmission line that interconnects them, referred to the receiver input. Often depicted as T_s or T_{sys}.

Thermal noise	The noise generated by thermal agitation of electrons in a conductor. The noise power, P, in watts, is given by $P = kT\Delta f$, where k is Boltzmann's constant in joules per kelvin, T is the conductor temperature in kelvins, and Δf is the bandwidth in hertz. The magnitude of the noise generated by an object is dependent upon the object's physical temperature. Thermal noise power, per hertz, is equal throughout the frequency spectrum, depending only on k and T. It is an undesired electrical disturbance in a circuit or communication channel.
Thrust	Unbalanced internal force exerted on a rocket during expulsion of its propellant mass.
Tracking	The process of continuously adjusting the orientation of an antenna so that its boresight follows the movements of the satellite about its nominal position. Used in earth stations equipped with large antennas and earth stations operating to satellites in inclined orbit, LEOs, MEOs, polar orbits.
Transmit data rate	Line rate speed of the digital signal being transmitted (including the FEC overhead).
Transponder	A transponder is a system of antennas and amplifiers.
Transponder intermodulation noise	Typically an amplifier creates noise when multiple carriers are present because signals tend to combine in frequency and create new signals (in the same way a mixer does), but at lower power levels. Since these signals are usually small they can be viewed as noise, and so they are characterized in the same way as uplink thermal noise. The noise from satellite intermodulation is presented as a ratio of the power of the carrier to the noise temperature, or C/T_{im} (intermodulation). (Note that this term is different from the uplink thermal noise contribution, uplink C/T.)
Traveling wave tube amplifier (TWTA)	A HPA that utilizes vacuum tube technology. Normally employed when high output power levels and wide bandwidths are required. Typically used on board the satellite and often in earth stations.
Trellis	A tree diagram where the branches following certain nodes are identical. These nodes (or states), when merged with similar states, form a graph that does not grow beyond 2k−1 where k is the constraint length [HEN200201].

Trellis coded modulation (TCM)	Method of coding multiple symbols to give improved symbol error rate performance in noisy conditions.
Turbo convolutional codes (TCC)	A coding and decoding scheme based in the use of two concatenated convolutional codes in parallel that achieves near capacity performance on additive white gaussian noise channel.
Turbo Product Code (TPC)	A forward error correction (FEC) coding scheme used to transmit digital data with the greatest efficiency and reliability of any coding scheme currently available.
Ultra HD	An emerging format that provides video quality that is the equivalent of 8-to-16 HDTV screens (33 million pixels, for the 7680×4320 resolution option), compared to a maximum 2 million pixels (1920×1080 resolution) for the current highest quality HDTV service. It clearly requires a lot more transmission/storage bandwidth, up to $16\times$ more; however the new H.265 high efficiency video coding allows one to double the resolution of the video for the same bitrate
Upconverter	A device for converting the frequency of a signal into a higher frequency.
Uplink Doppler frequency shift	See Downlink Doppler frequency shift.
Uplink signal power	Transmit power that emanates from an antenna.
Uplink thermal noise	Noise generated by the satellites own receiving system. The satellite itself has a receiving antenna and Low Noise Amplifier (LNA) and has a G/T value. The G/T value, losses, power transmitted from the earth station and other parameters are used to calculate the ratio of the signal (or carrier) power to the thermal noise of the uplink system. This component is referred to as uplink C/T.
Vertical polarization	Type of linear polarization where the electric field is approximately aligned with the local vertical plane at an on-ground transmission or reception point. Also see frequency reuse.
Very small aperture terminal (VSAT)	A complete terminal (typically with a small $4-5$ ft antenna) that is designed to interact with other terminals in a satellite delivered data network, commonly in a "star" configuration through a hub.

	The occupied bandwidth of the VSATs' transmitted carrier is typically only a few kHz wide. The VSAT terminal uses a special and often proprietary modulation, scrambling and coding algorithms; this allows the Hub or Network operator to control the system and present billing based on a data throughput, or other form of usage basis. Communications typically include the Network Layer of the protocol stack, that is, IP. VSATs are utilized in a variety of applications and are designed as low cost units. Commonly several VSATs networks are operated through the same hub (shared services) which reduces the initial installation/set up costs [BAR200101].
Viterbi algorithm (VA)	A technique for searching a decoding trellis to yield a path with the smallest distance. A maximum likelihood decoding algorithm devised by A. J. Viterbi in 1967. This is also known as maximum likelihood decoding.
	A widely-used forward Error Correction (FEC) scheme for satellite and other noisy communication channels. There are two key components of a channel using Viterbi encoding: the Viterbi encoder (at the transmitter) and Viterbi Decoder (at the receiver). A Viterbi encoder includes extra information in the transmitted signal to reduce the probability of errors in the received signal that may be corrupted by noise.
Viterbi decoder	Decoder that uses the Viterbi Algorithm (VA) for decoding a bitstream that has been encoded using FEC based on a Convolutional code.
Viterbi decoding	A decoding algorithm devised by A. J. Viterbi that uses a search tree or trellis structure and continually calculates the Hamming (or Euclidean) distance between received and valid code words within the constraint length [HEN200201].

Waveguide	A material medium (a transmission line) that confines and guides a propagating electromagnetic wave. At microwave frequencies, a waveguide normally consists of a hollow metallic conductor (pipe), usually rectangular, elliptical, or circular in cross-section. The cross-section area depends on the frequency: higher frequencies require smaller cross-sectional areas. This type of waveguide may, under certain conditions, contain a solid or gaseous dielectric material [ANS200001]. It is typically used in earth stations to connect the HPAs to the antenna.
Waveguide losses	Losses experienced between the antenna feed and the transmitting or receiving amplifier.

REFERENCES

[AER200801] The Aerospace Corporation, Satellite Communications Glossary, In *Crosslink* (ISSN 1527-5264 [print], ISSN 1527-5272 [Web]). Corporate Communications, P.O. Box 92957, M1-447, Los Angeles, CA 90009-2957. The Aerospace Press, P.O. Box 92957, Los Angeles, CA 90009-2957.

[AMS200001] T. S. Glickman Editor, *Glossary of Meteorology,* American Meteorological Society, Cambridge, Massachusetts, June 2000.

[ANS200001] ANS T1.523-2001, Telecom Glossary 2000, American National Standard (ANS), an outgrowth of the Federal Standard 1037 series, *Glossary of Telecommunication Terms,* 1996.

[BLO200701] Staff, "QAM Defined", Blonder Tongue Laboratories, Inc., One Jake Brown Road, Old Bridge, NJ 08857, 2007.

[BAR200101] M. Bartlett, Satellite FAQ, S+AS Limited, 6 The Walled Garden, Wallhouse, Torphichen, West Lothian, EH48 4NQ, SCOTLAND, July 2001.

[EUT200701] Eutelsat, Glossary, 70, rue Balard, F-75502 Paris Cedex 15, France, http://www.eutelsat.com/business/2_6_2.html.

[GEO200101] Staff, "Geostationary, LEO, MEO, HEO Orbits Including Polar and Sun-Synchronous Orbits with Example Systems and a brief section on Satellite History", 2001, http://www.geo-orbit.org.

[HEN200201] H. Hendrix, "Viterbi Decoding Techniques for the TMS320C54x DSP Generation", Texas Instruments, Application Report SPRA071A – January 2002. Texas Instruments.

[JAH200201] R. G. Jahn, E. Y. Choueiri, "Electric Propulsion", *Encyclopedia of Physical Science and Technology,* Third Edition, Academic Press, 2002. Volume 5.

[JOU199901] M. K. Juonolainen, Forward Error Correction in INSTANCE, Cand Scient Thesis, 1/2/1999, University of Oslo, Department of Informatics.

[LAU200701] J. E. Laube, HughesNet, "Introduction to the Satellite Mobility Support Network, HughesNet User Guide" 2005–2007.

[ORT200801] ORTEL/EMCORE, "RF and Microwave Fiber-Optic Design Guide", Application Note, March 7, 2003, 10420 Research Road, SE, Albuquerque, NM 87123.

[SAT200501] Satellite Internet Inc., "Satellite Physical Units & Definitions", http://www.satellite-internet.ro/satellite-internet-terminology-definitions.htm.

[SAT200801] http://www.satsig.net.

[USA199801] U.S. Army Information Systems Engineering Command Fort Huachuca, Arizona, "Automated Information Systems, Design Guidance, Commercial Satellite Transmission", August 1998.

[ZYR199801] J. Zyren, A. Petrick, "Tutorial on Basic Link Budget Analysis", Application Note AN9804.1, June 1998, Intersil Corporation, 1001 Murphy Ranch Road, Milpitas, CA 95035.

INDEX

---

*Innovations in Satellite Communications and Satellite Technology: The Industry Implications of DVB-S2X,
High Throughput Satellites, Ultra HD, M2M, and IP*, First Edition. Daniel Minoli.
© 2015 John Wiley & Sons, Inc. Published 2015 by John Wiley & Sons, Inc.